SCIENCE FRENCH COURSE

BY

C. W. PAGET MOFFATT, M.A., M.B., B.CHIR.

AUTHOR OF "SCIENCE GERMAN COURSE"

Revised by

NOEL CORCORAN, B.A.(COM.), B.SC., M.SC.TECH.

1966

CHEMICAL PUBLISHING CO., INC.

212 Fifth Avenue New York, N.Y.

Science French Course

ISBN: 978-0-8206-0038-3

Chemical Publishing Company:
www.chemical-publishing.com
www.chemicalpublishing.net

First Published by University Tutorial Press Ltd., 1951
Reprinted 1954, 1959

First American Edition:

Chemical Publishing Company, Inc. - New York 1966

Second Impression:

Chemical Publishing Company, Inc. - 2011

Printed in the United States of America

PREFACE

THE title of this book sufficiently indicates its purpose, namely to provide students who have to read French books on scientific subjects with the necessary minimum of grammar and a selection of extracts from which a little preliminary practice may be obtained.

The method followed will readily be seen from a glance at the Table of Contents. It may here be explained that Part I contains what must be learnt at the start and during the reading of the *Premières Lectures*, while Part II can be learnt later or kept for reference.

The Author's thanks are due to several teachers who have given him the benefit of their experience in the preparation of this book.

NOTE TO THE THIRD EDITION

IN preparing the third edition, certain parts of the Grammar Section have been omitted; the object has been to retain sufficient matter to enable the science student or practising scientist to read technical matter in French, without devoting too much space to the niceties of translation into French. In particular, the paragraphs dealing with the uses of the subjunctive have been greatly condensed.

The *Premières Lectures*, consisting of fairly simple reading matter, have been retained unchanged, but the remainder of the extracts have been replaced by more modern material, taken mainly from papers in scientific journals. A wide range of subjects is covered, and the material used is of the type which the scientist is often called upon to read.

The Reviser's thanks are due to the publishers of the journals from which the extracts have been selected, and to the Senate of the University of London for their kind permission to reproduce extracts from B.Sc. examination papers.

NOTE TO THE FOURTH EDITION

OPPORTUNITY has been taken to include in the Vocabulary some one hundred and fifty additional words, and thanks are due to all those who have made helpful suggestions in this connection.

CONTENTS

PART I.—ELEMENTS OF GRAMMAR

PART II.—NOTES ON SOME IMPORTANT CONSTRUCTIONS

PART III.—IRREGULAR ACCIDENCE AND PARADIGMS OF VERBS

PART IV

PART V

SCIENCE FRENCH COURSE

PART I

ELEMENTS OF GRAMMAR

SUBSTANTIVES, ADJECTIVES, AND ARTICLE

GENDER

1. The gender of a French substantive must be clearly distinguished from the sex of the object which it denotes. Whether this object is a person or a thing, the substantive is either masculine or feminine: there are no neuter substantives.

Certain general rules for ascertaining the gender of French substantives according to their meaning or form can be given, but they are subject to numerous exceptions and are of little use to the beginner whose object is to learn to read the language.

PLURAL OF SUBSTANTIVES

2. The regular method of forming the plural of a French substantive is to add **-s** to the singular.

arbre, *tree*	**arbres,** *trees*
fleur, *flower*	**fleurs,** *flowers*

For exceptions see §§ 195-9.

AGREEMENT

3. An adjective generally assumes different forms according as the substantive to which it refers is masculine or feminine, singular or plural. The regular method of forming the feminine singular of an adjective, *i.e.* the form to be used with a feminine singular substantive, is to add **-e** to the masculine singular. The plural of each gender is regularly formed by the addition of **-s** to the singular. Thus, with

arbre (masc.) and **fleur** (fem.) the forms of the adjective **petit,** *little,* are—

Sing.	**petit arbre**	**petite fleur**
Pl.	**petits arbres**	**petites fleurs**

Obs. In the case of some adjectives the feminine has the same form as the masculine (§ 206), and in others the masculine plural has the same form as the masculine singular (§ 196). For other irregularities see §§ 195-8, 206-11.

THE ARTICLE: DEFINITE, INDEFINITE, AND PARTITIVE

4. This principle of agreement applies not only to adjectives of quality, but also to pronominal adjectives, including the article.

The forms of the article, definite and indefinite, are—

		Masc.	Fem.	Masc.	Fem.
Sing.	Before a consonant	**le**	**la**	**un**	**une**
	Before a vowel or *h* mute	**l'**	**l'**		
Plural		**les**	**les**	(§ 6)	

Thus—

	Masc.	Fem.
Sing.	**le rocher,** *the rock*	**la pierre,** *the stone*
Pl.	**les rochers,** *the rocks*	**les pierres,** *the stones*
Sing.	**l'été,** *the summer*	**l'eau,** *the water*
	l'hiver, *the winter*	
Sing.	**un jour,** *a day*	**une heure,** *an hour*

5. The prepositions **à,** *to, at,* and **de,** *of, from,* coalesce with the masculine singular and with the plural of the definite article thus—

Sing.	for **à le**	**au**	for **de le**	**du**
Pl.	for **à les**	**aux**	for **de les**	**des**

au signe, *to the sign*	**du livre,** *of the book*
aux signes, *to the signs*	**des livres,** *of the books*

Obs. 1. Before the feminine singular, however, and when the definite article becomes **l'** by elision, **à** and **de** remain separate words and do not contract; *e.g.* **de la semaine,** *of the week;* **à l'or,** *to the gold.*

Obs. 2. Before a word beginning with a vowel, **de** becomes **d'.**

d'un arbre, *of a tree* **d'anciens livres,** *of ancient books*

6. The preposition **de** followed by the definite article—**du, de la, de l', des**—is used before substantives with the sense of *some, any* (expressed or implied), and is then known as the partitive article.

J'ai du pain, de la viande, de l'eau et des œufs	*I have (some) bread, meat, water, and eggs*

7. The article is repeated in French even where it would not be repeated in English. The same rule applies not only to the definite, indefinite, and partitive article, but also to the demonstrative (§ 91) and possessive (§ 86) adjectives, and to the prepositions **à, de,** and **en.**

Il me faut une aiguille et du fil	*I want a needle and cotton*
Ses frères et ses sœurs sont arrivés	*His brothers and sisters have come*
L'argent et l'or se trouvent en Australie et en Amérique	*Silver and gold are found in Australia and America*
Il a étudié aux universités de Paris et de Londres	*He has studied at the Universities of Paris and London*

PERSONAL PRONOUNS

8. Personal and relative pronouns are in French, as in English, the only words which vary in form according to the case in which they stand. Thus—

First Person

	Sing.	Plur.
Nom. (subject)	**je,** *I*	**nous,** *we*
Acc. (direct object)	**me,** *me*	**nous,** *us*
Dat. (indirect object)	**me,** (*to*) *me*	**nous,** (*to*) *us*

Second Person

	Sing.	Plur.
Nom. (subject)	**tu,** *thou*	**vous,** *ye, you*
Acc. (direct object)	**te,** *thee*	**vous,** *you*
Dat. (indirect object)	**te,** (*to*) *thee*	**vous,** (*to*) *you*

Third Person Masc.

	Sing.	Plur.
Nom. (subject)	**il,** *he, it*	**ils,** *they*
Acc. (direct object)	***le,** *him, it*	**les,** *them*
Dat. (indirect object)	**lui,** (*to*) *him,* (*to*) *it*	**leur,** (*to*) *them*

* There is also an invariable neuter pronoun **le** which often need not be translated, *e.g.* **comme on peut facilement le deviner,** *as we may easily guess.*

Third Person Fem.

	Sing.	Plur.
Nom. (subject)	**elle,** *she*	**elles,** *they*
Acc. (direct object)	**la,** *her, it*	**les,** *them*
Dat. (indirect object)	**lui,** (*to*) *her,* (*to*) *it*	**leur,** (*to*) *them*

9. The third person reflexive pronoun, in both singular and plural, accusative and dative, is **se.** For the first and second persons **me, te, nous,** and **vous** are used as reflexive pronouns.

10. The plural **vous** is used like the English *you* in addressing a single person, and words agreeing with it are in the singular; *e.g.* **vous êtes très aimable,** *you are very kind.* **Tu** is used in addressing members of one's family or very intimate friends.

11. When the following word begins with a vowel, **je** is elided to **j',** **me** to **m',** **te** to **t',** **le** and **la** to **l',** **se** to **s'.**

For examples illustrating the use of the personal pronouns see §§ 20, 21.

VERBS

12. The most important thing in French accidence is naturally the conjugation of verbs, regular and irregular. Of the former, four conjugations are recognised by grammarians according as the infinitive ends in (1) **-er,** (2) **-ir,** (3) **-oir,** (4) **-re.** Of the latter some are of such common occurrence that the beginner will do well to master them alongside of the regular verbs (p. 73).

13. The simple tenses of a French finite verb are—

Indicative	*Conditional*	*Subjunctive*	*Imperative*
Present	Present	Present	Present
Imperfect		Imperfect	
Past Definite			
Future			

14. A corresponding compound tense is formed by means of the appropriate tense of the auxiliary followed by the past participle. Thus—

Indicative	*Conditional*	*Subjunctive*
Past Indefinite	Perfect	Past Indefinite
Pluperfect		Pluperfect
Past Anterior		
Future Perfect		

There is also a perfect infinitive corresponding to the present infinitive, and a perfect participle corresponding to the present participle.

15. The passive voice (§ 57) is formed with the auxiliary **être,** *to be.* The only uncompounded part of a French verb that is passive is the past participle of a transitive verb, exactly as in English. This participle has masculine and feminine forms, both singular and plural, like an adjective; *e.g.*

	Masc.	Fem.
Sing.	**porté**	**portée**
Pl.	**portés**	**portées**

16. Negation is most commonly expressed by **ne** (before a vowel **n'**) . . . **pas, ne** preceding the verb and **pas** following it. In compound tenses **pas** is placed between the auxiliary and the participle; *e.g.* **je ne viens pas,** *I do not come* or *I am not coming;* **je n'ai pas vu,** *I have not seen.* See further § 54.

17. The following is the arrangement of the regular conjugations in this book:—

The conjugation of the two auxiliaries and of the six commonest irregular verbs is given tense by tense along with the regular verbs; for other irregular verbs see pp. 69-90.

18. Infinitive and Participles

	have	*be*	*carry*	*finish*	*receive*	*break*
Pr. Inf.	avoir	être	porter	finir	recevoir	rompre
Pr. Part.	ay-ant	ét-ant	port-**ant**	fin-**issant**	re-**cevant**	romp-**ant**
Past P.	eu	*été	port-é	fin-**i**	re-**çu**	romp-**u**

Conjugation of the Present Indicative

je, j'	ai	suis	port-e	fin-**is**	re-**çois**	romp-**s**
tu	as	es	**-es**	**-is**	**-çois**	**-s**
il	a	est	**-e**	**-it**	**-çoit**	**-t**
nous	av-ons	sommes	**-ons**	**-issons**	**-cevons**	**-ons**
vous	av-ez	êtes	**-ez**	**-issez**	**-cevez**	**-ez**
ils	ont	sont	**-ent**	**-issent**	**-çoivent**	**-ent**

* Invariable.

Infinitive and Participles

	go	*come*	*be able*	*wish*	*know*	*make, do*
Pr. Inf.	aller	venir	pouvoir	vouloir	savoir	faire
Pr. Part.	all-ant	ven-ant	pouv-ant	voul-ant	sach-ant	fais-ant
Past. P.	all-é	ven-u	pu	voul-u	su	fait

Conjugation of the Present Indicative

je, j'	vais	viens	peux *or* puis	veux	sais	fais
tu	vas	viens	peux	veux	sais	fais
il	va	vient	peut	veut	sait	fait
nous	all-ons	ven-ons	pouv-ons	voul-ons	sav-ons	fais-ons
vous	all-ez	ven-ez	pouv-ez	voul-ez	sav-ez	faites
ils	vont	viennent	peuvent	veulent	sav-ent	font

Past Indefinite.—The tense consists of the present tense of **avoir** (*or* **être,** § 60) followed by the past participle; *e.g.* **j'ai eu,** *I have had;* **j'ai été,** *I have been;* **j'ai porté,** *I have carried;* **je suis allé,** *I have gone.*

POSITION OF THE PERSONAL PRONOUNS

19. The nominative pronouns (§ 8) normally stand before the verb, except in questions (§ 51).

20. The object, direct or indirect, is placed before the verb or, in a compound tense, before the auxiliary.

Je leur parle	*I am speaking to them*
Je vous assure que le chien l'a mangé	*I assure you that the dog has eaten it*

Obs. For the position of object pronouns when the verb is in the imperative see § 48.

21. If two object pronouns, one direct the other indirect, precede the verb, the direct object must be of the third person. The indirect object is placed first if it is of the first or second person, last if it is of the third.

Ils me l'ont donné	*They have given it to me*
Je les lui apporte	*I am bringing them to him* or *to her*

THE PRESENT INDICATIVE

22. The present tense in French corresponds to two English forms, *I carry* and *I am carrying*.

Il porte un verre d'eau	*He is carrying a glass of water*

23. The present in French is also used, as in English, to express the immediate future.

Je vais demain au musée	*I am going to the museum to-morrow*

24. The present is used of an action or a state that has been going on for some time, though English idiomatically uses the present-perfect.

Je vous attends depuis midi	*I have been waiting for you since noon*
J'étudie les mathématiques depuis longtemps	*I have been studying mathematics for a long time*

THE PAST INDEFINITE

25. The past indefinite corresponds to the English present-perfect (*I have carried*). It is also the tense used in conversation where in English the simple past tense would be used.

J'ai fini mon thème ce matin	*I finished my exercise this morning*

26. The past indefinite is formed by the addition of the past participle to the present of the verb **avoir** or, in the case of some intransitive verbs, to the present of the verb **être** (§ 60). The past participle of a transitive verb agrees in gender and number with the direct object *if that direct object precedes.*

Nous les avons trouvés ici	*We found them here*
Avez-vous fini la leçon ? Oui, je l'ai finie	*Have you finished the lesson? Yes, I have finished it*
Ont-ils reçu les lettres ? Oui, ils les ont reçues	*Have they received the letters? Yes, they have received them*

27. Intransitive verbs, however, especially those denoting motion, form their past indefinite by adding the past participle to the present of **être.** The commonest of these verbs are **aller** and **venir** (see also § 60).

Il est allé voir le professeur	*He went to see the lecturer*
Nous sommes venus vous trouver	*We came to find you*

28. In this case the past participle is treated as an adjective and varies with the number and gender of the subject.

Elles sont venues à Paris pour assister aux conférences	*They came* (or *have come*) *to Paris to attend the lectures*

Read *Premières Lectures*, No. 1, p. 92.

29. CONJUGATION OF THE IMPERFECT INDICATIVE

	have	*be*	*carry*	*finish*	*receive*	*break*
	avoir	être	porter	finir	recevoir	rompre
je, j'	av-**ais**	ét-**ais**	port-**ais**	fin-**issais**	recev-**ais**	romp-**ais**
tu	**-ais**	**-ais**	**-ais**	**-issais**	**-ais**	**-ais**
il	**-ait**	**-ait**	**-ait**	**-issait**	**-ait**	**-ait**
nous	**-ions**	**-ions**	**-ions**	**-issions**	**-ions**	**-ions**
vous	**-iez**	**-iez**	**-iez**	**-issiez**	**-iez**	**-iez**
ils	**-aient**	**-aient**	**-aient**	**-issaient**	**-aient**	**-aient**

	go	*come*	*be able*	*wish*	*know*	*make, do*
	aller	venir	pouvoir	vouloir	savoir	faire
je, j'	all-ais	ven-ais	pouv-ais	voul-ais	sav-ais	fais-ais
tu	-ais	-ais	-ais	-ais	-ais	-ais
il	-ait	-ait	-ait	-ait	-ait	-ait
nous	-ions	-ions	-ions	-ions	-ions	-ions
vous	-iez	-iez	-iez	-iez	-iez	-iez
ils	-aient	-aient	-aient	-aient	-aient	-aient

PLUPERFECT INDICATIVE.—The tense consists of the imperfect indicative of **avoir** (or **être,** § 60) followed by the past participle; *e.g.* **j'avais eu,** *I had had;* **j'avais été,** *I had been;* **j'avais porté,** *I had carried;* **j'étais allé,** *I had gone.*

Obs. The personal endings of the imperfect indicative are the same in all verbs, but in most verbs of the second conjugation **-iss-** is inserted before them.

30. CONJUGATION OF THE PAST DEFINITE INDICATIVE

	have	*be*	*carry*	*finish*	*receive*	*break*
	avoir	être	porter	finir	recevoir	rompre
je, j'	eus	fus	port-**ai**	fin-**is**	reç-**us**	romp-**is**
tu	eus	fus	**-as**	**-is**	**-us**	**-is**
il	eut	fut	**-a**	**-it**	**-ut**	**-it**
nous	eûmes	fûmes	**-âmes**	**-îmes**	**-ûmes**	**-îmes**
vous	eûtes	fûtes	**-âtes**	**-îtes**	**-ûtes**	**-îtes**
ils	eurent	furent	**-èrent**	**-irent**	**-urent**	**-irent**

---

	go	*come*	*be able*	*wish*	*know*	*make, do*
	aller	venir	pouvoir	vouloir	savoir	faire
je, j'	all-ai	vins	p-us	voul-us	s-us	f-is
tu	-as	vins	-us	-us	-us	-is
il	-a	vint	-ut	-ut	-ut	-it
nous	-âmes	vînmes	-ûmes	-ûmes	-ûmes	-îmes
vous	-âtes	vîntes	-ûtes	-ûtes	-ûtes	-îtes
ils	-èrent	vinrent	-urent	-urent	-urent	-irent

PAST ANTERIOR.—The tense consists of the past definite indicative of **avoir** (or **être**, § 60) followed by the past participle; *e.g.* **j'eus eu,** *I had had;* **j'eus été,** *I had been;* **j'eus porté,** *I had carried;* **je fus allé,** *I had gone.*

THE REMAINING PAST TENSES OF THE INDICATIVE

31. The imperfect corresponds to three English forms, *e.g.* **je portais** means *I carried, I was carrying,* and *I used to carry.* Hence the imperfect is used in descriptions: it represents an action as in progress, or expresses a continuous state or an habitual action.

Je travaillais pendant que vous dormiez	*I was working* (or *worked*) *while you were asleep*
Les anciens Egyptiens embaumaient leurs morts	*The ancient Egyptians used to embalm their dead*

32. The imperfect is used of an action or state that at a certain point in the past had already been going on for some time; cp. § 24.

J'apprenais le français depuis deux ans quand j'allai à Paris	*I had been learning French for two years when I went to Paris*

33. The past definite corresponds to the English past, *e.g.* **je vis,** *I saw,* **j'allai,** *I went.* It is the historic tense and describes successive actions fully completed.

Une tuile lui tomba sur la tête et le tua	*A tile fell on his head and killed him*

Obs. In conversation the past indefinite is used for the past definite.

34. In narratives the past definite *and* imperfect are often both employed, the former to give the successive events, the latter to describe attendant circumstances, details, etc.

Il parut devant Narva à la tête de cette grande armée le 1er octobre. Le czar qui, dans de pareilles saisons, *faisait* quelquefois quatre cents lieues en poste, à cheval, pour voir quelque mine ou quelque canal, *n'épargnait* pas plus ses troupes que lui-même: il *savait* d'ailleurs que les Suédois faisaient la guerre au cœur de l'hiver comme dans l'été. Ainsi le czar *assiégeait* Narva à trente degrés du pôle et Charles XII *s'avançait* pour la secourir. Le czar ne *fut* pas plus tôt arrivé devant la place, qu'il se *hâta* de mettre en pratique tout ce qu'il venait d'apprendre dans ses voyages: il *traça* son camp, le *fit* fortifier de tous côtés, *éleva* des redoutes et *ouvrit* lui-même la tranchée.

Obs. The beginner may postpone the study of the above passage for the present.*

35. The pluperfect is used as in English.

Ils avaient fait tout ce qu'ils pouvaient	*They had done all they could*

36. The past anterior is only used in clauses beginning with **à peine,** *scarcely*, **aussitôt que,** *as soon as*, **lorsque,** *when*, and a few less common phrases.

A peine furent-ils revenus qu'ils virent le professeur	*Scarcely had they returned when they saw the master*

Obs. 1. Note here the inversion of subject and verb (§ 52, *Obs.*).

Obs. 2. The past anterior is not used of habitual or repeated action.

Read *Premières Lectures*, No. 2, p. 92.

37. CONJUGATION OF THE FUTURE INDICATIVE

	have	*be*	*carry*	*finish*	*receive*	*break*
	avoir	être	porter	finir	recevoir	rompre
je, j'	aurai	serai	port-**erai**	fin-**irai**	recev-**rai**	romp-**rai**
tu	auras	seras	**-eras**	**-iras**	**-ras**	**-ras**
il	aura	sera	**-era**	**-ira**	**-ra**	**-ra**
nous	aurons	serons	**-erons**	**-irons**	**-rons**	**-rons**
vous	aurez	serez	**-erez**	**-irez**	**-rez**	**-rez**
ils	auront	seront	**-eront**	**-iront**	**-ront**	**-ront**

---

* The translation is as follows:—

On October 1st he appeared before Narva at the head of this great army. The Tsar, who at such seasons would sometimes travel post or ride on horseback four hundred leagues in order to inspect some mine or some canal, spared his troops no more than he spared himself, for he knew that the Swedes were accustomed to make war in the depth of winter as well as in summer. So within thirty degrees of the Pole the Tsar was besieging Narva, and Charles XII was advancing to relieve it. No sooner had the Tsar arrived before the fortress, than he hastened to put into practice all that he had just learned in his travels: he marked out his encampment, had it fortified on all sides, erected redoubts, and himself worked at cutting the trenches.

	go	*come*	*be able*	*wish*	*know*	*make, do*
	aller	venir	pouvoir	vouloir	savoir	faire
je, j'	irai	viendrai	pourrai	voudrai	saurai	ferai
tu	iras	viendras	pourras	voudras	sauras	feras
il	ira	viendra	pourra	voudra	saura	fera
nous	irons	viendrons	pourrons	voudrons	saurons	ferons
vous	irez	viendrez	pourrez	voudrez	saurez	ferez
ils	iront	viendront	pourront	voudront	sauront	feront

FUTURE-PERFECT.—The tense consists of the future tense of **avoir** (or **être,** § 60) followed by the past participle; *e.g.* **j'aurai eu,** *I shall have had;* **j'aurai été,** *I shall have been;* **j'aurai porté,** *I shall have carried;* **je serai allé,** *I shall have gone.*

Obs. The personal endings of the future are the same in all verbs.

THE FUTURE

38. The future tense corresponds to two English forms; *e.g.* **je finirai** means *I shall finish* and *I shall be finishing.*

Elle ira à l'école dans huit jours	*She will be going to school in a week*
Nous partirons jeudi	*We shall leave on Thursday*

39. The future is used in dependent clauses (other than conditions) if future time is clearly referred to, though English idiomatically uses the present.

Quand tu auras fini l'expérience, tu pourras partir	*When you have finished the experiment, you will be able to go*
Je viendrai aussitôt qu'il sera ici	*I shall come as soon as he is here*

Obs. In conditional clauses referring to the future French idiom is the same as English; cp. the example in § 192.

THE FUTURE-PERFECT

40. The future-perfect refers to the time when an act still future will be completed, and its use is parallel to that of the future simple.

Quand le facteur sera venu, je serai prêt à travailler	*When the postman has come, I shall be ready to work*

Read *Premières Lectures*, No. 3, p. 93.

41. Conjugation of the Present Conditional

	have avoir	*be* être	*carry* porter	*finish* finir	*receive* recevoir	*break* rompre
je, j'	aurais	serais	port-**erais**	fin-**irais**	recev-**rais**	romp-**rais**
tu	aurais	serais	**-erais**	**-irais**	**-rais**	**-rais**
il	aurait	serait	**-erait**	**-irait**	**-rait**	**-rait**
nous	aurions	serions	**-erions**	**-irions**	**-rions**	**-rions**
vous	auriez	seriez	**-eriez**	**-iriez**	**-riez**	**-riez**
ils	auraient	seraient	**-eraient**	**-iraient**	**-raient**	**-raient**

	go aller	*come* venir	*be able* pouvoir	*wish* vouloir	*know* savoir	*make, do* faire
je, j'	irais	viendrais	pourrais	voudrais	saurais	ferais
tu	irais	viendrais	pourrais	voudrais	saurais	ferais
il	irait	viendrait	pourrait	voudrait	saurait	ferait
nous	irions	viendrions	pourrions	voudrions	saurions	ferions
vous	iriez	viendriez	pourriez	voudriez	sauriez	feriez
ils	iraient	viendraient	pourraient	voudraient	sauraient	feraient

Perfect Conditional.—The tense consists of the present conditional tense of **avoir** (or **être**, § 60) followed by the past participle; *e.g.* **j'aurais eu,** *I should have had;* **j'aurais été,** *I should have been;* **j'aurais porté,** *I should have carried;* **je serais allé,** *I should have gone.*

The Conditional Mood

42. The present conditional corresponds to two English forms; *e.g.* **je recevrais** means *I should receive* or *I should be receiving.*

Je le ferais encore, si j'avais à le faire	*I should do it again if I had to do it*
Si je faisais cela, je trahirais ma patrie	*If I did that, I should be betraying my country*

43. The "modest" conditional expresses diffidence or hesitation: it makes a statement for the truth of which the speaker does not vouch.

Cent personnes auraient péri dans le tremblement de terre	*A hundred persons are said to have lost their lives in the earthquake*
Les Norvégiens auraient découvert l'Amérique quelques siècles avant Colomb	*The Norwegians are said to have discovered America some centuries before Columbus*

44. The use of the perfect conditional is the same as that of the corresponding tense in English.

Read *Premières Lectures*, No. 4, p. 94.

The Subjunctive Mood

45. The subjunctive is much more widely used in French than in English. It is found chiefly in dependent clauses. In principal sentences it expresses a wish and is generally preceded by the conjunction **que,** *that.*

Qu'il vive heureux !	*May he live happily !*
Vive le roi !	*Long live the king !*

The second person present subjunctive of **vouloir** forms a polite imperative.

Veuillez m'indiquer le chemin de la ville	*Kindly tell me the way to the town*

46. Conjugation of the Present Subjunctive

	have	*be*	*carry*	*finish*	*receive*	*break*
	avoir	être	porter	finir	recevoir	rompre
je, j'	aie	sois	port-e	fin-isse	re-çoive	romp-e
tu	aies	sois	-es	-isses	-çoives	-es
il	ait	soit	-e	-isse	-çoive	-e
nous	ayons	soyons	-ions	-issions	-cevions	-ions
vous	ayez	soyez	-iez	-issiez	-ceviez	-iez
ils	aient	soient	-ent	-issent	-çoivent	-ent

	go	*come*	*be able*	*wish*	*know*	*make, do*
	aller	venir	pouvoir	vouloir	savoir	faire
je, j'	aille	vienne	puisse	veuille	sache	fasse
tu	ailles	viennes	puisses	veuilles	saches	fasses
il	aille	vienne	puisse	veuille	sache	fasse
nous	allions	venions	puissions	voulions	sachions	fassions
vous	alliez	veniez	puissiez	vouliez	sachiez	fassiez
ils	aillent	viennent	puissent	veuillent	sachent	fassent

PAST INDEFINITE SUBJUNCTIVE.—The tense consists of the present subjunctive of **avoir** (or **être**, § 60) followed by the past participle; *e.g.* **j'aie eu, j'aie été, j'aie porté, je sois allé.**

Read *Premières Lectures*, No. 5, p. 95.

47. CONJUGATION OF THE IMPERFECT SUBJUNCTIVE

	have	*be*	*carry*	*finish*	*receive*	*break*
	avoir	être	porter	finir	recevoir	rompre
je, j'	eusse	fusse	port-**asse**	fin-**isse**	reç-**usse**	romp-**isse**
tu	eusses	fusses	**-asses**	**-isses**	**-usses**	**-isses**
il	eût	fût	**-ât**	**-ît**	**-ût**	**-ît**
nous	eussions	fussions	**-assions**	**-issions**	**-ussions**	**-issions**
vous	eussiez	fussiez	**-assiez**	**-issiez**	**-ussiez**	**-issiez**
ils	eussent	fussent	**-assent**	**-issent**	**-ussent**	**-issent**

	go	*come*	*be able*	*wish*	*know*	*make, do*
	aller	venir	pouvoir	vouloir	savoir	faire
je, j'	allasse	vinsse	pusse	voulusse	susse	fisse
tu	allasses	vinsses	pusses	voulusses	susses	fisses
il	allât	vînt	pût	voulût	sût	fît
nous	allassions	vinssions	pussions	voulussions	sussions	fissions
vous	allassiez	vinssiez	pussiez	voulussiez	sussiez	fissiez
ils	allassent	vinssent	pussent	voulussent	sussent	fissent

PLUPERFECT SUBJUNCTIVE.—The tense consists of the imperfect subjunctive tense of **avoir** (or **être**, § 60) followed by the past participle; *e.g.* **j'eusse eu, j'eusse été, j'eusse porté, je fusse allé.**

THE IMPERATIVE MOOD

48. When the imperative (first person plural or second person) is used affirmatively, the governed pronouns *follow* it, though they precede all other forms of the verb, and the imperative also when it is used negatively. After an imperative an accusative pronoun always precedes a dative, and **moi** and **toi** take the place of **me** and **te.**

Ecoutez-moi	*Listen to me*
Ne m'écoutez pas	*Don't listen to me*
Donnez-les-nous	*Give them to us*
Disons-le-lui	*Let us tell him*
Ne le leur disons pas	*Let us not tell them*

Obs. Hyphens are used to connect the pronouns with the affirmative imperative.

Read *Premières Lectures*, No. 6, p. 96.

49. Conjugation of the Imperative

	have	*be*	*carry*	*finish*	*receive*	*break*
	avoir	être	porter	finir	recevoir	rompre
	aie	sois	porte	finis	reçois	romp-s
qu'il	ait	soit	porte	finisse	reçoive	rompe
	ayons	soyons	portons	finissons	recevons	rompons
	ayez	soyez	portez	finissez	recevez	rompez
qu'ils	aient	soient	portent	finissent	reçoivent	rompent

	go	*come*	*be able*	*wish*	*know*	*make, do*
	aller	venir	pouvoir	vouloir	savoir	faire
	va	viens	(none)	veuille	sache	fais
qu'il	aille	vienne	”	veuille	sache	fasse
	allons	venons	”	veuillons	sachons	faisons
	allez	venez	”	veuillez	sachez	faites
qu'ils	aillent	viennent	”	veuillent	sachent	fassent

The second singular imperative is the same as the second singular present indicative, except that in the first conjugation **s** is dropped. The first and second plural imperative are the same as the first and second plural present indicative without the pronoun. The third person both singular and plural is borrowed from the present subjunctive.

En and Y

50. To the list of pronouns given in §§ 8, 9 two more, **en** and **y**, may now be added.

En, *of it*, *of them*, *from it*, *some of it*, *some*, *any*, is used as a genitive of the third person pronoun referring to things.

Y is similarly used for the dative of the third person with reference to things.

En and **y** follow an affirmative imperative, but precede other forms of the verb; combined with other pronouns **en** and **y** come last.

J'en suis bien aise	*I am very glad of it*
Il a reçu votre lettre et y répondra	*He has received your letter and will reply to it*
Avez-vous tous mes livres ? J'en ai trois	*Have you all my books? I have three of them*
Si j'avais de l'argent, je lui en donnerais	*If I had any money, I would give him some (of it)*

Obs. In the last example the words *of it* are placed in parenthesis as unnecessary; **en** may often thus remain untranslated.

NOTE 1.—If after an affirmative imperative **en** is preceded by an indirect object in the first or second person singular, **me** or **te** is used, not **moi** or **toi**; *e.g.* **donnez-m'en,** *give me some (of it).*

NOTE 2.—**Ils y étaient,** *they were there.*

NOTE 3.—**En,** used with the article, is equivalent to the English possessive adjective: **ils en mangent la chair,** *they eat its flesh.*

INTERROGATIVE CONJUGATION

51. In an interrogative sentence the subject, if it is a personal pronoun, follows the verb. In a compound tense the pronoun immediately follows the auxiliary.

Où est-il ?	*Where is he?*
Chanté-je faux ?	*Am I singing out of tune?*
A-t-il apporté des fossiles ?	*Has he brought any fossils?*

Observe—

(1) a hyphen is inserted between verb and pronoun;

(2) in the first person singular present indicative of the first conjugation an acute accent is placed on the final **e**;

(3) if the verb ends and the pronoun begins with a vowel, **-t-** is inserted.

Obs. Inversion of subject and verb does not always indicate a question: it is found in affirmative sentences which begin with **ainsi, à peine, encore, en vain, peut-être**; *e.g.* **Peut-être voulait-il voir l'expérience,** *Perhaps he wanted to see the experiment.*

52. When the subject is a substantive, it comes first, then the verb, then (connected by a hyphen with the verb) a pronoun repeating the subject.

Les élèves savent-ils la leçon ?	*Do the pupils know the lesson?*
Cet accident arrive-t-il souvent ?	*Does this accident often happen?*
Combien d'expériences le professeur a-t-il fait ce matin ?	*How many experiments did the lecturer perform this morning?*

53. There is another form of interrogation in which no inversion of verb and pronoun takes place. It consists in prefixing to the verb the phrase **est-ce que,** *is it that?*

Est-ce que je chante faux ?
Est-ce que les élèves savent la leçon ?
Est-ce que cet accident arrive souvent ?

Negative Conjugation

54. The negative with verbs consists of two parts: **ne** and some strengthening word or "complement."

The **ne** immediately precedes the verb, and the second part of the negative follows the simple verb in simple tenses and is placed immediately before the past participle in compound tenses. With the infinitive both parts of the negative precede (§ 55).

The negatives are as follows:

ne . . . pas, *not*	**ne . . . jamais,** *never*
ne . . . point, *not at all*	**ne . . . guère,** *scarcely*
ne . . . que, *only*	**ne . . . rien,** *nothing*
ne . . . plus, *no more, no longer*	**ne . . . personne,** *nobody*
	ne . . . aucun, *not any, no* (adj.)

ne . . . ni . . . ni, *neither . . . nor.*

Examples:

Je n'ai pas d'encre (§ 173)	*I have no ink*
Je ne l'ai jamais reçu	*I have never received it*
Je n'ai reçu aucune lettre	*I have received no letter*
Nous n'avons que trois heures	*We have only three hours*
Elle n'a vu personne	*She has seen nobody*
Je n'admets point cela	*I can't admit that*
Ils n'ont rien vu	*They have seen nothing*
Je n'ai ni père ni mère	*I have neither father nor mother*

Obs. Certain of these complements may be combined.

Ils ont trouvé d'immenses régions dans lesquelles ne poussent guère que des mousses	*They found vast districts in which scarcely anything but moss grows*
Je n'avance plus que lentement	*I get on but slowly*

55. When the present infinitive is used negatively, **ne pas,** etc., *precede* the verb. With the perfect infinitive **pas,** etc., follow the auxiliary.

Voulez-vous bien ne pas briser le verre ?	*Will you please not break the glass?*
N'avoir pas aimé c'est n'avoir pas vécu	*Not to have loved is not to have lived*

Negative-Interrogative Conjugation

56. The negative-interrogative conjugation requires no explanation, as the rules for negative and for interrogative conjugation are combined. **Ne** comes in front of the whole compound and **pas** or other complementary word at the end.

N'avez-vous pas fini votre travail ?	*Have you not finished your work?*

or **Est-ce que vous n'avez pas fini votre travail ?**

Ne serez-vous jamais prêt ?	*Will you never be ready?*

or **Est-ce que vous ne serez jamais prêt ?**

Ne lui a-t-elle rien donné ?	*Has she not given him anything?*

Read *Premières Lectures,* No. 7, p. 97.

The Passive Voice

57. The passive voice is formed in French exactly as in English.

Il était aimé	*He was loved*
J'ai été loué	*I have been praised*

58. The past participle agrees in gender and number with the subject, but **été** remains invariable.

Elle était aimée	*She was loved*
Nous avons été loués	*We have been praised*

59. The passive voice is used less frequently than in English. Its place may be taken by (1) a reflexive verb (§ 70), or (2) **on** with an active verb (§ 123).

(1) **Il s'appelle Henri**	*He is called Henry* or *His name is Henry*
(2) **On dit qu'il a été malade**	*He is said to have been ill*

INTRANSITIVE VERBS

60. Most intransitive verbs form their compound tenses, like transitive verbs, with the auxiliary **avoir,** but the following verbs take **être**:—

aller, *to go*	**partir,** *to set out*
arriver, *to arrive*	**rester,** *to remain*
entrer, *to go* or *come in*	**retourner,** *to go back*
mourir, *to die*	**sortir,** *to go out*
naître, *to be born*	**tomber,** *to fall*
venir, *to come*	and their compounds

Certain other verbs are conjugated with **avoir** to denote action in progress and with **être** to denote a state or a completed action. These forms are usually translated literally into English.

Il a passé par Paris	*He passed through Paris*
L'été est passé	*Summer is over*

61. The past participle of these verbs agrees in gender and number with the subject.

Trois étudiants étaient entrés dans la salle de lecture	*Three students had entered the reading-room*

62. Some actions are denoted by transitive verbs in English but by intransitive verbs in French. Such are—

entrer dans, *to enter*	**succéder à,** *to succeed (come after)*
nuire à, *to injure*	**survivre à,** *to survive*
obéir à, *to obey*	**jouir de,** *to enjoy*
résister à, *to resist*	

63. Conversely some intransitive English verbs correspond to transitive verbs in French.

attendre, *to wait for*	**chercher,** *to look for*
demander, *to ask for*	**regarder,** *to look at*

Reflexive Verbs

64. To conjugate reflexive verbs insert **me, te, se, nous, vous, se** immediately before the verb in simple tenses and immediately before the auxiliary in compound tenses. The auxiliary is **être,** not **avoir.**

65. Some verbs are naturally reflexive (*i.e.* cannot be used without the reflexive pronoun), as for instance **se repentir;** others are accidentally reflexive (*i.e.* can be used with or without the reflexive pronoun), as for instance **se laver,** *to wash oneself,* **laver** being a transitive verb meaning *to wash.*

Among the latter class of verbs some are directly reflexive, *i.e.* the reflexive pronoun denotes the direct object of the verb's action, *e.g.* **se laver;** others are indirectly reflexive, *i.e.* the reflexive pronoun denotes the indirect object of that action, *e.g.* **s'acquérir,** *to obtain for oneself.* Verbs indirectly reflexive may take a direct object (§ 68).

66. In compound tenses of reflexive verbs the past participle agrees with the direct object if it precedes the verb, but not with the indirect object.

Nous nous sommes blessés	*We have hurt ourselves*
Elle s'était tuée	*She had killed herself*
Ils s'étaient gagné un grand renom	*They had won great fame*
Elles se sont nui	*They have injured* (lit. *done harm to*) *themselves*

67. Se may be reciprocal.

Ils se ressemblent fort	*They are very much alike*

68. In the imperative used affirmatively the reflexive pronoun follows the verb.

Levez-vous vite	*Get up quickly*
Ne vous levez pas de trop bonne heure	*Don't get up too early*

69. In English many verbs can be used either transitively or intransitively without change of form; such are *to turn, to stop, to wash.* In French these become reflexive, as the object (unexpressed in English) refers to the same person as the subject.

Après la mort le corps se décompose	*After death the body decomposes*
Les anciens Gaulois se nourrissaient de viande, de lait, et de pain	*The ancient Gauls lived on meat, milk, and bread*

70. The French reflexive is often equivalent to the English passive.

Le sel se produit au moyen de l'évaporation	*Salt is produced by evaporation*
A une température élevée il se produit un gaz	*At a high temperature a gas is produced* (lit. *there is produced a gas*)

71. Note **se servir de,** *to use;* **se passer de,** *to do without.*

Read *Premières Lectures*, No. 9, p. 100.

Impersonal Verbs

72. Impersonal verbs are only used in the third person singular. They may be (1) naturally, (2) accidentally impersonal. Naturally impersonal verbs are never used otherwise than impersonally, while some ordinary intransitive verbs may be used as accidentally impersonal verbs.

Il faut travailler pour réussir	*It is necessary to work in order to succeed*
Dans ce cas il s'y produit un mouvement circulaire	*In this case a circular movement is produced* (§ 59)
Il en sortait une fumée noire et épaisse	*There was issuing from it a thick black smoke*

73. The commonest impersonal verb is **il y a,** *there is, there are* (lit. *it has there*).

Il y aura beaucoup de monde à l'exposition	*There will be a great many people at the exhibition*

74. Il y a is equivalent to the English *ago* in such phrases as

Je les ai vus il y a deux heures	*I saw them two hours ago*

Read *Premières Lectures*, No. 10, p. 101.

75. Il fait (3rd singular present of **faire** used impersonally) is used in expressions about the weather.

Il a fait beau ce matin, mais maintenant il fait du brouillard	*It was fine this morning, but now it is foggy*

76. Impersonal constructions are common in French. The logical subject may be plural.

Il tombe de la pluie	*Rain is falling*
Il en sort des rayons lumineux	*Bright rays issue from it*

77. The verbs **avoir** and **être** are used as auxiliaries as follows:—

Avoir	**Etre**
All transitive verbs, active voice.	All transitive verbs, passive voice.
Most intransitive verbs.	Some intransitive verbs.
All true impersonal verbs.	All reflexive verbs.

Idiomatic Uses of Certain Verbs as Auxiliaries

78. Devoir as a notional verb means *to owe.* As an auxiliary with the present infinitive it means—

(1) In the conditional, *ought;*

(2) In other moods, *must, have to, am to.*

Vous devriez sortir tous les jours	*You ought to go out every day*
Il a dû abandonner l'essai	*He has had to* (or *been obliged to*) *give up the attempt*
L'expérience devait aboutir à une grande découverte	*The experiment was* (*destined*) *to lead to a great discovery*
C'est pourquoi le crocodile doit ramper	*That is why the crocodile is obliged to crawl*

79. The perfect conditional of **devoir** or **pouvoir** with the present infinitive corresponds to the idiomatic English *ought to have, could have.*

Vous auriez dû finir l'expérience	*You ought to have finished the experiment*
Il aurait pu travailler mieux	*He could have worked better*

80. Aller followed by an infinitive is used of an act in the immediate future.

Je vais finir l'expérience	*I am going to finish the experiment*

81. Venir de with the infinitive means *to have just*.

Nous venions de rencontrer le grand chimiste	*We had just met the great chemist*

82. Il faut, *it is necessary*, is best translated by *must*, *be obliged to*.

Il fallait absolûment partir ce soir même	*We were absolutely obliged to leave that very evening*
Il faut que vous fassiez cela *or* **Il vous faut faire cela**	*You must do that*

83. Faire is used in the sense of to cause a thing to be done and is variously rendered.

Je ferai venir l'étudiant	*I will send for the student*
Il a fait faire les instruments à Paris	*He has had the instruments made in Paris*

When the infinitive that follows **faire** has its own direct object, the object of **faire** itself becomes indirect.

Le professeur lui avait fait finir l'expérience	*The lecturer had made him finish the experiment*
Il lui fera dire la vérité	*He will make him tell the truth*

84. Laisser, entendre, voir have the same construction as **faire.**

Il a laissé faire	*He has left things alone*
J'ai entendu détonner l'hydrogène	*I heard the hydrogen explode*
Je vois venir le professeur de chimie	*I see the chemistry teacher coming*

Obs. Note the word-order in French: in affirmative sentences the infinitive is not separated from the verb used as an auxiliary.

85. **Savoir** with an infinitive means *to know how to.* Note also **je ne saurais** (cp. § 43), *I cannot.*

Je ne saurais vous dire s'il sait patiner	*I can't tell you whether he knows how to skate*

Read *Premières Lectures*, No. 11, p. 102.

§§ 200-205, on the Comparison of Adjectives and Adverbs, may now be read for a first time.

POSSESSIVE ADJECTIVES AND PRONOUNS

Possessive Adjectives

86. The possessive adjectives are—

		Singular Masc.	Singular Fem.	Plural For both genders	
Sing.	1st Person	**mon**	**ma**	**mes**	*my*
	2nd ,,	**ton**	**ta**	**tes**	*thy*
	3rd ,,	**son**	**sa**	**ses**	*his, her, its*
Pl.	1st ,,	**notre**		**nos**	*our*
	2nd ,,	**votre**		**vos**	*your*
	3rd ,,	**leur**		**leurs**	*their*

87. The possessive adjective, like any other adjective, agrees in gender and number with the substantive to which it refers.

mon mari, *my husband* **ta femme,** *your wife*
son parapluie, *his umbrella* or *her umbrella*
sa sœur, *his sister* or *her sister*
ses sœurs, *his sisters* or *her sisters*

Obs. **Son** and **sa** mean *his* or *her* indifferently, just as **mon** and **ton,** **ma** and **ta,** are used without regard to the sex of the possessor.

88. When a feminine word begins with a vowel or **h** mute, the forms **mon, ton, son** are used instead of **ma, ta, sa,** in order to avoid the retention of an unelided vowel before a word beginning with a vowel.

mon âme (instead of **ma âme**), *my soul*
ton horloge (instead of **ta horloge**), *your clock*

Possessive Pronouns

89. The possessive pronouns are—

	Singular		Plural		
	Masc.	Fem.	Masc.	Fem.	
1st Pers. Sing.	**le mien**	**la mienne**	**les miens**	**les miennes**	*mine*
2nd „ „	**le tien**	**la tienne**	**les tiens**	**les tiennes**	*thine*
3rd „ „	**le sien**	**la sienne**	**les siens**	**les siennes**	*his, hers*
1st Pers. Pl.	**le nôtre**	**la nôtre**	**les nôtres**		*ours*
2nd „ „	**le vôtre**	**la vôtre**	**les vôtres**		*yours*
3rd „ „	**le leur**	**la leur**	**les leurs**		*theirs*

90. Possessive pronouns take the gender and number of the substantive to which they refer.

Mes tables ne sont pas si précises que les vôtres ou les siennes	*My tables are not so exact as yours or his* (or *hers*)

DEMONSTRATIVE ADJECTIVES AND PRONOUNS

The Demonstrative Adjective

91. The demonstrative adjective in French has four forms.

	Singular		Plural
Masc.	**ce, cet**	*this* or *that*	**ces,** *these* or *those*
Fem.	**cette**		

92. In the masculine singular **cet** is used before a substantive or adjective beginning with a vowel or **h** mute.

ce garçon, *this boy* or *that boy* **cet ami,** *this friend* or *that friend*

ce héros, *this hero* or *that hero* **cet homme,** *this man* or *that man*

cet heureux événement, *this* or *that happy event*

cette horloge, *this* or *that clock* **ces horloges,** *these* or *those clocks*

93. When *this* and *that* are used in the same sentence in English, the distinction is preserved in French by **ci** (shortened from **ici**), *here*, and **là,** *there*, placed after the substantives and connected therewith by hyphens.

ce cheval-ci et ce chien-là, *this horse and that dog*

Demonstrative Pronouns

94. The demonstrative pronouns are—

Singular.

Masc.	**celui,** *he* } *that,*	**celui-ci** } *this one,*	**celui-là** } *that one,*
Fem.	**celle,** *she* } *the one*	**celle-ci** } *the latter*	**celle-là** } *the former*

Plural.

Masc.	**ceux** } *those*	**ceux-ci** } *these,*	**ceux-là** } *those,*
Fem.	**celles** }	**celles-ci** } *the latter*	**celles-là** } *the former*

Singular.

Neut.	**ce,** *it, that,* etc.	**ceci,** *this*	**cela,** *that*

95. Celui, celle, and their plurals, **ceux, celles,** must always be followed by the preposition **de** or a relative pronoun. They usually refer to a substantive preceding; when they do not, they can only be used of persons.

Voici mon thermomètre et voilà celui de mon frère	*Here is my thermometer and there is my brother's*
Ceux qui font des heureux sont les vrais conquérants	*Those who make men happy are the true conquerors*

96. Celui-ci, etc., **celui-là,** etc., are used as real demonstratives, *i.e.* to point out or indicate, **celui-ci** referring to the nearer object, **celui-là** to the more remote.

Vos livres sont mieux reliés que ceux-là	*Your books are better bound than those*
Achetez des prunes et des marrons; mangez celles-là et gardez ceux-ci	*Buy some plums and chestnuts; eat the former and keep the latter*

97. The demonstrative pronoun **ce** is used (i) as the subject of the verb **être,** (ii) in conjunction with the relative pronouns.

98. Ce, as the subject of **être,** may take the place of the pronouns **il, ils, elle, elles.**

C'est un farceur	*He is a humbug*
Ce sont de bons mathématiciens	*They are good mathematicians*

99. C'est is also used impersonally for **il est,** *it is,* before a proper noun, a common noun preceded by any determining word, or a pronoun.

C'est Charles	*It is Charles*
C'est mon médecin	*It is my doctor*
C'est nous	*It is we*

NOTE.—**Ce sont** must be used before a third person plural.

Ce sont les enfants	*It is the children*

100. Ceci and **cela** are used as the equivalents of the English *this* and *that* when not referring to a substantive already mentioned.

Ceci est la maison rouge	*This is the red house*
Pourquoi vous a-t-il donné cela ?	*Why did he give you that?*

RELATIVE PRONOUNS

101. The relative pronouns are **qui,** *who, which, that;* **lequel,** *who, which;* **ce qui,** *what* (*that which*).

102. Qui is declined as follows, the singular and plural being identical in form:—

	Of Persons	Of Animals and Things
Nom.	**qui,** *who*	**qui,** *which, that*
Acc.	**que,** *whom*	**que,** *which, that*
Gen.	**de qui, dont,** *of whom*	**dont,** *of which*
Dat.	**à qui,** *to whom*	(**auquel,** etc., *to which*)

103. The relative pronoun **lequel** is declined as follows:—

	Singular		Plural		
	Masc.	Fem.	Masc.	Fem.	
Nom.	**lequel**	**laquelle**	**lesquels**	**lesquelles**	*which*
Acc.	**lequel**	**laquelle**	**lesquels**	**lesquelles**	*which*
Gen.	**duquel**	**de laquelle**	**desquels**	**desquelles**	*of which*
Dat.	**auquel**	**à laquelle**	**auxquels**	**auxquelles**	*to which*

Note the relative use of **où,** *where,* in place of a preposition and **lequel.**

Voici l'hôtel où il est descendu	*This is the hotel at which he put up*

104. Qui, after a preposition, can only refer to persons; in reference to animals or things one of the various forms of **lequel** must be used.

Les assassins par qui il fut tué ont été pendus	*The assassins by whom he was killed have been hung*
C'est l'étude à laquelle je consacre mes loisirs	*This is the study to which I devote my leisure*

105. Et qui, et que are equivalent to the English *and one which,* etc.

La pomme de terre est une plante très commune maintenant et que chacun a vue	*The potato is now a very common plant, and one which everybody has seen*
La baleine a de petits yeux, pas plus grands que ceux d'un bœuf et qui sont placés de chaque côté de la tête	*The whale has small eyes, no larger than those of an ox; they are placed one on each side of its head*

106. The genitive **dont** may be used of persons or things, but must always come next to its antecedent.

Alexandre fut un héros dont nous admirons les exploits	*Alexander was a hero whose exploits we admire*
Le cheval dont je parlais est mort	*The horse of which I was speaking is dead*

Obs. **Dont** must sometimes be translated *with which, e.g.*

Son poil, dont on remplit les coussins, est long et fin	*Its hair, with which cushions are filled, is long and fine*

107. When **dont** cannot come next to its antecedent, its place is taken by **de qui, duquel,** etc.

Le monsieur dans la famille de qui (*or* **duquel) il avait été domestique**	*The gentleman in whose family he had been a servant*

108. Ce qui is the demonstrative pronoun **ce** in conjunction with the relative **qui,** and is used as the equivalent of the English relative *what;* it may be thus declined:—

Nom.	**ce qui**	*that which, what*
Acc.	**ce que**	*that which, what*
Gen.	**ce dont**	*that of which*
Dat.	**ce à quoi**	*that to which*

Il aime ce qui est juste et équitable	*He likes what is just and fair*
J'aime ce que vous aimez	*I like what you like*
Je ne comprends pas ce dont vous parlez	*I do not understand what you are talking about*
Ce à quoi elle pense est un secret	*What she is thinking of is a secret*

109. For the relative adjective **quel que,** *whatever,* see § 120.

INTERROGATIVE ADJECTIVE AND PRONOUNS

The Interrogative Adjective

110. The interrogative adjective has four forms:—

	Singular	Plural	
Masc.	**quel**	**quels**	*which?*
Fem.	**quelle**	**quelles**	*what?*

111. Quel may be either directly or indirectly interrogative.

Direct question—

Quelle route dois-je suivre ?	*Which road must I follow?*
Quel est votre numéro d'ordre à l'examen ?	*What is your examination number?*

Indirect question—

Je vous demande quelle route je dois suivre	*I ask you which road I must follow*
L'examinateur m'a demandé quel était mon numéro	*The examiner asked me what my number was*

112. Quel also corresponds to the English *What a . . . !* used in exclamations.

Quel dommage !	*What a pity!*

INTERROGATIVE PRONOUNS

113. The interrogative pronouns are **qui ?** *who?* **que ? quoi ?** *what?* **lequel ?** *which?* **Lequel** is declined in the same way as the relative pronoun.

	Singular and Plural	Singular
Nom.	**qui ?** *who?*	**que ?** *what?*
Acc.	**qui ?** *whom?*	**que ?** *what?*
Gen.	**de qui ?** *of whom?*	**de quoi ?** *of what?*
Dat.	**à qui ?** *to whom?*	**à quoi ?** *to what?*

114. For the interrogative pronoun **qui,** in its various cases, the following locution is often used:—

Nom.	**qui est-ce qui ?**	*who (is it that)?*
Acc.	**qui est-ce que ?**	*whom?*
Gen.	**de qui est-ce que ?**	*of whom?*
Dat.	**à qui est-ce que ?**	*to whom?*

Qui vous a dit de sortir ?
Qui est-ce qui vous a dit de sortir ? } *Who has told you to go out?*

Qui cherchent-ils ?
Qui est-ce qu'ils cherchent ? } *For whom are they looking?*

De qui parle-t-il ?
De qui est-ce qu'il parle ? } *Of whom is he speaking?*

A qui parlez-vous ?
A qui est-ce que vous parlez ? } *To whom are you speaking?*

115. Similar substitution may be made for the various cases of **que** or **quoi.** In the nominative the simple form **que** is rarely used.

Nom.	**qu'est-ce qui ?**	*what (is it that)?*
Acc.	**qu'est-ce que ?**	*what?*
Gen.	**de quoi est-ce que ?**	*of what?*
Dat.	**à quoi est-ce que ?**	*to what?*

Qu'est-ce qui vous empêche de sortir ? — *What prevents you from going out?*

Que dit-il ?
Qu'est-ce qu'il dit ? } *What does he say?*

De quoi parle-t-il ?
De quoi est-ce qu'il parle ? } *Of what is he speaking?*

A quoi travaille-t-il ?
A quoi est-ce qu'il travaille ? } *At what is he working?*

116. Lequel, in its various forms, is used, like the English *which*, when there is a choice of several persons or things.

Lequel de ces garçons est votre fils ?	*Which of these boys is your son?*
Voici deux plumes ; laquelle préférez-vous ?	*Here are two pens; which do you prefer?*

INDEFINITE ADJECTIVES AND PRONOUNS

117. The following words can be used either as indefinite adjectives or as indefinite pronouns:—

Singular		Plural		
Masc.	Fem.	Masc.	Fem.	
aucun	**aucune**	**aucuns**	**aucunes**	*any, no, none*
autre		**autres**		*other*
l'un et l'autre	**l'une et l'autre**	**les uns et les autres**	**les unes et les autres**	*both*
l'un ou l'autre	**l'une ou l'autre**	**les uns ou les autres**	**les unes ou les autres**	*either*
ni l'un ni l'autre	**ni l'une ni l'autre**	**ni les uns ni les autres**	**ni les unes ni les autres**	*neither*
nul	**nulle**	**nuls**	**nulles**	*no, none*
		plusieurs		*several*
tel	**telle**	**tels**	**telles**	*such*
tout	**toute**	**tous**	**toutes**	*all, every*

***Aucun** (or **nul**) **espoir ne nous reste**	*No hope is left to us*
***Je n'ai vu aucun de vos amis**	*I have seen none of your friends*
Prenez l'autre éprouvette	*Take the other test-tube*
Les autres sont partis	*The others are gone*
Ni l'une ni l'autre maison n'est à vendre	*Neither house is for sale*
Plusieurs ont été tués	*Several have been killed*
Plusieurs soldats ont été tués	*Several soldiers have been killed*
Telle fut sa fin	*Such was his end*
Toute la terre est habitée	*The whole earth is inhabited*
Tous ont péché	*All have sinned*

* **Aucun** (negative) and **nul** when used with a verb require the latter to be preceded by **ne.**

118. Tout may be an adjective meaning in the singular *any*, in the plural *all.*

Toute femme saurait le faire	*Any woman could do it*
Tous les vases sont pleins	*All the vessels are full*

119. Tout as an adverb is invariable, except when it precedes a feminine adjective beginning with a consonant or **h** mute, in which case it agrees.

Elle était tout étonnée, toute charmée	*She was quite astonished, quite delighted*

Le tout is a noun and means *the whole.*

Indefinite Adjectives

120. The following words can be used as indefinite adjectives only:—

Singular Masc.	Singular Fem.	Plural Masc.	Plural Fem.	
chaque				*each, every*
même		**mêmes**		*self, same*
quelque		**quelques**		*some, a few*
quel . . . que	**quelle . . . que**	**quels . . . que**	**quelles . . . que**	*whatever*
quelconque		**quelconques**		*any . . . whatever*

Les professeurs mêmes font la même chose	*The lecturers themselves do the same thing*
Nous avons quelques heures à attendre	*We have some hours to wait*
***Quelle qu'ait été son intention, le résultat a été bon**	*Whatever his intention may have been, the result was good*
Prenez deux points quelconques	*Take any two points*

Obs. **Même** is also an adverb, meaning *even.*

Les plus sages même le font	*Even the wisest do it*

Indefinite Pronouns

121. The following words are used as indefinite pronouns only. They may be divided into variable and invariable.

---

* **Quel . . . que** is always followed by the subjunctive.

122. Variable Indefinite Pronouns

Singular Masc.	Singular Fem.	Plural Masc.	Plural Fem.
chacun	**chacune**		
each one			
l'un	**l'une**	**les uns**	**les unes**
(the) one			
l'un l'autre	**l'une l'autre**	**les uns les autres**	**les unes les autres**
one another			
quelqu'un	**quelqu'une**	**quelques-uns**	**quelques-unes**
someone			

Remettez ces éprouvettes chacune à sa place	*Put each of these test-tubes back into its place*
L'une d'elles est brisée	*One of them is broken*
Ces deux hommes se haïssent l'un l'autre	*These two men hate each other*
Tous ces gens se haïssent les uns les autres	*All these people hate one another*
Achetez quelques-uns de ces chronomètres	*Buy some of these chronometers*

123. Invariable Indefinite Pronouns

on, l'on, *one, people, they* — **quelque chose,** *something*

personne, *anyone, no one* — **rien,** *anything, nothing*

On trouve des diamants en Afrique	*Diamonds are found in Africa*
Je viendrai si l'on* m'appelle	*I shall come if I am called*
Personne† n'est assez sot pour le croire	*Nobody is so foolish as to believe him*
Qui vous l'a dit ? Personne	*Who told you so? No one*
Y a-t-il personne d'assez courageux ?	*Is there anybody courageous enough?*
J'ai appris quelque chose de fâcheux	*I have learnt something annoying*
Qui ne risque rien, n'a rien	*Nothing venture, nothing have*
Qu'avez-vous ? Rien	*What is the matter with you? Nothing*
Y a-t-il rien de plus beau ?	*Is there anything more beautiful?*

---

* **L'on** is often substituted for **on** after **et, ou, où, que, si,** for the sake of euphony.

† The masculine pronoun **personne** must be distinguished from the feminine substantive **une personne,** *a person.*

DISJUNCTIVE PERSONAL PRONOUNS

124. The forms of the personal pronouns given in §§ 8, 9 are used, as was explained in §§ 19-21, when the pronoun stands in conjunction with a verb as its subject or object, and they are hence known as conjunctive personal pronouns.

125. The disjunctive pronouns are not thus used in direct combination with a verb. They have only one form for the nominative and accusative; the dative is formed by prefixing **à**.

	Singular	Plural
First Person (both genders)	**moi,** *I, me*	**nous,** *we, us*
Second Person ,, ,,	**toi,** *thou, thee*	**vous,** *ye, you*
Third Person (Masculine)	**lui,** *he, him*	**eux,** *they, them*
,, ,, (Feminine)	**elle,** *she, her*	**elles,** *they, them*

126. A disjunctive personal pronoun may stand alone in answer to a question.

Qui va là ? Moi *Who is there? I*

127. A disjunctive personal pronoun is used as the second term of a comparison.

Nous sommes plus âgés qu'eux *We are older than they*

128. Disjunctive personal pronouns are used when there are two or more subjects, direct objects, or indirect objects to the same verb, one or more of such subjects or objects being pronouns.

Lui et elle sont arrivés *He and she have arrived*

129. A disjunctive personal pronoun is used as the antecedent to a relative pronoun.

Moi, qui suis innocent, serai condamné *I, who am innocent, shall be condemned*

130. A disjunctive personal pronoun is used after a preposition.

A qui parlez-vous ? A eux *To whom are you speaking? To them*

131. A disjunctive personal pronoun may stand as the complement of the impersonal **c'est, ce sont.**

Est-ce vous, mon cher ?	*Is it you, my dear fellow?*
Ce sont eux	*It is they*

132. The disjunctive personal pronouns are used with the adjective **même** to form the emphatic personal pronouns.

	Singular	Plural
First Person	**moi-même,** *myself*	**nous-mêmes,** *ourselves*
Second ,,	**toi-même,** *thyself* **vous-même,** *yourself*	**vous-mêmes,** *yourselves*
Third ,,	**lui-même,** *himself* **elle-même,** *herself*	**eux-mêmes** **elles-mêmes** *themselves*

Je l'ai fait moi-même	*I did it myself*
Elle l'a dit elle-même	*She said so herself*

REFLEXIVE PRONOUNS

133. The reflexive conjunctive pronouns are **me, te, se, nous, vous, se.** See § 64 on reflexive verbs.

For disjunctive reflexives the ordinary disjunctive pronouns are used, with or without **même**; there is also a special form **soi,** used in reference to an indefinite pronoun such as **on** or **chacun.**

Il a de l'argent sur lui	*He has money on him*
Chacun pour soi	*Each for himself*

THE NUMERALS

For the Table of Numerals see pp. 60, 61.

134. The notation is the same in French as in English, but vulgar fractions are but little employed, except those in low terms, owing to the universality of the decimal system.

NOTE 1.—Ordinals are abbreviated thus:—**1er, 2e, 3e,** etc.

NOTE 2.—Vulgar fractions are expressed as in English, by using a cardinal number as numerator, an ordinal as denominator; *e.g.* **trois-cinquièmes ($\frac{3}{5}$), trois millièmes ($\frac{3}{1000}$).**
But note **un demi ($\frac{1}{2}$), deux-tiers ($\frac{2}{3}$), trois-quarts ($\frac{3}{4}$).**

NOTE 3.—*A vulgar fraction* is **une fraction ordinaire.** *A decimal fraction* is **une fraction décimale;** *a decimal point* is **une virgule décimale.**

NOTE 4.—No elision takes place before **huit** and **onze. Au onzième siècle,** *in the* 11*th century;* **le huit mars,** *March* 8*th.*

NOTE 5.—In French **un billion** means *a thousand millions,* **un trillion,** *a million millions.*

NOTE 6.—The nouns **vingtaine,** *a score,* **centaine,** *a hundred,* etc., denote round numbers, *about twenty, about a hundred,* etc.

NOTE 7.—*Units, tens, hundreds,* . . . are respectively **unités, dizaines, centaines, . . .**

135. Vingt and **cent** take the sign of the plural when they are multiplied, but not followed, by another number.

Quatre-vingts morceaux	80 *pieces*
Huit cent un ans	801 *years*

136. Mille, *a thousand,* never takes an *s.*

Cinq mille milles	5,000 *miles*

137. Mil is used for **mille** in dates of the Christian era

Newton mourut en l'an mil sept cent vingt-sept	*Newton died in* 1727

ORDINALS

138. The ordinal numbers, with the exception of **premier,** *first,* and **second** (the alternative for **deuxième**), *second,* are formed by adding **-ième** to the cardinals, final **-e** mute being elided.

e.g. **quatre** **quatrième**

But note orthographical modifications in **cinquième** (from **cinq**), and **neuvième** (from **neuf**).

139. Unième and **deuxième,** NOT **premier** and **second,** are used in compounds.

e.g. **vingt et un** **vingt et unième**

140. Cardinals are used instead of ordinals in the titles of monarchs and in naming the days of the month except **premier.**

Le premier juin	*The first of June*
Louis seize fut décapité le vingt-et-un janvier 1793	*Louis XVI was beheaded January* 21*st,* 1793

Time of Day

141. The following examples show the French method of stating the hour. The word **minutes,** *minutes*, may be omitted, as in English:—

Quelle heure est-il ?	*What o'clock is it?*
Il est une heure	*It is one o'clock*
Il est midi	*It is twelve o'clock (noon)*
Il est minuit	*It is twelve o'clock (midnight)*
Il est deux heures et demie	*It is half-past two*
Il est midi et (un) quart	*It is a quarter-past twelve*
Il est midi moins un quart	*It is a quarter to twelve*
Il est une heure cinq (minutes)	*It is five minutes past one*
Il est dix heures vingt-cinq (minutes)	*It is five-and-twenty past ten*
Le train partira à midi moins vingt (minutes)	*The train will leave at twenty (minutes) to twelve*

Note.—The abbreviation **m.** (**matin,** *morning*) is used for *a.m.*, and **s.** (**soir,** *evening*) for *p.m.*

Dimension

142. There are three ways of expressing in French *This courtyard is a hundred feet long and eighty feet wide*.

(1) **Cette cour est longue de cent pieds et large de quatre-vingts pieds.**

(2) **Cette cour a cent pieds de long et** (*or* **sur**) **quatre-vingts pieds de large.**

(3) **Cette cour a cent pieds de longueur sur quatre-vingts pieds de largeur.**

Read *Premières Lectures,* No. 12, p. 104.

QUASI-IRREGULAR VERBS OF THE FIRST CONJUGATION

143. In the first conjugation some verbs present irregularities which are mainly orthographical.

(1) Verbs with **e** (unaccented) in the last syllable but one change this **e** mute to **è** before a mute **e.**

mener, *to lead* **je mène,** *I lead*

(2) Verbs of which the present infinitive ends in **-eler** or **-eter** double the consonant **l** or **t** before the **e** mute instead of taking the grave accent.

appeler, *to call* **j'appelle,** *I call*
jeter, *to throw* **je jette,** *I throw*

A few of these verbs, instead of doubling the consonant, take a grave accent on the **e** preceding the consonant.

acheter, *to buy* **j'achète,** *I buy*

(3) Verbs with **é** in the last syllable but one have a grave accent, instead of an acute, before **e** mute, except in the future indicative and present conditional, where the acute accent is retained.

préférer, *to prefer* **je préfère,** *I prefer*
je préférerai, *I shall prefer*

Verbs ending in **-éer** form an exception to this rule, and retain the acute accent throughout.

créer, *to create* **je crée,** *I create*

(4) Verbs ending in **-yer** change the **y** into **i** before **e** mute.

ployer, *to bend* **je ploie,** *I bend*

Verbs ending in **-ayer** may retain the **y** throughout, but usually change it to **i** before **e** mute.

payer, *to pay* **je paie,** *I pay*

(5) In verbs ending in **-ger** or **-cer** an **e** is inserted after the **g** of the former, and a cedilla is put under the **c** of the latter, whenever the inflexional suffix begins with **a** or **o**.

manger, *to eat* **nous mangeons,** *we eat*
tracer, *to trace* **nous traçons,** *we trace*

PREPOSITIONS

N.B.—The object of this chapter is to point out the chief differences of idiom between the two languages as regards this part of speech.

144. A (1) means *to, at* (Latin *ad*); (2) denotes manner, means, purpose, possession, standard (of judging).

Nous vivons à l'anglaise	*We live English fashion*
Peignez-vous à l'huile ou à l'aquarelle ?	*Do you paint in oils or in water-colours ?*
Vous trouverez le pot au lait dans la salle à manger	*You will find the milk-jug in the dining-room*
Ces crayons sont à vous, non pas à moi	*The pencils are yours, not mine*
Ce métal est reconnaissable à sa couleur	*This metal may be recognised by its colour*

Note also—

Peu à peu	*Little by little*
Quant à lui	*As for him*
A part	*Aside*
Au moyen de	*By means of*
Au menton	*On the chin*
Au mois de	*In the month of*
A ce sujet	*On this subject*

145. Acheter, demander, and verbs with a similar meaning take the dative of the person and accusative of the thing, as also do verbs of "taking away."

J'ai demandé une éprouvette à l'étudiant	*I asked the student for a test-tube*
J'ai dû emprunter de l'argent à un ami	*I had to borrow some money from a friend*

146. Avec and **de** may both be used to express *with* (instrumental). **Avec** denotes mere instrumentality, **de** gives in addition the idea of cause or means.

On écrit avec un crayon ou avec une plume	*One writes with a pencil or a pen*
Il a écrit cette lettre de sa propre main	*He has written the letter with his own hand*

147. Chez, lit. *at the house of,* is sometimes to be translated *among, in.*

Parfois on rencontre chez les oiseaux aquatiques un long cou et des pattes courtes	*Sometimes long necks and short legs are to be found in aquatic birds*

148. Dans means *into, in.* It is more definite than **en,** and is used before an article or demonstrative adjective. With such verbs as **boire,** *to drink,* **prendre,** *to take,* **puiser,** *to draw,* **dans** means *out of.*

Il a pris cela dans sa poche	*He has taken it out of his pocket*

Obs. In speaking of time **dans** means *at the end of,* **en,** *in.*

Il fera cet ouvrage en trois jours	*He will take three days to do that piece of work* (whenever he begins)
Il terminera cet ouvrage dans trois jours	*He will finish that piece of work in three days* (from now)

149. De means *from* or *of.* It is used in a figurative sense of the agent or instrument. After **de** expressing instrument the partitive article is omitted.

Il brûle de rage	*He is burning with fury*
La montagne est couverte de neige	*The mountain is covered with snow*

150. De may precede an infinitive which is not governed by a finite verb.

Il est possible de calculer	*It is possible to calculate*
C'est bien le cas de le dire	*One may well say so*

Note also—

De façon à	*In such a way as*
De beaucoup	*By far*
De tels	*Such*
D'autres	*Others*
De nos jours	*In our day*
Du moins	*At least*
Autour de	*Around*
De tous temps	*In every age*

151. Devant, *before* (of place only); **avant,** *before* (of time).

Il s'est tenu devant la porte	*He stood in front of the door*
Avant la guerre de Cent Ans, les paysans étaient prospères	*Before the Hundred Years' War the peasants were prosperous*

152. En means *in*, in a wider and vaguer sense than **dans.** It never stands before **le, la,** or **les.**

With the names of countries it means both *in* and *to.*

Dans huit jours j'irai en France: je suis resté assez longtemps en Angleterre	*In a week I shall go to France: I have stayed long enough in England*

Obs. **Au** is used with masculine singular and **aux** with plural names of countries.

Au Japon, au Portugal, au Groënland, aux Indes	*To* (or *in*) *Japan, Portugal, Greenland, India*

En also denotes material.

Une boîte en bois	*A wooden box*
Une maison en pierre	*A stone house*

En also means *as a, in the manner of a.*

Vivre en savant	*To live like a learned man*

Obs. 1. In **de ville en ville,** *from town to town,* **de mal en pis,** *from bad to worse,* **de plus en plus,** *more and more,* **en** is used to express motion, as in **aller en Allemagne.**

Obs. 2. Note **en moyenne,** *on an average;* **en vie,** *alive;* **consister en,** *to consist of.*

153. Entre.

Entre les mains de . . .	*In the hands of . . .*

154. Faute de.

Faute d'argent	*For lack of money*

155. Le long de.

Ils errèrent le long du chemin	*They wandered along the road*

156. Par, *by,* especially to express the agent after a passive verb.

Jeter par la fenêtre	*To throw out of* (i.e. *through*) *the window*

Obs. 1. When **par** is used of weather it is translated *in* or *on.*

Par un si beau jour	*On such a fine day*
Par le temps qu'il fait	*In this weather*

Obs. 2. **Par** is used distributively in **six francs par jour,** *six francs a day.*

Par with **commencer, finir,** is usually rendered *by* or *at.*

Tu me ferais plaisir si tu voulais commencer par le commencement	*I should be obliged if you would begin at the beginning*
Il a fini par réussir	*He succeeded in the end*

Note also—

Par où va-t-on à Laon ?	*How does one go to Laon? Which is the way to Laon?*
Je suis arrivé par là	*I came that way*

157. Pour corresponds with the English *for,* except in **trois pour cent,** *three per cent.*

Pour ce qui est de cela, ne vous en souciez pas	*As for that, don't bother about it*

158. The adjective or adverb modified by **trop** and **assez** must be followed by **pour** with—

Either (*a*) an infinitive—

Il est assez riche pour être indépendant	*He is rich enough to be independent*

Or (*b*) a **que** clause with the indicative of that which is regarded as a fact, with the subjunctive if the idea of purpose is dwelt on.

Il est assez riche pour qu'on puisse s'adresser à lui	*He is rich enough to justify one in applying to him*

159. Sur is translated *from* in such phrases as **prendre du pain sur la table, prendre un drapeau sur l'ennemi.**

160. Sur also corresponds with *by* in dimensions.

Cette salle a trente pieds de long sur vingt de large	*This hall is thirty feet long by twenty feet wide*

161. A travers, au travers de, *through, across.*

Il va à travers le champ (au travers du champ)	*He crosses the field* (diagonally)

162. Many present participles are used as prepositions. Such are **pendant,** *during;* **suivant,** *according to*, etc.

Suivant moi, il est impossible de faire cela	*In my opinion it is impossible to do that*

163. Prepositions enter largely into the formation of compound nouns in French; *e.g.*—

L'hôtel de ville, *the town-hall;* **un verre à vin,** *a wine-glass;* **un fer à cheval,** *a horse-shoe;* **un cheval de course,** *a race-horse;* **l'Afrique du Sud,** *South Africa;* **le voyage de retour,** *the return journey.*

164. The process here illustrated necessitates altering the English order, as in French the name of the thing determined, not of its determinant, must come first; *e.g.*—

La télégraphie sans fil	*Wireless telegraphy*

Obs. To translate a French double compound noun, begin at the *end.*

Une fabrique de poudre à canon	*A gunpowder manufactory*
La compagnie de bateaux à vapeur du Rhin	*The Rhine Steamship Company*
Des gâteaux de farine d'avoine	*Oatmeal cakes*

165. The negative adverb **ne** must in the following cases be left untranslated:—

(i) After comparatives—

Ces fruits sont plus mûrs que je ne l'attendais	*This fruit is riper than I expected*

(ii) After verbs of fearing—

Je crains que les mathématiques ne soient une étude difficile	*I fear that mathematics is a difficult study*

(iii) After verbs of hindering—

Cela empêche que l'extrémité inférieure ne soit recourbée vers le haut	*That prevents the lower extremity from being bent back towards the top*

(iv) After verbs of denial or doubt used negatively or interrogatively—

Je ne nie pas que ce gaz ne se dégage facilement	*I do not deny that this gas is evolved easily*

PART II

NOTES ON SOME IMPORTANT CONSTRUCTIONS

THE ARTICLE AND THE SUBSTANTIVE

The Definite Article

166. The definite article is employed much more widely in French than in English. The following are some of the additional uses.

167. The definite article is used in French before any substantive taken in a general sense, *i.e.* before—

(1) an abstract noun, *e.g.* **la vérité,** *truth;*

(2) the name of a material, *e.g.* **le fer,** *iron;*

(3) the name of a class, *e.g.* **les étudiants,** *students;*

(4) the name of a country, *e.g.* **l'Angleterre,** *England;*

(5) the name of a mountain, *e.g.* **le mont Blanc,** *Mont Blanc.*

Les expériences nous prouvent que la nature de l'électricité n'est pas encore comprise par les savants	*Experiments prove to us that the nature of electricity is not yet understood by scientists*

168. The definite article is also used before titles, and when an adjective precedes a proper name.

Les Français ne se défient plus de la perfide Albion, grâce à l'habilité du roi Edouard sept	*The French no longer distrust perfidious Albion, thanks to the cleverness of King Edward VII*

Obs. The article is omitted after the preposition **en.**

Voyager de France en Espagne	*To travel from France into Spain*

169. The definite article is used in place of the English possessive adjective where the possessor is sufficiently indicated by the context.

Il s'est brûlé la main en faisant cette expérience	*He burnt his hand in performing this experiment*

170. French uses the definite, not the indefinite, article in speaking of price.

Cela m'a coûté trois francs la livre	*That cost me three francs a pound*

The Indefinite Article

171. The indefinite article is omitted in the following cases in which it is used in English:—

(1) in apposition;

(2) with the complement of such verbs as **devenir,** *to become;* **être,** *to be;* **rester,** *to remain.*

(1) **La Russie, pays très peu accidenté, a beaucoup de fleuves aux cours longs et lents**	*Russia, a flat country, has many rivers with long slow courses*
(2) **Il est devenu médecin très célèbre, mais il ne sera jamais savant**	*He has become a very famous doctor, but he will never be a scientist*

The Partitive Article

172. A substantive taken in a partitive sense is preceded in French by **du, de la, de l', des.**

Les plantes puisent de l'eau dans le sol	*Plants draw water from the soil*
Ces arbres porteront des fruits	*Those trees will bear fruit*

173. There are two cases, however, in which **de** only is used instead of the partitive article—

(1) with a substantive preceded by an adjective;

(2) after a substantive or adverb denoting quantity.

Je veux de grands flacons pour mesurer ce fluide	*I require large flasks to measure this fluid*
Donnez-moi beaucoup de charbon	*Give me plenty of coal*
Nous n'avons besoin que de peu d'outils	*We need but few instruments*

NOTE.—The negative adverbs **pas de, point de,** are included under this heading.

Ils n'ont pas de charbon	*They have no coal*

174. In French the article, whether definite or partitive, is repeated before each substantive to which it applies. So also with pronouns, personal, demonstrative or possessive.

L'or, le cuivre et l'argent se trouvent en Australie	*Gold, copper, and silver are found in Australia*
J'ai besoin du soufre et du phosphore pour cette expérience	*I need sulphur and phosphorus for this experiment*

175. Note the possibility of combining prepositions.

Je fais une visite à des amis	*I am visiting some friends*

POSITION OF THE ADJECTIVE

176. In French a qualifying adjective sometimes precedes and sometimes follows the substantive. Certain adjectives change in meaning according to their position, *i.e.* before or after the noun—

Un brave homme, *a worthy man*	**Un homme brave,** *a courageous man*
Un grand homme, *a great man*	**Un homme grand,** *a tall man*
Mon propre chapeau, *my own hat*	**Mon chapeau propre,** *my clean hat*
Un honnête homme, *an honest man*	**Un homme honnête,** *a polite man*

177. Sometimes the position of the adjective is determined by euphony, a short adjective preceding and a long adjective following the substantive.

The adjectives enumerated below normally precede the substantive:—

beau, *beautiful, fine*	**mauvais,** *bad*
bon, *good*	**méchant,** *wicked*
cher, *dear*	**meilleur,** *better*
digne, *worthy*	**moindre,** *less*
grand, *great, tall*	**petit,** *small*
gros, *big*	**saint,** *holy*
haut, *high*	**sot,** *foolish*
jeune, *young*	**vaste,** *vast*
joli, *pretty*	**vieux,** *old*
long, *long*	**vilain,** *ugly*

Study the following examples:—

Le nerf optique	*The optic nerve*
Le grand physicien allemand	*The great German physicist*
Des rayons jaunes et verts	*Yellow and red rays*
Un joli petit singe africain	*A pretty little African monkey*

DEPENDENT CLAUSES

178. The verb in a dependent clause may be in (*a*) the indicative, (*b*) the subjunctive, (*c*) the conditional.

179. The indicative is used when the idea is that of certainty, fact, or even probability; the subjunctive when the idea is one of assumption, doubt, or desire.

C'est un fait certain que je suis ici	*It is an indisputable fact that I am here*
Que pensez-vous de l'idée que nous passions nos vacances en Suisse ?	*What do you think of the idea of our spending our holidays in Switzerland?*
Il faut faire tomber de l'eau goutte à goutte sur le carbure, de manière que le dégagement gazeux se puisse régler	*You must pour some water drop by drop on to the carbide so that the evolution of the gas can be regulated*
Il dégagea le gaz de manière que le professeur était content	*He disengaged the gas in such a way that the teacher was pleased*

Obs. In the last two examples the indicative implies "with the result that," the subjunctive "with the intention that."

180. Si, *if* (conditional), takes the indicative, except when the condition is represented as unfulfilled in the past, in which case the pluperfect subjunctive may be used for the pluperfect indicative.

THE SUBJUNCTIVE

181. The subjunctive is used in French dependent clauses according to certain definite rules, usually after verbs denoting doubt, wishing, feeling, etc. It is almost always translated by the indicative in English, the one regular exception being final clauses expressing purpose. In these English also uses the subjunctive or else, very frequently, a simple present infinitive.

Faites ceci afin qu'on soit content de vous	*Do this in order that you may give (in order to give) satisfaction*
Je désire qu'il le fasse	*I desire that he does it* (or, *I wish him to do it*)
Il ne croit pas que nous ayons raison	*He does not believe that we are right*

THE CONDITIONAL

182. The conditional is used with **quand même.**

Quand même le problème serait difficile, je l'essayerais	*Even if the problem were difficult, I should attempt it*

183. Si, *whether,* introducing an indirect question may be followed by any tense, including the conditional.

Il se demandait si elle y réussirait	*He wondered whether she would succeed in the attempt*

THE INFINITIVE

184. The infinitive, being by nature a verbal substantive, can be used—

(i) as the subject of the verb **être** and some impersonal verbs;

(ii) without a preposition as the direct object of certain verbs;

(iii) with the prepositions **à** and **de** after certain verbs and adjectives;

(iv) after the prepositions **après, par, pour, sans;** in this case the infinitive is generally translated by the English gerund or verbal substantive in *-ing*.

(i)	**Travailler n'est pas toujours un plaisir**	*Work* (or *Working*) *is not always a pleasure*
	Il ferait beau voir cela	*It would be a fine thing to see that*
(ii)	**Je désire rester ici**	*I wish to stay here*
	A peine sait-il lire	*He can hardly read*
(iii)	**Il consentit à nous montrer sa bibliothèque**	*He consented to show us his library*
	Il est difficile à convaincre	*He is difficult to convince*
	Le temps est venu de me reposer	*The time has come for me to rest*
	Je serai heureux de connaître un tel homme	*I shall be happy to know such a man*
(iv)	**Après avoir dit cela, il se tut**	*After saying that, he was silent*
	Il est trop paresseux pour travailler	*He is too lazy to work*

185. The infinitive with **à** is also used with some substantives, and often corresponds to the English verbal substantive in *-ing* used as an adjective.

La pierre à bâtir	*Building stone*
Une salle à manger	*A dining-room*
Un cheval à vendre	*A horse for sale*

186. The infinitive is preferable to a subordinate clause introduced by a conjunction whenever the subject of that clause would be the same as that of the principal clause. Contrast—

Il partit sans me voir	*He departed without seeing me*
Je partis sans qu'il me vît	*I left without his seeing me*
Je n'irai pas à moins de ne pas être accompagné	*I will not go unless unaccompanied*

187. De is often used with the infinitive—

(i) as the mark of a genitive,

(ii) as a mere link,

(iii) with verbs of *entreating*, *permitting*, etc.

Je me souvins d'avoir vu cet homme	*I remembered having seen that man*
Il est mieux d'échouer que de n'avoir pas essayé	*It is better to fail than not to have made an attempt*
Il m'a prié d'y aller	*He begged me to go there*

THE PRESENT PARTICIPLE AND GERUND (OR VERBAL)

188. The verbal form in **-ant** may be used as a participle or as a gerund. In both cases it is invariable. The latter is usually preceded by the preposition **en,** which often represents the English *in*, *by*, or *while*.

L'argent cristallise, en formant ce qu'on nomme l'arbre de Diane	*Silver crystallises, forming what is called the* arbor Dianae
En employant cette formule on constate la quantité de déplacement	*By using this formula we ascertain the amount of displacement*
Au fond, se détachant à l'horizon, on voyait quelques dômes étincelant au soleil	*In the distance, standing out on the horizon, could be seen a few domes sparkling in the sun*
En entrant il salua toute la compagnie	*As he came in, he greeted all those present*

189. Many present participles may be used, as in English, as pure adjectives.

La neige étincelante	*The sparkling snow*
Des ressemblances frappantes	*Striking resemblances*

THE PAST PARTICIPLE

190. The past participle used as an adjective agrees with its substantive in gender and number.

L'argent déplacé cristallise	*The silver displaced crystallises*

Rules have already been given as to the agreement or invariability of the past participle when conjugated with **avoir** (§ 26) and when conjugated with **être** in intransitive (§ 27), passive (§ 58), and reflexive (§ 66) verbs.

191. The past participle of impersonal verbs is always invariable.

Les chaleurs qu'il a fait ce mois-ci ont été insupportables	*The heat we have had this month has been unbearable*
Malgré les grands frais qu'il leur a fallu faire, ils gagneront beaucoup d'argent	*In spite of the great expenses to which they have been put, they will make a good deal of money*

Obs. English idiom seldom admits of a literal rendering of a French impersonal expression.

CONJUNCTIONS

192. Conjunctions are of two classes—

(i) Co-ordinating conjunctions, which connect two phrases or clauses without making one subordinate to the other. These are—

et, *and*	**ni,** *nor*
car, *for*	**ou,** *or*
mais, *but*	**que,** *than*

soit . . . soit, *either . . . or*

(ii) Subordinating conjunctions, which make the clause they introduce subordinate to some other clause. Such conjunctions are—

(*a*) simple;

(*b*) compound.

193. There are only four simple subordinating conjunctions—

que, *that, than*	**si,** *if*
quand, *when*	**comme,** *as*

194. All the other subordinating conjunctions are compounds of **que.** Some of these take the indicative, some the subjunctive.

A. Subordinating Conjunctions followed by the Indicative

à mesure que, *in proportion as*
ainsi que, *as*
après que, *after*
aussitôt que, *as soon as*
dès que, *as soon as*
depuis que, *since*
lorsque, *when*
parce que, *because*
pendant que, *while*
tandis que, *while*
puisque, *since*
selon que, *according as*

B. Subordinating Conjunctions followed by the Subjunctive

The following is a list of all the conjunctions in frequent use that take the subjunctive:—

afin que, **pour que** } *in order that*
de façon que, **de manière que**, **de sorte que** } *so that*
pourvu que, *provided that*
avant que, *before*
quoique, **bien que** } *although*
sans que, *without that*
non que, *not that*
jusqu'à ce que, *until*
où que, *wherever*
supposé que, *supposing*
à moins que . . . (ne), *unless*
quelque . . . que, *however*
que (to avoid repetition of another conjunction), *if*

C. Subordinating Conjunctions followed by the Indicative or Subjunctive (§ 179)

de façon que, **de manière que**, **de (or en) sorte que** } *in such a way that*
tellement que, *so that*

PART III

IRREGULAR SUBSTANTIVES AND ADJECTIVES
COMPARISON OF ADJECTIVES AND ADVERBS
TABLE OF NUMERALS
PARADIGMS OF REGULAR AND IRREGULAR VERBS

IRREGULAR PLURALS

195. The rule that substantives and adjectives form their plural by the addition of **-s** to the singular is subject to certain exceptions, the more important of which are given below.

196. If the singular ends in **-s, -x,** or **-z,** the same form serves also for the plural; *e.g.* **os,** *bone* or *bones;* **gros** (masc.), *big;* **prix,** *prize* or *prizes;* **gaz,** *gas* or *gases.*

197. Words ending in **-au** or **-eu** and some ending in **-ou** add **-x** to form the plural, and words ending in **-al** change **-al** into **-aux.**

Singular	Plural
un beau château, *a fine castle*	**de beaux châteaux**
le métal principal, *the chief metal*	**les métaux principaux**

198. The following plurals are irregular:—

bleu	*blue*	**bleus**
le travail	*the work*	**les travaux**
l'œil	*the eye*	**les yeux**
ciel	*the sky*	**les cieux**

PLURAL OF COMPOUND SUBSTANTIVES

199. The guiding principle in the formation of the plural of compound substantives is that of the several elements in a compound only substantives and adjectives can take the sign of the plural.

There are six methods of composition—

1. Substantive + substantive. Both take the sign of the plural.

Singular	Plural
un chou-fleur, *a cauliflower*	**des choux-fleurs**
un oiseau-mouche, *a humming-bird*	**des oiseaux-mouches**

2. Substantive + adjective. Both take the sign of the plural.

Singular	Plural
un cerf-volant, *a kite*	**des cerfs-volants**

3. Substantive + preposition + substantive. The first or determining substantive alone takes the sign of the plural.

Singular	Plural
un ver-à-soie, *a silkworm*	**des vers-à-soie**
un arc-en-ciel, *a rainbow*	**des arcs-en-ciel**

4. Invariable word + substantive. The substantive only takes the sign of the plural.

Singular	Plural
une arrière-pensée, *an afterthought*	**des arrière-pensées**

5. Verb + substantive. Here the substantive is the object of the verb, which may be considered as an imperative. Consequently neither component takes the sign of the plural.

Singular	Plural
une tire-bouchon, *a cork-screw*	**des tire-bouchon**

6. Miscellaneous compounds with no syntactical relation between their components.

Singular	Plural
un passe-partout, *a master-key*	**des passe-partout**

COMPARISON

200. The comparative of adjectives is formed by prefixing the adverb **plus,** *more,* to the positive.

	Singular	Plural	
Masc.	**plus grand**	**plus grands**	*greater, larger*
Fem.	**plus grande**	**plus grandes**	

Moins, *less,* can be prefixed in the same way.

201. *Than* in a comparison expressing quality is translated by **que,** but in a comparison involving number **de** is used.

L'or est plus pesant que l'argent, et moins pesant que le platine	*Gold is heavier than silver and less heavy than platinum*
Cette plaque a moins d'un centimètre d'épaisseur	*That plate is less than a centimetre thick*

202. To form the "comparative of equality" prefix the adverb **aussi** to the adjective. The second *as* (conjunction) is translated by **que.**

Le platine est aussi rare que l'or	*Platinum is as rare as gold*

203. The superlative is formed by prefixing the definite article **le, la,** or **les** to the comparative.

	Singular	Plural	
Masc.	**le plus grand**	**les plus grands**	*the greatest*
Fem.	**la plus grande**	**les plus grandes**	

La télégraphie sans fil est la plus grande découverte de notre époque	*Wireless telegraphy is the greatest invention of our age*
Les aborigènes d'Australie sont la race la moins civilisée de la terre	*The aborigines of Australia are the least civilised race in the world*

204. Adverbs are compared in the same way as adjectives, except that the superlative of an adverb is preceded always by the article **le** (invariable).

Ils ont agi plus courageusement que vous	*They have acted more courageously than you*
C'est elle qui a agi le plus courageusement	*It is she who has acted most courageously*

Note—

Plus on a, plus on veut	*The more we have, the more we want*
Ni moi non plus	*Nor I either*

205. Three adjectives, with their corresponding adverbs, have an irregular comparative.

Positive	Comparative	Superlative
bon, *good*	**meilleur**	**le meilleur**
mauvais, *bad*	**pire** *or* **plus mauvais**	**le pire** *or* **le plus mauvais**
petit, *little*	**moindre** *or* **plus petit**	**le moindre** *or* **le plus petit**

Positive	Comparative	Superlative
bien, *well*	**mieux**	**le mieux**
mal, *badly*	**pis** *or* **plus mal**	**le pis** *or* **le plus mal**
peu, *little*	**moins**	**le moins**

IRREGULAR INFLECTION OF ADJECTIVES

206. Adjectives that end in **-e** in the masculine singular do not change for the feminine.

Masculine	Feminine	
belge	**belge**	*Belgian*
jeune	**jeune**	*young*

207. Adjectives ending in **-er** and some adjectives in **-et** take the grave accent over the last **e** but one in the feminine.

premier	**première**	*first*
complet	**complète**	*complete*

208. Adjectives ending in **-c** in the masculine generally change the **c** into **que** for the feminine.

public	**publique**	*public*

But—

blanc	**blanche**	*white*
grec	**grecque**	*Greek*

209. Adjectives in **-f** and **-x** change this into **-ve** and **-se** respectively in the feminine.

actif	**active**	*active*
heureux	**heureuse**	*happy*

But—

doux	**douce**	*sweet*

210. Adjectives in **-el, -eil, -en, -on,** and some in **-et** double the final consonant before the **e** mute of the feminine.

cruel	**cruelle**	*cruel*
pareil	**pareille**	*like*
ancien	**ancienne**	*ancient*
bon	**bonne**	*good*
muet	**muette**	*dumb*

211. Other Irregular Adjectives

Masculine	Feminine	
beau, bel	**belle**	*beautiful, fine*
dissous	**dissoute**	*dissolved*
épais	**épaisse**	*thick*
exprès	**expresse**	*express*
fou, fol	**folle**	*mad*
frais	**fraîche**	*fresh, cool*
jumeau	**jumelle**	*twin*
long	**longue**	*long*
mou, mol	**molle**	*soft*
nouveau, nouvel	**nouvelle**	*new*
nul	**nulle**	*no*
vieux, vieil	**vieille**	*old*

Obs. In the case of the five adjectives which have two forms in the masculine singular, **beau, fou, mou, nouveau, vieux** are used when the following substantive begins with a consonant, **bel, fol, mol, nouvel, vieil** when it begins with a vowel or **h** mute.

Un bel été	*A beautiful summer*
Un beau jour	*A fine day*

212. For irregular plurals see §§ 196-198.

ADVERBS

213. Adverbs of manner are formed by the addition of **-ment** to the feminine of the adjective.

Masc.	Fem.		Adverb
doux	**douce**	*gentle*	**doucement,** *gently*
frais	**fraîche**	*fresh*	**fraîchement,** *freshly*
heureux	**heureuse**	*happy*	**heureusement,** *happily*

214. If the adjective ends in a vowel, **-ment** is added directly to the masculine.

sage-ment, *wisely* **poli-ment,** *politely* **absolu-ment,** *absolutely*

215. Adjectives ending in **-ant, -ent,** change **-nt** into **-mment.**

courant, *fluent*	**couramment,** *fluently*
savant, *learned*	**savamment,** *learnedly*
récent, *recent*	**récemment,** *lately*

216. From the following adjectives adverbs are formed irregularly:—

bref	**brièvement**	*briefly*
commode	**commodément**	*conveniently*
commun	**communément**	*commonly*
conforme	**conformément**	*conformably*
confus	**confusément**	*confusedly*
continu	**continûment**	*continuously*
diffus	**diffusément**	*diffusedly*
énorme	**énormément**	*enormously*
exprès	**expressément**	*expressly*
gentil	**gentiment**	*prettily, nicely*
immense	**immensément**	*immensely*
lent	**lentement**	*slowly*
obscur	**obscurément**	*obscurely*
précis	**précisément**	*precisely*
présent	**présentement**	*presently*
profond	**profondément**	*deeply*
traître	**traitreusement**	*treacherously*
uniforme	**uniformément**	*uniformly*

217. Some adjectives are used without change of form as adverbs.

parler bas, *to speak low*	**marcher vite,** *to walk quickly*
parler haut, *to speak loud*	**chanter faux,** *to sing out of tune*

TABLE OF NUMERALS

Cardinals and Ordinals

218. The cardinals and ordinals are given in the following tables:—

	Cardinals	Ordinals
1	**un** (masc.), **une** (fem.)	**premier** (masc.), **première** (fem.)
2	**deux**	**second** (masc.), **seconde** (fem.) deuxième *
3	**trois**	troisième
4	**quatre**	**quatrième**
5	**cinq**	**cinquième**
6	**six**	sixième *or* sizième †
7	**sept**	septième
8	**huit**	huitième
9	**neuf**	**neuvième**
10	**dix**	dixième *or* dizième
11	**onze**	onzième
12	**douze**	douzième
13	**treize**	treizième
14	**quatorze**	quatorzième
15	**quinze**	quinzième
16	**seize**	seizième
17	**dix-sept**	dix-septième
18	**dix-huit**	dix-huitième
19	**dix-neuf**	dix-neuvième
20	**vingt**	vingtième
21	**vingt et un**	**vingt et unième**
22	**vingt-deux**	**vingt-deuxième**
23	vingt-trois	vingt-troisième
24	vingt-quatre	vingt-quatrième
25	vingt-cinq	vingt-cinquième
26	vingt-six	vingt-sixième
27	vingt-sept	vingt-septième
28	vingt-huit	vingt-huitième
29	vingt-neuf	vingt-neuvième

* **Second** denotes *second* of two; **deuxième,** *second* among a larger number.

† To be distinguished from **seizième.**

	CARDINALS	ORDINALS
30	**trente**	trentième
31	trente et un	trente et unième
32	trente-deux	trente-deuxième
40	**quarante**	quarantième
41	quarante et un	quarante et unième
42	quarante-deux	quarante-deuxième
50	**cinquante**	cinquantième
51	cinquante et un	cinquante et unième
52	cinquante-deux	cinquante-deuxième
60	**soixante**	soixantième
61	soixante et un	soixante et unième
62	soixante-deux	soixante-deuxième
70	**soixante-dix**	soixante-dixième
71	**soixante et onze**	soixante et onzième
72	**soixante-douze**	soixante-douzième
73	**soixante-treize**	soixante-treizième
74	**soixante-quatorze**	soixante-quatorzième
75	**soixante-quinze**	soixante-quinzième
76	**soixante-seize**	soixante-seizième
77	**soixante-dix-sept**	soixante-dix-septième
78	**soixante-dix-huit**	soixante-dix-huitième
79	**soixante-dix-neuf**	soixante-dix-neuvième
80	**quatre-vingts**	quatre-vingtième
81	**quatre-vingt-un**	quatre-vingt-unième
82	quatre-vingt-deux	quatre-vingt-deuxième
90	**quatre-vingt-dix**	quatre-vingt-dixième
91	**quatre-vingt-onze**	quatre-vingt-onzième
92	**quatre-vingt-douze**	quatre-vingt-douzième
99	**quatre-vingt-dix-neuf**	quatre-vingt-dix-neuvième
100	**cent**	centième
101	**cent un**	**cent unième**
102	cent deux	cent deuxième
200	**deux cents**	**deux centième**
300	trois cents	trois centième
500	cinq cents	cinq centième
580	cinq cent quatre-vingts	cinq cent quatre-vingtième
586	cinq cent quatre-vingt-six	cinq cent quatre-vingt-sixième
1000	**mille**	millième
2000	**deux mille**	deux millième
10,000	dix mille	dix millième

PARADIGMS OF VERBS

THE FOUR REGULAR CONJUGATIONS

219. French verbs are divided into four conjugations according to the endings of the infinitive, which are (1) **-er,** (2) **-ir,** (3) **-oir,** (4) **-re.**

220. From the five **principal parts** of a French verb all the other parts may be formed. The principal parts are (1) the **present infinitive;** (2) the **present participle** stem, *i.e.* the stem left after cutting off the ending **-ant;** (3) the **past participle;** (4) the **present indicative** (first person singular); (5) the **past definite indicative** (first person singular).

221. The various parts of a verb are thus formed, with the addition of the suffixes shown in the paradigms.

Indicative	Formed from	To form the first person add
Present, 2nd and 3rd pers. sing.	present indicative (1st sing.)	
,, plural	present participle stem	**-ons**
Imperfect	,, ,, ,,	**-ais**
Past definite (throughout)	past definite (1st sing.)	
Future	infinitive	**-ai**
Subjunctive		
Present (except in the 3rd conj.)	present participle stem	**-e**
Imperfect	past definite (2nd sing.)	**-se**
Conditional, present	infinitive	**-ais**
Imperative, present (except 3rd person)	present indicative	

Note.—In forming the **future indicative** and **present conditional** from the infinitive, **oi** must be dropped in the case of verbs of the third conjugation and **e** in the case of verbs of the fourth

REGULAR VERBS

222. FIRST CONJUGATION (INFINITIVE IN -ER)

MODEL: **porter,** *to carry.*

INDICATIVE MOOD

PRESENT

je **port-e**
tu port-**es**
il port-**e**

nous port-**ons**
vous port-**ez**
ils port-**ent**

IMPERFECT

je **port-ais**
tu port-**ais**
il port-**ait**

nous port-**ions**
vous port-**iez**
ils port-**aient**

PAST DEFINITE

je **port-ai**
tu port-**as**
il port-**a**

nous port-**âmes**
vous port-**âtes**
ils port-**èrent**

FUTURE

je **port-er-ai**
tu port-**er-as**
il port-**er-a**

nous port-**er-ons**
vous port-**er-ez**
ils port-**er-ont**

CONDITIONAL PRESENT

je **port-er-ais**
tu port-**er-ais**
il port-**er-ait**

nous port-**er-ions**
vous port-**er-iez**
ils port-**er-aient**

IMPERATIVE MOOD

port-e
(qu'il port-**e**)

port-**ons**
port-**ez**
(qu'ils port-**ent**)

SUBJUNCTIVE MOOD

PRESENT

je **port-e**
tu port-**es**
il port-e

nous port-**ions**
vous port-**iez**
ils port-**ent**

IMPERFECT

je **port-asse**
tu port-**asses**
il port-**ât**

nous port-**assions**
vous port-**assiez**
ils port-**assent**

INFINITIVE PRESENT: **port-er,** *to carry.*

PARTICIPLE PRESENT: **port-ant,** *carrying.*

PARTICIPLE PAST (PASSIVE): **port-é,** *carried.*

Examples of the Use of the Simple Tenses

il porte ma valise,	*he (generally) carries my bag.* *he is (now) carrying my bag.*
il portait ma valise,	*he carried my bag (at various times).* *he was carrying my bag (at that time).*
il porta ma valise,	*he carried my bag (on one occasion).*
il portera ma valise,	*he will carry my bag (to-morrow).*
il porterait ma valise si je le lui demandais,	*he would carry my bag if I were to ask him* or *if I asked him.*
j'espérais qu'il porterait ma valise,	*I was hoping that he would carry my bag,* i.e. *in the past I hoped for what was then future.*
portez-la pour moi,	*carry it for me.*
il faut que vous la portiez,	*you must carry it.*
je marchais vite, quoique je portasse une grande valise,	*I was walking quickly, although I was carrying a big bag.*

223. SECOND CONJUGATION (INFINITIVE IN **-IR**)

Model: **finir,** *to finish*

INDICATIVE MOOD

Present

je **fin-is**
tu fin-**is**
il fin-**it**
nous fin-**iss-ons**
vous fin-**iss-ez**
ils fin-**iss-ent**

Imperfect

je **fin-iss-ais**
tu fin-**iss-ais**
il fin-**iss-ait**
nous fin-**iss-ions**
vous fin-**iss-iez**
ils fin-**iss-aient**

Past Definite

je **fin-is**
tu fin-**is**
il fin-**it**
nous fin-**îmes**
vous fin-**îtes**
ils fin-**irent**

Future

je **fin-ir-ai**
tu fin-**ir-as**
il fin-**ir-a**
nous fin-**ir-ons**
vous fin-**ir-ez**
ils fin-**ir-ont**

Conditional Present

je **fin-ir-ais**
tu fin-**ir-ais**
il fin-**ir-ait**
nous fin-**ir-ions**
vous fin-**ir-iez**
ils fin-**ir-aient**

IMPERATIVE MOOD

fin-is
(qu'il fin-iss-**e**)
fin-iss-ons
fin-iss-ez
(qu'ils fin-**iss-ent**)

SUBJUNCTIVE MOOD

PRESENT	IMPERFECT
je **fin-iss-e**	je **fin-iss-e**
tu fin-**iss-es**	tu fin-**iss-es**
il fin-**iss-e**	il **fin-ît**
nous fin-**iss-ions**	nous fin-**iss-ions**
vous fin-**iss-iez**	vous fin-**iss-iez**
ils fin-**iss-ent**	ils fin-**iss-ent**

INFINITIVE PRESENT: **fin-ir,** *to finish.*
PARTICIPLE PRESENT: **fin-iss-ant,** *finishing.*
PARTICIPLE PAST (PASSIVE): **fini,** *finished.*

EXAMPLES OF THE USE OF THE SIMPLE TENSES

je finis mon déjeuner à huit heures,	*I finish my breakfast at eight o'clock.*
maintenant je finis mon dîner,	*now I am finishing my dinner.*
notre classe finissait à midi,	*our class used to finish at twelve.*
à ce moment-là je finissais le livre,	*at that moment I was finishing the book.*
il finit sa vie à l'âge de soixante-dix ans,	*he ended his life at the age of seventy.*
nous finirons notre travail demain,	*we shall finish our work to-morrow.*
je croyais que vous le finiriez plus tôt,	*I thought you would finish it sooner.*
si vous ne disiez rien, nous le finirions plus tôt,	*if you said* (or *if you were to say*) *nothing, we should finish it sooner.*
finissons-le tout de suite,	*let us finish it at once.*
il faut que nous finissions la partie,	*we must finish the game.*
ils demandaient qu'il finît à temps,	*they asked that he should finish in good time.*

224. THIRD CONJUGATION (INFINITIVE IN -OIR)

MODEL: **recevoir,** *to receive*

INDICATIVE MOOD

PRESENT	FUTURE
je **reç-ois**	je **recevr-ai**
tu reç-**ois**	tu recevr-**as**
il reç-**oit**	il recevr-**a**
nous **recev-ons**	nous recevr-**ons**
vous recev-**ez**	vous recevr-**ez**
ils **reçoiv-ent**	ils recevr-**ont**

IMPERFECT

je **recev-ais**
tu recev-**ais**
il recev-**ait**

nous recev-**ions**
vous recev-**iez**
ils recev-**aient**

PAST DEFINITE

je **reç-us**
tu reç-**us**
il reç-**ut**

nous reç-**ûmes**
vous reç-**ûtes**
ils reç-**urent**

CONDITIONAL PRESENT

je **recevr-ais**
tu recevr-**ais**
il recevr-**ait**

nous recevr-**ions**
vous recevr-**iez**
ils recevr-**aient**

IMPERATIVE MOOD

reç-ois
(qu'il reç-**oive**)

recev-ons
recev-**ez**
(qu'ils **reç-oivent**)

SUBJUNCTIVE MOOD

PRESENT

je **reçoiv-e**
tu reçoiv-es
il reçoiv-e

nous **recev-ions**
vous recev-**iez**
ils **reçoiv-ent**

IMPERFECT

je **reç-usse**
tu reç-**usses**
il reç-**ût**

nous reç-**ussions**
vous reç-**ussiez**
ils reç-**ussent**

INFINITIVE PRESENT: **recev-oir,** *to receive.*

PARTICIPLE PRESENT: **recev-ant,** *receiving.*

PARTICIPLE PAST (PASSIVE): **reç-u,** *received.*

EXAMPLES OF THE USE OF THE SIMPLE TENSES

je reçois des lettres tous les jours,	*I receive letters every day.*
je reçois ce cadeau en vous remerciant,	*I receive this present with thanks.*
ma mère recevait le jeudi,	*my mother used to be at home on Thursdays.*
je recevais l'argent quand il entra,	*I was receiving the money when he came in.*
il reçut un coup mortel,	*he was mortally wounded.*
vous recevrez le colis demain,	*you will receive the package to-morrow.*
je croyais que vous le recevriez aujourd'hui,	*I thought you would receive it to-day.*
si je l'expédiais demain, vous le recevriez jeudi,	*if I sent* (or *were to send*) *it off to-morrow, you would receive it on Thursday.*

recevez-le pour moi,	*receive it for me.*
il ne pense pas que je reçoive tant d'argent,	*he does not think that I receive so much money.*
il fallait que je reçusse les visiteurs,	*I had to receive the visitors.*

NOTE.—Verbs conjugated like **recevoir** are—

devoir (past participle **dû**), *to have to.*
s'apercevoir, *to perceive.*
concevoir, *to conceive.*
décevoir, *to deceive, to disappoint.*
percevoir, *to levy* (taxes).

The other verbs of this conjugation, about fifteen in number, are in one way or another irregular (§§ 262-272).

225. FOURTH CONJUGATION (INFINITIVE IN -RE)

MODEL: **rendre,** *to render, give back*

INDICATIVE MOOD

PRESENT	FUTURE
je **rend-s**	je **rendr-ai**
tu rend-**s**	tu rendr-**as**
il rend*	il rendr-**a**
nous rend-**ons**	nous rendr-**ons**
vous rend-**ez**	vous rendr-**ez**
ils rend-**ent**	ils rendr-**ont**
IMPERFECT	CONDITIONAL PRESENT
je **rend-ais**	je **rendr-ais**
tu rend-**ais**	tu rendr-**ais**
il rend-**ait**	il rendr-**ait**
nous rend-**ions**	nous rendr-**ions**
vous rend-**iez**	vous rendr-**iez**
ils rend-**aient**	ils rendr-**aient**
PAST DEFINITE	IMPERATIVE MOOD
je **rend-is**	**rend-s**
tu rend-**is**	(qu'il rend-**e**)
il rend-**it**	
nous rend-**îmes**	**rend-ons**
vous rend-**îtes**	**rend-ez**
ils rend-**irent**	(qu'ils rend-**ent**)

* Verbs of which the stem does not end in *d* or *t* add *t*; *e.g.* **rompre, il rompt.**

SUBJUNCTIVE MOOD

PRESENT	IMPERFECT
je **rend-e**	je **rend-isse**
tu rend-**es**	tu rend-**isses**
il rend-**e**	il rend-**ît**
nous rend-**ions**	nous rend-**issions**
vous rend-**iez**	vous rend-**issiez**
ils rend-**ent**	ils rend-**issent**

INFINITIVE PRESENT: **rend-re,** *to render, give back.*

PARTICIPLE PRESENT: **rend-ant,** *rendering, giving back.*

PARTICIPLE PAST (PASSIVE): **rend-u,** *rendered, given back.*

EXAMPLES OF THE USE OF THE SIMPLE TENSES

il vous rend le salut,	*he is returning your greeting.*
il rend son compte tous les mois,	*he sends in his account every month.*
à cette heure il me rendait visite,	*at this time of day he used to call on me.*
il me rendait visite quand ma femme entra,	*he was calling on me when my wife came in.*
le général rendit la forteresse,	*the general surrendered the fortress.*
je m'y rendrai dans une heure,	*I shall go there in an hour's time.*
je m'y rendrais, s'il m'invitait,	*I shall go there, if he were to invite me* (or *if he invited me*).
il croyait que vous lui rendriez justice,	*he believed you would do him justice.*
rendez-moi mon argent,	*give me back my money.*
il est impossible que je vous le rende,	*it is impossible for me to give it you back.*
il était impossible que je vous le rendisse,	*it was impossible for me to give it you back.*

QUASI-IRREGULAR VERBS OF THE FIRST CONJUGATION

226. A number of verbs of the first conjugation present certain peculiarities in the formation of some tenses, but as these peculiarities result from the laws of French pronunciation and follow fixed rules, such verbs cannot be classed as irregular.

Obs. In the tables of conjugation those tenses only are given in which some peculiarity occurs.

227. These quasi-irregular verbs fall into five classes—

I.—Verbs with **e** mute (not followed by **l** or **t**) in the last syllable of the stem.

II.—Verbs of which the present infinitive ends in **-eler** or **-eter.**

III.—Verbs with close **e** (**é**) in the last syllable of the stem.

IV.—Verbs of which the present infinitive ends in **-yer.**

V.—Verbs of which the present infinitive ends in **-ger** or **-cer.**

I.—VERBS WITH E MUTE (NOT FOLLOWED BY L OR T) IN THE LAST SYLLABLE OF THE STEM

228. It is a rule in French that mute **e** cannot occur in two successive syllables. Hence if the last syllable of the verb-stem has **e** mute, as in **men-er, appel-er,** and the ending also has **e** mute (*e.g.* **-e, -ent, -erai**), the **e** mute of the stem takes the sound of open **e,** and this is indicated either by a grave accent (§ 229) or by the doubling of the final consonant of the stem (§ 230); *e.g.* **je mène, ils mèneront, tu appelles, vous appelleriez.**

Obs. Theoretically **e** mute makes a syllable; thus **mène** consists of two syllables, **appelle** of three.

229. MODEL: men-er, *to lead*

PRINCIPAL PARTS: **men-er, men-ant, men-é, je mèn-e, je men-ai**

INDICATIVE, PRESENT	je **mèn-e** nous men-ons	tu **mèn-es** vous men-ez	il **mèn-e** ils **mèn-ent**
INDICATIVE, FUTURE	je **mèn-erai** nous **mèn-erons**	tu **mèn-eras** vous **mèn-erez**	il **mèn-era** ils **mèn-eront**
CONDITIONAL, PRESENT	je **mèn-erais** nous **mèn-erions**	tu **mèn-erais** vous **mèn-eriez**	il **mèn-erait** ils **mèn-eraient**
IMPERATIVE, PRESENT	 men-ons	**mèn-e** men-ez	
SUBJUNCTIVE, PRESENT	je **mène** nous menions	tu **mèn-es** vous meniez	il **mèn-e** ils **mèn-ent**

II.—VERBS OF WHICH THE PRESENT INFINITIVE ENDS IN -ELER OR -ETER

230. Verbs ending in **-eler, -eter,** usually **double the consonant l** or **t** before **e** mute instead of taking the grave accent.

MODEL: **appel-er,** *to call*

PRINCIPAL PARTS: **appel-er, appel-ant, appel-é, j'appel-l-e, j'appel-ai**

INDICATIVE, PRESENT	**j'appel-l-e** nous appel-ons	tu **appel-l-es** vous appel-ez	il **appel-l-e** ils **appel-l-ent**
INDICATIVE, FUTURE	**j'appel-l-erai** nous **appel-l-erons**	tu **appel-l-eras** vous **appel-l-erez**	il **appel-l-era** ils **appel-l-eront**
CONDITIONAL, PRESENT	**j'appel-l-erais** nous **appel-l-erions**	tu **appel-l-erais** vous **appel-l-eriez**	il **appel-l-erait** ils **appel-l-eraient**
SUBJUNCTIVE, PRESENT	**j'appel-l-e** nous appel-ions	tu **appel-l-es** vous appel-iez	il **appel-l-e** ils **appel-l-ent**
IMPERATIVE, PRESENT	appel-ons	**appel-l-e** appel-ez	

231. MODEL: **jet-er,** *to throw*

PRINCIPAL PARTS: **jet-er, jet-ant, jet-é, je jet-t-e, je jet-ai**

INDICATIVE, PRESENT	je **jet-t-e** nous jet-ons	tu **jet-t-es** vous jet-ez	il **jet-t-e** ils **jet-t-ent**
INDICATIVE, FUTURE	je **jet-t-erai** nous **jet-t-erons**	tu **jet-t-eras** vous **jet-t-erez**	il **jet-t-era** ils **jet-t-eront**
CONDITIONAL, PRESENT	je **jet-t-erais** nous **jet-t-erions**	tu **jet-t-erais** vous **jet-t-eriez**	il **jet-t-erait** ils **jet-t-eraient**
SUBJUNCTIVE, PRESENT	je **jet-t-e** nous jet-ions	tu **jet-t-es** vous jet-iez	il **jet-t-e** ils **jet-t-ent**
IMPERATIVE, PRESENT	jet-ons	**jet-t-e** jet-ez	

III.—VERBS WITH CLOSE E (É) IN THE LAST SYLLABLE OF THE STEM

232. Verbs with **é** in the last syllable of the stem have a **grave accent,** instead of an acute, before **e** mute, except in the future indicative and present conditional, where the acute accent is retained.

233. Model: **préfér-er,** *to prefer*

Principal Parts: **préfér-er, préfér-ant, préfér-é, je préfér-e, je préfér-ai**

Indicative, Present	je **préfèr-e** nous préfér-ons	tu **préfèr-es** vous préfér-ez	il **préfèr-e** ils **préfèr-ent**
Indicative, Future	je préférer-ai nous préférer-ons	tu préférer-as vous préférer-ez	il préférer-a ils préférer-ont
Conditional, Present	je préférer-ais nous préférer-ions	tu préférer-ais vous préférer-iez	il préférer-ait ils préférer-aient
Subjunctive, Present	je **préfèr-e** nous préfér-ions	tu **préfèr-es** vous préfér-iez	il **préfèr-e** ils **préfèr-ent**
Imperative, Present	préfér-ons	**préfèr-e** préfér-ez	

Obs. The future indicative and present conditional in this class of verbs present no irregularity.

IV.—VERBS OF WHICH THE PRESENT INFINITIVE ENDS IN -YER

234. Verbs of the first conjugation of which the stem ends in **y** change this letter to **i** before **e** mute.

235. Model: **ploy-er,** *to bend*

Principal Parts: **ploy-er, ploy-ant, ploy-é, je ploi-e, je ploy-ai**

Indicative, Present	je **ploi-e** nous ploy-ons	tu **ploi-es** vous ploy-ez	il **ploi-e** ils **ploi-ent**
Indicative, Future	je **ploi-erai** nous **ploi-erons**	tu **ploi-eras** vous **ploi-erez**	il **ploi-era** ils **ploi-eront**
Conditional, Present	je **ploi-erais** nous **ploi-erions**	tu **ploi-erais** vous **ploi-eriez**	il **ploi-erait** ils **ploi-eraient**
Subjunctive, Present	je **ploi-e** nous ploy-ions	tu **ploi-es** vous ploy-iez	il **ploi-e** ils **ploi-ent**
Imperative, Present	ploy-ons	**ploi-e** ploy-ez	

Note.—Verbs in **-uyer** are conjugated like **ployer,** *e.g.* **essuyer,** *to wipe,* **j'essuie, j'essuierai, j'essuierais, j'essuie.** Verbs in **-ayer** usually retain the **y,** but may change it to **i.** Thus, of **payer,** *to pay,* the present indicative is **je paye** or **paie,** and the future indicative is **je payerai** or **paierai.**

V.—VERBS OF WHICH THE PRESENT INFINITIVE ENDS IN -GER OR -CER

236. By the rules of French pronunciation the letters **g** and **c** are hard before **a, o,** and **u,** and soft before **e** and **i.** To preserve the soft sound in verbs ending in **-ger** and **-cer,** an **e** is inserted after the **g** and a **cedilla** is put under the **c** (ç) whenever these letters are followed by **a** or **o** (**u** does not occur in the endings of this conjugation).

237. Model: **mang-er,** *to eat*

Principal Parts: **mang-er, mang-e-ant, mang-é, je mang-e, je mang-e-ai**

Indicative, Present	je mang-e	tu mang-es	il mang-e
	nous **mang-e-ons**	vous mang-ez	ils mang-ent
Indicative, Imperfect	je **mang-e-ais**	tu **mang-e-ais**	il **mang-e-ait**
	nous mang-ions	vous mang-iez	ils **mang-e-aient**
Indicative, Past Definite	je **mang-e-ai**	tu **mang-e-as**	il **mang-e-a**
	nous **mang-e-âmes**	vous **mang-e-âtes**	ils mang-èrent
Subjunctive, Past	je **mang-e-asse**	tu **mang-e-asses**	il **mang-e-ât**
	nous **mang-e-assions**	vous **mang-e-assiez**	ils **mang-e-assent**
Imperative, Present		mang-e	
	mang-e-ons	mang-ez	

Model: **trac-er,** *to trace*

Principal Parts: **trac-er, traç-ant, trac-é, je trac-e, je traç-ai**

Indicative, Present	je trac-e	tu trac-es	il trac-e
	nous **traç-ons**	vous trac-ez	ils trac-ent
Indicative, Imperfect	je **traç-ais**	tu **traç-ais**	il **traç-ait**
	nous trac-ions	vous trac-iez	ils **traç-aient**
Indicative, Past Definite	je **traç-ai**	tu **traç-as**	il **traç-a**
	nous **traç-âmes**	vous **traç-âtes**	ils trac-èrent
Subjunctive, Past	je **traç-asse**	tu **traç-asses**	il **traç-ât**
	nous **traç-assions**	vous **traç-assiez**	ils **traç-assent**
Imperative, Present		trace-e	
	traçons	trac-ez	

Obs. All verbs in **-ger** and **-cer,** without exception, are thus conjugated.

IRREGULAR VERBS

238. In the following tables the irregular verbs are arranged in four conjugations, and are grouped as far as possible according to points of similarity.

239. Four tenses are omitted altogether from the tables, viz. the imperfect indicative and imperfect subjunctive, the present conditional and the present imperative. These four tenses (as was stated in § 221) can always be formed as follows:—

The **imperfect indicative** is found by changing the final **-ant** of the present participle into **-ais.**

The **imperfect subjunctive** is found by adding **-se** to the second person singular of the past definite.

The **present conditional** is found by adding **-s** to the first person singular of the future indicative.

The **imperative** is identical with the corresponding persons of the present indicative, omitting **-s** in the second person singular of the first conjugation.

The above tenses are conjugated regularly in all persons and both numbers.

NOTE.—**Avoir, être,** and **savoir** are the only verbs which do not follow the above rules. For **vouloir** see § 272.

240. The **future** indicative is usually formed according to rule from the present infinitive, but is too frequently irregular to be omitted from the tables in this chapter; when regular it is not printed in dark type.

241. In addition to the four tenses mentioned in § 239 the **past definite** and **future** indicative can always be regularly conjugated throughout when the first person singular is known. The only tenses which can be irregular in conjugation are the present indicative and the present subjunctive. These are therefore always given in full in the tables, as also is the past definite in a few cases in which peculiar collocations of letters occur. Those parts of the present subjunctive and of the plural of the present indicative which are regularly formed from the present participle (§ 221) are not printed in dark type.

VERBS OF THE FIRST CONJUGATION

Inf.	Participles	Pres. Ind.	
242. **all-er** *to go*	 all-ant all-é	**vais** **vas** **va**	all-ons all-ez **vont**
243. **envoy-er** *to send*	 envoy-ant envoy-é	**envoi-e** **envoi-es** **envoi-e**	envoy-ons envoy-ez **envoi-ent**

VERBS OF THE SECOND CONJUGATION

244. **acquér-ir** *to acquire*	 **acquér-ant** **acquis**	**acquier-s** **acquier-s** **acquier-t**	acquér-ons acquér-ez **acquièr-ent**
245. **assaill-ir** *to assail*	 **assaill-ant** assaill-i	**assaill-e** **assaill-es** **assaill-e**	assaill-ons assaill-ez assaill-ent
246. **bouill-ir,** *to boil* (intrans.)	 **bouill-ant** bouill-i	**bou-s** **bou-s** **bou-t**	bouill-ons bouill-ez bouill-ent
247. **cour-ir** *to run*	 **cour-ant** **cour-u**	**cour-s** **cour-s** **cour-t**	cour-ons cour-ez cour-ent
248. **couvr-ir** *to cover*	 **couvr-ant** **couver-t**	**couvr-e** **couvr-es** **couvr-e**	couvr-ons couvr-ez couvr-ent
249. **cueill-ir** *to pluck*	 **cueill-ant** cueill-i	**cueill-e** **cueill-es** **cueill-e**	cueill-ons cueill-ez cueill-ent

Compounds, etc.

§ **242. s'en aller,** *to go away.*

§ **243. renvoyer,** *to send back.*

§ **244. conquérir,** *to conquer;* **s'enquérir,** *to inquire;* **requérir,** *to require.*

§ **245. tressaillir,** *to start, to shudder.*

§ **246. ébouillir,** *to boil away;* **rebouillir,** *to boil again; to boil* (transitive) is rendered by **faire bouillir.**

VERBS OF THE FIRST CONJUGATION

PAST DEF.	FUTURE	PRES. SUBJ.	
all-ai	**i-rai**	**aill-e** **aill-es** **aill-e**	all-ions all-iez **aill-ent**
envoy-ai	**enver-rai**	**envoi-e** **envoi-es** **envoi-e**	envoy-ions envoy-iez **envoi-ent**

VERBS OF THE SECOND CONJUGATION

acqu-is	**acquer-rai**	**acquièr-e** **acquièr-es** **acquièr-e**	acquér-ions acquér-iez **acquièr-ent**
assaill-is	assaill-irai	assaill-e assaill-es assaill-e	assaill-ions assaill-iez assaill-ent
bouill-is	bouill-irai	bouill-e bouill-es bouill-e	bouill-ions bouill-iez bouill-ent
cour-us	**cour-rai**	cour-e cour-es cour-e	cour-ions cour-iez cour-ent
couvr-is	couvr-irai	couvr-e couvr-es couvr-e	couvr-ions couvr-iez couvr-ent
cueill-is	**cueill-erai**	cueill-e cueill-es cueill-e	cueill-ions cueill-iez cueill-ent

COMPOUNDS, ETC.

§ 247. **accourir,** *to hasten;* **concourir,** *to concur;* **discourir,** *to discourse;* **parcourir,** *to traverse;* **secourir,** *to help.*

§ 248. **découvrir,** *to discover;* **recouvrir,** *to cover up again;* **offrir,** *to offer;* **ouvrir,** *to open;* **souffrir,** *to suffer.*

§ 249. **accueillir,** *to welcome;* **recueillir,** *to reap;* **se recueillir,** *to collect one's thoughts.*

Inf.	Participles	Pres. Ind.	
250. **dorm-ir** *to sleep*	 **dorm-ant** **dorm-i**	**dor-s** **dor-s** **dor-t**	dorm-ons dorm-ez dorm-ent
251. **fu-ir** *to flee*	 **fuy-ant** fu-i	fu-is fu-is fu-it	fuy-ons fuy-ez **fui-ent**
252. **ment-ir** *to lie, to tell a lie*	 **ment-ant** ment-i	**men-s** **men-s** **men-t**	ment-ons ment-ez ment-ent
253. **mour-ir** *to die*	 **mour-ant** **mort**	**meur-s** **meur-s** **meu-t**	mour-ons mour-ez **meur-ent**
254. **part-ir** *to depart*	 **part-ant** part-i	**par-s** **par-s** **par-t**	part-ons part-ez part-ent
255. **se repent-ir** *to repent*	 **se repent-ant** repent-i	**me repen-s** **te repen-s** **se repen-t**	nous repent-ons vous repent-ez se repent-ent
256. **sent-ir** *to feel, to smell*	 **sent-ant** sent-i	**sen-s** **sen-s** **sen-t**	sent-ons sent-ez sent-ent
257. **serv-ir** *to serve*	 **serv-ant** serv-i	**ser-s** **ser-s** **ser-t**	serv-ons serv-ez serv-ent
258. **sort-ir** *to go out*	 **sort-ant** sort-i	**sor-s** **sor-s** **sor-t**	sort-ons sort-ez sort-ent
259. **ten-ir** *to hold*	 **ten-ant** **ten-u**	**tien-s** **tien-s** **tien-t**	ten-ons ten-ez **tien-n-ent**

Compounds, etc.

§ **250. endormir,** *to lull to sleep;* **s'endormir,** *to fall asleep;* **se rendormir,** *to go to sleep again.*

§ **251. s'enfuir,** *to flee.*

§ **252. démentir,** *to give the lie.*

§ **254. départir,** *to divide;* **repartir,** *to reply, to start again.*

§ **256. consentir,** *to consent;* **pressentir,** *to forbode;* **ressentir,** *to resent;* **se ressentir,** *to feel.*

PAST DEF.	FUTURE	PRES. SUBJ.	
dorm-is	dorm-irai	dorm-e dorm-es dorm-e	dorm-ions dorm-iez dorm-ent
fu-is	fu-irai	**fui-e** **fui-es** **fui-e**	fuy-ions fuy-iez **fui-ent**
ment-is	ment-irai	ment-e ment-es ment-e	ment-ions ment-iez ment-ent
mour-us	**mour-rai**	**meur-e** **meur-es** **meur-e**	mour-ions mour-iez **meur-ent**
part-is	part-irai	part-e part-es part-e	part-ions part-iez part-ent
me repent-is	me repent-irai	me repent-e te repent-es se repent-e	nous repent-ions vous repent-iez se repent-ent
sent-is	sent-irai	sent-e sent-es sent-e	sent-ions sent-iez sent-ent
serv-is	serv-irai	serv-e serv-es serv-e	serv-ions serv-iez serv-ent
sort-is	sort-irai	sort-e sort-es sort-e	sort-ions sort-iez sort-ent
tin-s*	**tien-d-rai**	**tien-n-e** **tien-n-es** **tien-n-e**	ten-ions ten-iez **tien-n-ent**

* PAST DEF. of **tenir: tin-s, tin-s, tin-t, tîn-mes, tîn-tes, tin-rent.**

COMPOUNDS, ETC.

§ 257. **desservir,** *to clear the table.*

§ 258. **ressortir,** *to go out again.*

§ 259. **s'abstenir,** *to abstain;* **appartenir,** *to belong;* **contenir,** *to contain;* **détenir,** *to detain;* **entretenir,** *to keep up;* **maintenir,** *to maintain;* **obtenir,** *to obtain;* **retenir,** *to retain;* **soutenir,** *to assert.*

Inf.	Participles	Pres. Ind.	
260. **ven-ir** *to come*	 **ven-ant** **ven-u**	**vien-s** **vien-s** **vien-t**	ven-ons ven-ez **vien-n-ent**
261. **vêt-ir** *to clothe*	 **vêt-ant** **vêt-u**	**vêt-s** **vêt-s** **vêt**	vêt-ons vêt-ez vêt-ent

VERBS OF THE THIRD CONJUGATION

262. **asse-oir** *to seat, to set*	 **assey-ant** **assis**	**assied-s** **assied-s** **assied**	assey-ons assey-ez assey-ent
263. **dev-oir** *to owe*	 **dev-ant** **dû***	**doi-s** **doi-s** **doi-t**	dev-ons dev-ez **doiv-ent**
264. **fall-oir** *to be necessary*	**fall-u**	**faut**	
265. **mouv-oir** *to move*	 **mouv-ant** **mû†**	**meu-s** **meu-s** **meu-t**	mouv-ons mouv-ez **meuv-ent**
266. **pleuv-oir,** *to rain* (impersonal)	 **pleuv-ant** **pl-u**	**pleu-t**	
267. **pouv-oir** *to be able*	 **pouv-ant** **pu**	**peu-x** or **pui-s** **peu-x** **peu-t**	pouv-ons pouv-ez **peuv-ent**
268. **recev-oir** *to receive*	 **recev-ant** **reç-u**	**reçoi-s** **reçoi-s** **reçoi-t**	recev-ons recev-ez **reçoiv-ent**

* Past Participle of **devoir:** sing. f. due, pl. m. dus, pl. f. dues.

† Past Participle of **mouvoir:** sing. f. mue, pl. m. mus, pl. f. mues.

Compounds, etc.

§ 260. **circonvenir,** *to circumvent;* **convenir,** *to suit, to agree;* **devenir,** *to become;* **intervenir,** *to intervene;* **parvenir,** *to reach;* **revenir,** *to return;* **se souvenir de,** *to remember.*

§ 261. **dévêtir,** *to divest ;* **revêtir,** *to cover.*

PAST DEF.	FUTURE	PRES. SUBJ.	
vin-s*	**vien-d-rai**	**vien-n-e** **vien-n-es** **vien-n-e**	ven-ions ven-iez **vien-n-ent**
vêt-is	vêt-irai	vêt-e vêt-es vêt-e	vêt-ions vêt-iez vêt-ent

VERBS OF THE THIRD CONJUGATION

ass-is	**assié-rai** or **assey-erai**	assey-e assey-es assey-e	assey-ions assey-iez assey-ent
d-us	dev-rai	**doiv-e** **doiv-es** **doiv-e**	dev-ions dev-iez **doiv-ent**
fall-ut	**fau-d-ra**	**faill-e**	
m-us	mouv-rai	**meuv-e** **meuv-es** **meuv-e**	mouv-ions mouv-iez **meuv-ent**
pl-ut	pleuv-ra	pleuv-e	
p-us	**pour-rai**	**puiss-e** **puiss-es** **puiss-e**	**puiss-ions** **puiss-iez** **puiss-ent**
reç-us	recev-rai	**reçoi-ve** **reçoi-ves** **reçoiv-e**	recev-ions recev-iez **reçoiv-ent**

* PAST DEF. of **venir: vin-s, vin-s, vin-t, vîn-mes, vîn-tes, vin-rent.**

COMPOUNDS, ETC.

§ 262. **s'asseoir,** *to sit down;* **rasseoir,** *to put back;* **se rasseoir,** *to sit down again.*

§ 263. **redevoir,** *to owe still.*

§ 268. **apercevoir,** *to perceive;* **concevoir,** *to conceive;* **décevoir,** *to deceive, disappoint.*

INF.	PARTICIPLES	PRES. IND.	
269.*		**sai-s**	**sav-ons**
sav-oir	**sach-ant**	**sai-s**	**sav-ez**
to know	**su**	**sai-t**	**sav-ent**
270.		**vau-x**	val-ons
val-oir	**val-ant**	**vau-x**	val-ez
to be worth	**val-u**	**vau-t**	val-ent
271.		**voi-s**	voy-ons
v-oir	**voy-ant**	**voi-s**	voy-ez
to see	**vu**	**voi-t**	**voi-ent**
272.†		**veu-x**	voul-ons
voul-oir	**voul-ant**	**veu-x**	voul-ez
to wish	**voul-u**	**veu-t**	**veul-ent**

VERBS OF THE FOURTH CONJUGATION

273.	**absol-v-ant**	**absou-s**	absol-v-ons
absou-d-re	**absou-s**	**absou-s**	absol-v-ez
to absolve	(f. **absou-te**)	**absou-t**	absol-v-ent
274.		**bat-s**	batt-ons
batt-re	batt-ant	**bat-s**	batt-ez
to beat	batt-u	**bat**	batt-ent
275.		boi-s	bu-v-ons
boi-re	**bu-v-ant**	boi-s	bu-v-ez
to drink	**bu**	boi-t	**boi-v-ent**
276.		conclu-s	conclu-ons
conclu-re	conclu-ant	conclu-s	conclu-ez
to conclude	**conclu**	conclu-t	conclu-ent

* IMPERATIVE of **savoir: sache, sachons, sachez.**

† IMPERATIVE of **vouloir: veuille, veuillez.**

COMPOUNDS, ETC.

§ **270. équivaloir,** *to be equivalent;* **prévaloir,** *to prevail,* has pres. subj. **prévale.**

§ **271. entrevoir,** *to catch a glimpse of;* **revoir** *to see again;* **prévoir,**

PAST DEF.	FUTURE	PRES. SUBJ.	
s-us	**sau-rai**	sach-e sach-es sach-e	sach-ions sach-iez sach-ent
val-us	**vau-d-rai**	**vail-l-e** **vail-l-es** **vail-l-e**	val-ions val-iez **vail-l-ent**
v-is	**ver-rai**	**voi-e** **voi-es** **voi-e**	voy-ions voy-iez **voi-ent**
voul-us	**vou-d-rai**	**veuil-l-e** **veuil-l-es** **veuil-l-e**	voul-ions voul-iez **veuil-l-ent**

VERBS OF THE FOURTH CONJUGATION

	absou-d-rai	absol-v-e absol-v-es absol-v-e	absol-v-ions absol-v-iez absol-v-ent
batt-is	batt-rai	batt-e batt-es batt-e	batt-ions batt-iez batt-ent
b-us	boi-rai	**boi-v-e** **boi-v-es** **boi-v-e**	bu-v-ions bu-v-iez **boi-v-ent**
concl-us	conclu-rai	conclu-e conclu-es conclu-e	conclu-ions conclu-iez conclu-ent

COMPOUNDS, ETC.

to foresee, and **pourvoir,** *to provide,* have fut. **-voirai,** and **pourvoir** has past def. **pourvus.**

§ **273. dissoudre,** *to dissolve.*

§ **274. se battre,** *to fight;* **abattre,** *to beat down;* **combattre,** *to fight;* **débattre,** *to dispute.*

§ **276. exclure,** *to exclude.*

INF.	PARTICIPLES	PRES. IND.	
277.		condui-s	condui-s-ons
condui-re	**condui-s-ant**	condui-s	condui-s-ez
to conduct	**condui-t**	condui-t	condui-s-ent
278.		confi-s	confi-s-ons
confi-re	**confi-s-ant**	confi-s	confi-s-ez
to pickle	**confi-t**	confi-t	confi-s-ent
279.		**connai-s**	connai-ss-ons
connaî-t-re	**connai-ss-ant**	**connai-s**	connai-ss-ez
to know	**conn-u**	**connaî-t**	connai-ss-ent
280.		cou-d-s	cou-s-ons
cou-d-re	**cou-s-ant**	cou-d-s	cou-s-ez
to sew	**cou-s-u**	**cou-d**	cou-s-ent
281.		**crain-s**	craign-ons
crain-d-re	**craign-ant**	**crain-s**	craign-ez
to fear	**crain-t**	**crain-t**	craign-ent
282.		croi-s	croy-ons
croi-re	**croy-ant**	croi-s	croy-ez
to believe	**cru**	croi-t	**croi-ent**
283.		**croî-s**	croi-ss-ons
croî-t-re	**croi-ss-ant**	**croî-s**	croi-ss-ez
to grow	**crû***	**croî-t**	croi-ss-ent
284.		cui-s	cui-s-ons
cui-re	**cui-s-ant**	cui-s	cui-s-ez
to cook	**cui-t**	cui-t	cui-s-ent
285.		di-s	di-s-ons
di-re	**di-s-ant**	di-s	**di-t-es**
to say	**di-t**	di-t	di-s-ent

* PAST PARTICIPLE of **croître:** s. f. crue, pl. m. crus, pl. f. crues.

COMPOUNDS, ETC.

§ **277. déduire,** *to deduct;* **éconduire,** *to show out;* **introduire,** *to intro-uce;* **produire,** *to produce;* **réduire,** *to reduce;* **séduire,** *to seduce;* **raduire,** *to translate.*

§ **278. déconfire,** *to discomfit.*

§ **279. méconnaître,** *not to recognise;* **reconnaître,** *to recognise.*

PAST DEF.	FUTURE	PRES. SUBJ.	
condui-s-is	condui-rai	condui-s-e condui-s-es condui-s-e	condui-s-ions condui-s-iez condui-s-ent
conf-is	confi-rai	confi-s-e confi-s-es confi-s-e	confi-s-ions confi-s-iez confi-s-ent
conn-us	connaî-t-rai	connai-ss-e connai-ss-es connai-ss-e	connai-ss-ions connai-ss-iez connai-ss-ent
cou-s-is	cou-d-rai	cou-s-e cou-s-es cou-s-e	cou-s-ions cou-s-iez cou-s-ent
craign-is	crain-d-rai	craign-e craign-es craign-e	craign-ions craign-iez craign-ent
cr-us	croi-rai	**croi-e** **croi-es** **croi-e**	croy-ions croy-iez **croi-ent**
cr-ûs	croî-t-rai	croi-ss-e croi-ss-es croi-ss-e	croi-ss-ions croi-ss-iez croi-ss-ent
cui-s-is	cui-rai	cui-s-e cui-s-es cui-s-e	cui-s-ions cui-s-iez cui-s-ent
di-s	di-rai	di-s-e di-s-es di-s-e	di-s-ions di-s-iez di-s-ent

COMPOUNDS, ETC.

§ **280. découdre,** *to unsew:* **recoudre,** *to sew again.*

§ **281. plaindre,** *to pity;* **se plaindre,** *to complain;* **contraindre,** *to compel.*

§ **282. mécroire,** *to disbelieve.*

§ **284. recuire,** *to cook again.*

§ **285. redire,** *to say again.* Other compounds have pres. ind. 2nd person pl. in **-disez;** *e.g.* from **médire,** *to slander,* **vous médisez.**

Inf.	Participles	Pres. Ind.	
286. **écri-re** *to write*	 **écri-v-ant** **écri-t**	écri-s écri-s écri-t	écri-v-ons écri-v-ez écri-v-ent
287. **fai-re** *to make, to do*	 **fai-s-ant** **fai-t**	fai-s fai-s fai-t	fai-s-ons **fai-t-es** **font**
288. **instrui-re** *to instruct*	 **instrui-s-ant** **instrui-t**	instrui-s instrui-s instrui-t	instrui-s-ons instrui-s-ez instrui-s-ent
289. **join-d-re** *to join*	 **joign-ant** **join-t**	**join-s** **join-s** **join-t**	joign-ons joign-ez joign-ent
290. **li-re** *to read*	 **li-s-ant** **lu**	li-s li-s li-t	li-s-ons li-s-ez li-s-ent
291. **maudi-re** *to curse*	 **maudi-ss-ant** **maudi-t**	maudi-s maudi-s maudi-t	maudi-ss-ons maudi-ss-ez maudi-ss-ent
292. **mett-re** *to put*	 mett-ant **mis**	**met-s** **met-s** **met**	mett-ons mett-ez mett-ent
293. **mou-d-re** *to grind*	 **mou-l-ant** **mou-l-u**	mou-d-s mou-d-s **mou-d**	mou-l-ons mou-l-ez mou-l-ent
294. **naî-t-re** *to be born*	 **nai-ss-ant** **né**	**nai-s** **nai-s** **naî-t**	nai-ss-ons nai-ss-ez nai-ss-ent

Compounds, etc.

§ **287. contrefaire,** *to feign;* **défaire,** *to undo;* **se défaire de,** *to get rid of;* **refaire,** *to do again;* **satisfaire,** *to satisfy;* **surfaire,** *to overcharge.*

§ **288. construire,** *to construct;* **détruire,** *to destroy.*

§ **289. conjoindre,** *to unite;* **déjoindre,** *to separate;* **enjoindre,** *to enjoin;* **rejoindre,** *to rejoin;* **oindre,** *to anoint.*

Past Def.	Future	Pres. Subj.	
écri-v-is	écri-rai	écri-v-e écri-v-es écri-v-e	écri-v-ions écri-v-iez écri-v-ent
f-is	**fe-rai**	**fa-ss-e** **fa-ss-es** **fa-ss-e**	**fa-ss-ions** **fa-ss-iez** **fa-ss-ent**
instrui-s-is	instrui-rai	instrui-s-e instrui-s-es instrui-s-e	instrui-s-ions instrui-s-iez instrui-s-ent
joign-is	join-d-rai	joign-e joign-es joign-e	joign-ions joign-iez joign-ent
l-us	li-rai	li-s-e li-s-es li-s-e	li-s-ions li-s-iez li-s-ent
maud-is	maudi-rai	maudi-ss-e maudi-ss-es maudi-ss-e	maudi-ss-ions maudi-ss-iez maudi-ss-ent
m-is	mett-rai	mett-e mett-es mett-e	mett-ions mett-iez mett-ent
moul-us	mou-d-rai	mou-l-e mou-l-es mou-l-e	mou-l-ions mou-l-iez mou-l-ent
naqu-is	naî-t-rai	nai-ss-e nai-ss-es nai-ss-e	nai-ss-ions nai-ss-iez nai-ss-ent

Compounds, etc.

§ **290. relire,** *to read again.*

§ **292. admettre,** *to admit;* **commettre,** *to commit;* **compromettre,** *to compromise;* **omettre,** *to omit;* **permettre,** *to permit;* **soumettre,** *to subdue;* **transmettre,** *to transmit.*

§ **293. émoudre,** *to sharpen;* **rémoudre,** *to re-sharpen;* **remoudre,** *to grind again.*

INF.	PARTICIPLES	PRES. IND.	
295. **nui-re** *to harm*	 **nui-s-ant** **nui**	nui-s nui-s nui-t	nui-s-ons nui-s-ez nui-s-ent
296. **paraî-t-re** *to appear*	 **parai-ss-ant** **par-u**	**parai-s** **parai-s** **paraî-t**	parai-ss-ons parai-ss-ez parai-ss-ent
297. **pein-d-re** *to paint*	 **peign-ant** **pein-t**	**pein-s** **pein-s** **pein-t**	peign-ons peign-ez peign-ent
298. **plai-re** *to please*	 **plai-s-ant** **plu**	plai-s plai-s **plaî-t**	plai-s-ons plai-s-ez plai-s-ent
299. **prend-re** *to take*	 **pren-ant** **pris**	prend-s prend-s **prend**	pren-ons pren-ez **pren-n-ent**
300.* **résou-d-re** *to resolve*	 **résol-v-ant**	**résou-s** **résou-s** **résou-t**	résol-v-ons résol-v-ez résol-v-ent
301. **ri-re** *to laugh*	 **ri-ant** **ri**	**ri-s** ri-s ri-t	**ri-ons** ri-ez ri-ent
302. **suffi-re** *to suffice*	 **suffi-s-ant** **suffi**	suffi-s suffi-s suffi-t	suffi-s-ons suffi-s-ez suffi-s-ent
303. **suiv-re** *to follow*	 **suiv-ant** **suiv-i**	**sui-s** **sui-s** **sui-t**	suiv-ons suiv-ez suiv-ent
304. **tai-re** *to keep secret*	 **tai-s-ant** **tu**	tai-s tai-s tai-t	tai-s-ons tai-s-ez tai-s-ent

* PAST PART.: **résou-s** (*dissolved*), f. **résou-te**; **résol-u** (*determined*).

COMPOUNDS, ETC.

§ **295. luire,** *to shine;* **reluire,** *to gleam;* these have no past definite.

§ **296. apparaître,** *to appear;* **comparaître,** *to appear in court;* **disparaître,** *to disappear.*

§ **297. atteindre,** *to reach;* **astreindre,** *to compel;* **ceindre,** *to gird;* **enfreindre,** *to infringe;* **éteindre,** *to extinguish;* **feindre,** *to feign;* **restreindre,** *to restrict;* **teindre,** *to dye.*

PAST DEF.	FUTURE	PRES. SUBJ.	
nui-s-is	nui-rai	nui-s-e nui-s-es nui-s-e	nui-s-ions nui-s-iez nui-s-ent
par-us	paraî-t-rai	parai-ss-e parai-ss-es parai-ss-e	parai-ss-ions parai-ss-iez parai-ss-ent
peign-is	pein-d-rai	peign-e peign-es peign-e	peign-ions peign-iez peign-ent
pl-us	plai-rai	plai-s-e plai-s-es plai-s-e	plai-s-ions plai-s-iez plai-s-ent
pr-is	prend-rai	**pren-n-e** **pren-n-es** **pren-n-e**	pren-ions pren-iez **pren-n-ent**
résol-us	résou-d-rai	résol-v-e résol-v-es résol-v-e	résol-v-ions résol-v-iez résol-v-ent
r-is	ri-rai	ri-e ri-es ri-e	ri-ions ri-iez ri-ent
suff-is	suffi-rai	suffi-s-e suffi-s-es suffi-s-e	suffi-s-ions suffi-s-iez suffi-s-ent
suiv-is	suiv-rai	suiv-e suiv-es suiv-e	suiv-ions suiv-iez suiv-ent
tu-s	tai-rai	tai-s-e tai-s-es tai-s-e	tai-s-ions tai-s-iez tai-s-ent

COMPOUNDS, ETC.

§ **298. complaire,** *to humour;* **déplaire,** *to displease.*

§ **299. apprendre,** *to learn;* **comprendre,** *to understand;* **entreprendre,** *to undertake;* **se méprendre,** *to be mistaken;* **reprendre,** *to resume;* **surprendre,** *to surprise.*

§ **301. sourire,** *to smile.*

§ **303. s'ensuivre** (impersonal), *to result;* **poursuivre,** *to pursue.*

§ **304. se taire,** *to be silent.*

Inf.	Participles	Pres. Ind.	
305. **trai-re** *to milk*	**tray-ant** **trai-t**	trai-s trai-s trai-t	tray-ons tray-ez **trai-ent**
306. **vainc-re** *to conquer*	**vainqu-ant** **vainc-u**	vainc-s vainc-s **vainc**	vainqu-ons vainqu-ez vainqu-ent
307. **viv-re** *to live*	**viv-ant** **véc-u**	**vi-s** **vi-s** **vi-t**	viv-ons viv-ez viv-ent

Compounds

§ 305. **abstraire,** *to abstract;* **distraire,** *to distract;* **extraire,** *to extract;* **soustraire,** *to subtract.*

---

308. CONJUGATION OF AVOIR, *TO HAVE*

INDICATIVE MOOD

Present

j'**ai**,	*I have*
tu **as**,	*thou hast*
il **a**, elle **a**,	*he* (or *she*) *has*
nous **av-ons**,	*we have*
vous **av-ez**,	*you have*
ils **ont**, elles **ont**,	*they have*

Imperfect

j'**av-ais**,	*I had*
tu av-**ais**,	*thou hadst*
il av-**ait**,	*he had*
nous av-**ions**,	*we had*
vous av-**iez**,	*you had*
ils av-**aient**,	*they had*

Past Definite

j'**eu-s**,	*I had*
tu eu-**s**,	*thou hadst*
il eu-**t**,	*he had*
nous **eû-mes**,	*we had*
vous **eû-tes**,	*you had*
ils eu-**rent**,	*they had*

Future

j'**aur-ai**,	*I shall have*
tu aur-**as**,	*thou wilt have*
il aur-**a**,	*he will have*
nous aur-**ons**,	*we shall have*
vous aur-**ez**,	*you will have*
ils aur-**ont**,	*they will have*

Conditional Present

j'**aur-ais**,	*I should*
tu aur-**ais**,	*thou wouldst*
il aur-**ait**,	*he would*
nous aur-**ions**,	*we should*
vous aur-**iez**,	*you would*
ils aur-**aient**,	*they would*

} *have*

IMPERATIVE MOOD

ai-e,	*have*
(qu'il **ai-t**),	*let him have*
ay-ons,	*let us have*
ay-ez,	*have*
(qu'ils **ai-ent**),	*let them have*

PAST DEF.	FUTURE	PRES. SUBJ.	
	trai-rai	**trai-e** **trai-es** **trai-e**	tray-ions tray-iez **trai-ent**
vainqu-is	vainc-rai	vainqu-e vainqu-es vainqu-e	vainqu-ions vainqu-iez vainqu-ent
véc-us	viv-rai	viv-e viv-es viv-e	viv-ions viv-iez viv-ent

COMPOUNDS

§ 306. **convaincre,** *to convince.*

§ 307. **revivre,** *to revive;* **survivre,** *to survive.*

SUBJUNCTIVE MOOD

PRESENT	IMPERFECT
j'**ai-e**	j'**eu-sse**
tu ai-**es**	tu eu-**sses**
il ai-**t**	il eû-**t**
nous **ay-ons**	nous eu-**ssions**
vous **ay-ez**	vous eu-**ssiez**
ils **ai-ent**	ils eu-**ssent**

INFINITIVE PRESENT: **av-oir,** *to have.*

PARTICIPLE PRESENT: **ay-ant,** *having.*

PARTICIPLE PAST (PASSIVE): **eu,** *had.*

309. CONJUGATION OF ÊTRE, *TO BE*

INDICATIVE MOOD

Present

je **suis**,	*I am*
tu **es**,	*thou art*
il **est**,	*he is*
nous **sommes**,	*we are*
vous **êtes**,	*you are*
ils **sont**,	*they are*

Imperfect

j'ét-**ais**,	*I was*
tu ét-**ais**,	*thou wast*
il ét-**ait**,	*he was*
nous ét-**ions**,	*we were*
vous ét-**iez**,	*you were*
ils ét-**aient**,	*they were*

Past Definite

je **fu-s**,	*I was*
tu **fu-s**,	*thou wast*
il **fu-t**,	*he was*
nous **fû-mes**,	*we were*
vous **fû-tes**,	*you were*
ils **fu-rent**,	*they were*

Future

je **ser-ai**,	*I shall be*
tu ser-**as**,	*thou wilt be*
il ser-**a**,	*he will be*
nous ser-**ons**,	*we shall be*
vous ser-**ez**,	*you will be*
ils ser-**ont**,	*they will be*

Conditional Present

je **ser-ais**	*I should be*
tu ser-**ais**,	*thou wouldst be*
il ser-**ait**,	*he would be*
nous ser-**ions**,	*we should be*
vous ser-**iez**,	*you would be*
ils ser-**aient**,	*they would be*

IMPERATIVE MOOD

soi-s,	*be*
(qu'il **soi-t**),	*let him be*
soy-ons,	*let us be*
soy-ez,	*be*
(qu'ils **soi-ent**),	*let them be*

SUBJUNCTIVE MOOD

Present

je **soi-s**
tu soi-s
il soi-**t**
nous **soy-ons**
vous **soy-ez**
ils **soi-ent**

Imperfect

je **fu-sse**
tu fu-**sses**
il **fû-t**
nous fu-**ssions**
vous fu-**ssiez**
ils fu-**ssent**

Infinitive Present: **êt-re**, *to be.*

Participle Present: **ét-ant**, *being.*

Participle Past: **ét-é** (invariable), *been.*

SUBSTANTIVES WITH TWO GENDERS

310. The following is a list of common substantives, the meaning of which varies with the gender.

	Masculine	Feminine
carpe	*wrist*	*carp*
caustique	*caustic substance*	*caustic curve*
faune	*faun*	*fauna*
garde	*guardian*	*guardianship, the Guards (a regiment)*
greffe	*chancery*	*graft*
livre	*book*	*pound*
manche	*handle*	*sleeve, English Channel*
mémoire	*memoir, note*	*memory*
merci	*thanks*	*mercy*
mort	*dead man*	*death*
moule	*mould*	*mussel*
mousse	*cabin-boy*	*moss, froth*
page	*page-boy*	*page of a book*
parallèle	*comparison*	*parallel line*
pendule	*pendulum*	*clock*
période	*conclusion, culmination*	*period (in mathematics and punctuation)*
somme	*sleep*	*sum*
souris	*smile*	*mouse*
tour	*turn, tour, trick*	*tower*
vapeur	*steamer*	*heat*
vase	*vessel*	*mud*
voile	*veil*	*sail*

PART IV

PREMIÈRES LECTURES

1

Les saisons

Nous divisions l'année en quatre parties. Nous appelons ces parties les quatre saisons. Ce sont: le printemps, l'été, l'automne et l'hiver. Chaque saison dure environ trois mois, mais l'hiver est généralement plus long que l'été. Le printemps est une belle saison. La neige ne tombe plus et les fleurs poussent. Les oiseaux qui passent l'hiver dans les pays chauds arrivent en Angleterre et commencent à chanter. Sur les arbres, nous apercevons des feuilles vertes et l'été approche. En été les jours sont longs et les nuits courtes. Le soleil brille et ses rayons réchauffent la terre et font paraître les fleurs. Nous ne voyons plus de neige que sur les hautes montagnes où elle reste éternellement. Le mois d'août est habituellement le plus chaud de l'année. L'automne est la troisième saison. Le temps peut varier beaucoup en automne. Parfois nous avons un bel automne: les fruits mûrissent et les fleurs égaient les jardins. Les vignobles sont pleins de gens qui récoltent les raisins. En hiver, le temps froid arrive, le vent souffle, la pluie et la neige tombent. Beaucoup de gens aiment l'hiver. Les enfants patinent, glissent sur l'étang et font des boules de neige dans le jardin.

2

Le système métrique

Le mètre est l'unité des mesures de longueur. Le décimètre est la dixième partie du mètre. Le mètre est divisé en cent parties, et ces parties sont nommées centimètres. Dix mètres font un décamètre, cent mètres font un hectomètre et mille mètres un kilomètre. Le kilomètre est plus court que le mille anglais. Huit kilomètres correspondent environ à cinq milles. Un décimètre cube d'eau pure à 4° centigrade pèse un kilogramme. Le kilogramme a mille grammes, cent décagrammes et dix hectogrammes. Le gramme a dix décigrammes, cent centigrammes et mille milligrammes. Ces mesures sont très petites. Un kilogramme d'eau porte le nom de litre. Dix litres font un décalitre, cent litres un hectolitre et mille litres un kilolitre. Nous avons divisé le litre en décilitres, centilitres et millilitres. Nous employons aussi le demi-litre et le quart de litre. Toutes les mesures du système métrique sont très utiles, et généralement faciles à manier. Les anciennes mesures françaises ont été abandonnées. Plusieurs pays ont adopté le système métrique. Ce système a été inventé en France à la fin du dix-huitième siècle. L'Allemagne emploie le système métrique. L'Angleterre a des mesures spéciales, très compliquées comme les anciennes mesures françaises. Beaucoup de gens préfèrent le système métrique.

3

LE BŒUF ET LA VACHE

Le bœuf et la vache sont des animaux domestiques. Ce sont des mammifères et des ruminants. Le bœuf est un animal assez grand; il a le museau carré, le front plat, les yeux grands et doux. Il a le corps couvert de poils courts et une longue queue. Deux cornes pointues ornent son front. La vache est plus petite que le bœuf. Ces deux animaux sont bruns rougeâtres ou blancs. Ils mangent de l'herbe et d'autres végétaux. Anciennement on employait le bœuf pour battre le blé, mais à présent les machines l'ont remplacé. Les bœufs tiraient aussi la charrue, mais nous employons maintenant des chevaux. En Allemagne les bœufs doivent labourer et en France on emploie même quelquefois les vaches pour cela. De tous temps les hommes ont mangé la chair du bœuf. La viande de cet animal est bonne et nourrissante. De la vache nous recevons du lait. Ce lait est meilleur que le lait de chèvre. On en fait du beurre et du fromage. Dans certaines parties du monde on ne peut pas obtenir de lait. Alors on importe du lait condensé et ce lait est utilisé aussi par les exploreurs et sur mer. Les buffalos étaient autrefois très nombreux dans l'Amérique du Nord. Ce sont des animaux très forts. Leurs cornes sont plus longues que les cornes du bœuf. On a tué beaucoup de ces bœufs sauvages, et ils ont diminué énormément. Les anciens Egyptiens aimaient beaucoup les bœufs et les vaches. Ils les adoraient et leur bâtirent des temples. Ils eurent comme dieu un bœuf spécial qu'ils appelèrent Apis.

4

Le crocodile

Le crocodile est un reptile. C'est un amphibie, c'est-à-dire qu'il peut vivre dans l'eau et sur terre. Ses membres sont courts et ne peuvent pas supporter le poids énorme de son corps. C'est pourquoi le crocodile doit, pour ainsi dire, ramper sur la partie inférieure de son corps. Il y a trois sortes de crocodiles: le gavial, le crocodile proprement dit, et l'alligator. On trouve le gavial aux Indes et il peut avoir dix-sept pieds de longueur. Le crocodile habite certaines parties de l'Asie, de l'Afrique et de l'Amérique. Les alligators n'habitent que l'Amérique. Les crocodiles étaient autrefois abondants dans toute la vallée du Nil. Ils y étaient au temps des vieux Egyptiens, qui en firent des dieux. On trouve à Memphis des temples que ces gens leur érigèrent. Cependant, les crocodiles ont disparu du delta du Nil et ont remonté ce fleuve. Les hommes les ont chassés des endroits habités. Les naturels de l'Egypte mangent la chair du crocodile.

Les crocodiles aiment la boue et y passent la saison sèche. Quand la pluie arrive, ils quittent leur retraite. La femelle du crocodile place ses œufs dans le sable. La nourriture des jeunes crocodiles consiste exclusivement en poisson, mais les grands crocodiles mangent des animaux et même des hommes. Une fois, des voyageurs attaquèrent un crocodile qui était sur le sable. Le crocodile resta immobile et les voyageurs ne pensaient pas qu'il était encore en vie. Ils allaient le laisser. Mais à peine eurent-ils tourné la tête que le crocodile remua et alla chercher refuge dans une rivière voisine.

5

Le cheval

Si vous regardez un cheval, vous remarquerez que son poil est court et généralement d'une seule couleur. Mais parfois il est de couleurs différentes et alors le cheval a des taches, jamais de raies comme le zèbre. Vous pourrez observer que la queue de cet animal est longue et qu'il possède aussi une longue crinière souvent très épaisse. Si on place devant vous trois ou quatre chevaux, vous remarquerez qu'ils sont différents. Les uns sont d'une espèce qui est très grande; ces animaux sont très lourds et très forts; ce sont eux qui tirent les fardeaux les plus pesants. Si vous examinez ensuite un autre cheval vous trouverez qu'il est différent. Il est moins grand et semble moins fort; c'est cette espèce qui est attelée aux voitures. Le troisième cheval, qui est très élégant et a des jambes fines, sera un cheval de course. Son allure est très rapide.

Vous aurez peut-être décidé après avoir regardé ces chevaux que leur force et leur capacité varient selon leur aspect. Mais tous sont utiles à l'homme, qui les a rendus domestiques depuis les temps les plus reculés. Après avoir parlé de l'extérieur du cheval, on pourra vous dire quelque chose de ses habitudes. Il est herbivore et on le nourrit d'herbe, de foin et d'avoine. A l'état sauvage on trouve les chevaux en troupes. A l'état domestique le cheval a pour compagnons et amis d'autres animaux. Par exemple, on vous citera l'histoire d'un cheval dont l'ami était un petit chat noir et qui était tout heureux quand son camarade était perché sur son dos.

Vous aurez peut-être entendu parler de l'intelligence du cheval, laquelle est remarquable ainsi que sa mémoire. C'est un des animaux domestiques les plus intéressants et un des meilleurs amis de l'homme. Même après sa mort, il sera utile. Il nous donnera alors son poil, dont on remplit les coussins. Avec sa peau, on fera du cuir. On vendra sa chair pour la nourriture des chats, et sa graisse sera utilisée dans la fabrication du savon.

Le cheval existe encore à l'état sauvage, et en Tartarie on trouve encore de grandes troupes de chevaux. Les Tartares les attrapent pour les faire travailler et ils en mangent aussi la chair. Du lait de la jument ils font du fromage. Dans les prairies de l'Amérique on trouve des mustangs qui sont proches parents du cheval. Les plus beaux représentants de la race chevaline sont en Arabie; nos chevaux de course sont souvent les descendants de chevaux arabes.

6

L'AUTRUCHE

L'autruche est le plus grand des oiseaux actuels. Le mâle adulte a presque huit pieds de hauteur et pèse trois cents livres. C'est donc un oiseau fort majestueux. Les autruches sont moins nombreuses qu'autrefois, et dans un certain nombre d'années on n'en trouvera peut-être plus. Nos descendants auront sans doute des oiseaux plus petits et penseront que les nôtres étaient immenses.

On rencontre les autruches en sociétés de quatre ou cinq et parfois même en troupes de trente à cinquante. Elles ont de longues jambes, un long cou et une petite tête. Leur queue, qui est formée de longues plumes, est très belle. Elles ne peuvent pas voler, parce que leurs ailes ne le leur permettent pas, étant trop petites. Toutefois, elles courent avec une rapidité considérable.

Les femelles déposent leurs œufs sur le sable. Ces œufs sont naturellement très grands et ont une coquille épaisse et dure. Autour du nid, les autruches mettent d'autres œufs qu'elles cassent et que les petits oiseaux mangeront quand ils sortiront de l'œuf. Généralement, c'est le soleil qui joue le rôle de poule couveuse; mais quelquefois les autruches passent quelque temps sur le nid. Dans ce cas, le mâle y reste la nuit et les femelles pendant la journée.

On trouve ces oiseaux dans les régions désertes et surtout en Afrique. Cependant, ils ne peuvent rester sans eau comme les chameaux. Dans l'Afrique du Sud, il y a des fermes où on élève les autruches pour avoir leurs plumes.

Ces plumes, qui sont blanches, grises ou noires, sont l'objet d'un grand commerce.

Parmi les indigènes, l'autruche remplace parfois le cheval et on a vu des nègres ayant pour monture un de ces oiseaux. C'est un spectacle très drôle. On ne mange pas la chair de l'autruche, qui est coriace.

7

La pomme de terre

La pomme de terre est une plante herbacée très commune maintenant et que chacun a vue. On ne pourrait pas trouver en Europe une personne à qui cette plante est inconnue. Ses feuilles sont simples, vert-foncé, arrondies, et de grandeurs différentes. La fleur est assez petite, et le fruit est une baie, d'abord verte, puis noire.

Les gens qui penseraient que la pomme de terre est un fruit seraient dans l'erreur. C'est au contraire un tubercule, c'est-à-dire une excroissance qui pousse sur une tige souterraine. Il y a plusieurs variétés de pommes de terre, dont le feuillage et les fleurs peuvent varier; la grandeur, la forme et la couleur des tubercules diffèrent aussi plus ou moins.

Une pelure brunâtre entoure la pomme de terre; cette pelure présente ce que nous appelons des yeux. Ces yeux peuvent produire chacun une nouvelle plante. Il n'est donc pas nécessaire de planter une pomme de terre tout entière; un morceau ayant un œil suffit. Le moment de semer les pommes de terre varie selon l'espèce et la localité.

La pomme de terre est utilisée comme article d'alimentation. C'est surtout l'amidon qu'elle renferme qui lui donne sa valeur comme tel, et ensuite la potasse et les autres sels qui y sont. L'azote n'y est présent qu'en petites quantités, et environ soixante-cinq pour cent de la pomme de terre est de l'eau. La composition du tubercule change selon le sol, l'engrais, la saison et la variété de la plante.

La pomme de terre a un ennemi qui lui cause une certaine maladie. C'est une sorte de champignon qui attaque d'abord les feuilles. Ces dernières périssent et peu à peu la maladie arrive dans les tubercules. Ce champignon est petit, même microscopique.

La pomme de terre serait originaire de l'Amérique. Les Espagnols l'auraient rencontrée près de Quito, où elle aurait été cultivée par les naturels du pays. Un moine l'aurait ensuite amenée en Espagne et de là elle aurait passé en Italie et en Belgique.

Sir Walter Raleigh la trouva dans la Virginie, d'où il l'apporta en Angleterre. Il la cultiva dans son domaine près de Cork. Elle ne passa dans la consommation générale que beaucoup plus tard. Sous Jacques Ier c'était un mets recherché. Peu à peu, on commença à la cultiver dans les comtés d'York et de Lancaster ainsi qu'en Ecosse, et enfin dans le reste du royaume britannique.

8

Les oiseaux

Les oiseaux sont des êtres ailés qui habitent toutes les parties du globe. Comme c'est aussi le cas chez les autres espèces animées, plusieurs sortes d'oiseaux, autrefois communes, n'existent plus aujourd'hui. Dans la période antédiluvienne, des oiseaux ont du exister de taille colossale auprès desquels nos grandes autruches sembleraient petites. Il est possible de calculer approximativement la hauteur et la grosseur de ces oiseaux préhistoriques au moyen d'empreintes et de fossiles.

On trouve beaucoup d'espèces d'oiseaux, mais toutes ont de grandes ressemblances. Le cou des oiseaux est généralement plus long que celui des quadrupèdes et la longueur en est déterminée par celle des pattes, quoique parfois on rencontre chez les oiseaux aquatiques un long cou et des pattes courtes. Les os sont formés de façon à unir la force et la légèreté; chez les oiseaux adultes la moelle est remplacée par de l'air; mais ceci ne concerne que les oiseaux qui volent beaucoup, les aigles, les hirondelles, etc. La température du corps est plus élevée chez les oiseaux que chez les hommes. Cela leur permet de résister à de très grands froids et explique a présence d'une certaine sorte d'oiseaux dans la Terre de Feu.

Les plumes qui couvrent le corps des oiseaux sont de formes variées. Si on regardait un aigle et une autruche, on remarquerait vite que les plumes de l'un ne ressemblent pas du tout à celles de l'autre. Un aigle a des plumes fermes et serrées, tandis que l'autruche a des plumes frisées. Les oiseaux aquatiques possèdent des glandes contenant de l'huile. Cette huile va dans leur plumage pour le rendre imperméable à l'eau tout en leur tenant chaud.

Les sens les plus développés chez les oiseaux sont la vue, l'ouïe et l'odorat. Ils ont trois paupières; la troisième, qui est transversale, couvre l'œil d'une substance à demi transparente. Ce serait grâce à elle que l'aigle pourrait regarder le soleil. L'ouïe des oiseaux est excellente; et l'odorat, surtout celui des oiseaux de proie, est très fin. Bien que certains oiseaux ne puissent pas voler, la majorité vole; beaucoup d'eux sont extrêmement forts et peuvent parcourir des distances énormes.

Une chose très intéressante a lieu dans le monde des oiseaux: nous voulons parler de la migration, qu'on ne doit pas confondre avec la visite plus ou moins accidentelle d'oiseaux égarés. Ce phénomène est tout entouré de mystère et attira l'attention des écrivains les plus anciens. Même de nos jours il n'est pas possible de l'expliquer d'une façon satisfaisante, quoique beaucoup de naturalistes aient essayé de le faire.

Nous savons que beaucoup d'oiseaux changent de demeure selon la saison. D'autres nous quittent en hiver, laissant quelques-uns de leur espèce passer cette saison dans notre pays. Dans cette catégorie on classe les rouges-gorges. On peut dire que tous les oiseaux habitant l'hémisphère nord sont plus ou moins migrateurs. Vers la fin de l'été nous remarquons en Angleterre une grande augmentation du nombre de nos oiseaux migrateurs; la cause de ce renforcement est l'arrivée des oiseaux du nord qui cherchent d'autres logis.

Donc, le mouvement de migration commence au nord et continue vers le sud. Il serait causé, selon la croyance populaire et l'opinion de certains naturalistes, par le froid. Selon d'autres, la raison de ce déplacement serait la rareté ou l'absence de nourriture, et nous devons avouer que cela semble plus probable. La disette commence à se faire sentir dans le nord et c'est pour cela que la migration va du nord au sud.

Le voyage de retour est encore plus difficile à expliquer. Probablement, les pays du sud sont tellement peuplés au moment de la migration que le manque de nourriture pourrait en chasser les oiseaux vers le nord à la fin de l'hiver. D'ailleurs, les oiseaux aiment retourner à leurs anciens nids et y élever leur petite famille.

9

La baleine

La baleine est un cétacé, elle est parente du marsouin et du dauphin. Bien qu'elle vive dans l'eau, ce n'est pas un poisson, elle est membre de la classe des mammifères. C'est le plus grand des animaux que nous ayons; une baleine adulte de l'espèce du Groënland peut mesurer jusqu'à 50 pieds de longueur. C'est donc un animal très imposant.

La forme d'un poisson a été donnée à la baleine pour qu'elle puisse nager facilement et parce que c'est la forme habituelle des habitants de l'eau. Cependant, quelle que soit la forme de la baleine, elle est tout de même un mammifère avec tous les signes caractéristiques d'un membre de cette classe. Elle a le sang chaud, tandis que les poissons ont le sang froid.

Chez la baleine le pelage que possèdent les mammifères est représenté par quelques poils au menton et à la lèvre supérieure, et souvent même ces poils sont seulement présents chez les jeunes animaux. Les membres antérieurs, quoiqu'ils ne soient pas très différents des nageoires, du moins à l'extérieur, possèdent intérieurement les os, les jointures et même une grande partie des muscles, des artères et des nerfs du bras de l'homme. Des traces de membres postérieurs existent dans le corps de l'animal.

La baleine est noire, quelquefois tachetée de gris. Sa tête est le tiers de la longueur totale de son corps. Sa lèvre supérieure est percée de deux trous par lesquels elle jette de l'eau. Sa gueule est énorme; elle n'a pas de dents, mais des fanons qui ne lui permettent d'avaler que de petits poissons et des mollusques. Sa langue est de nature spongieuse et consiste en une substance dont on retire de l'huile.

La gorge de la baleine est fort étroite, n'excédant que rarement quatre pouces de diamètre; ceci semble très étrange chez un animal de si grande taille et explique pourquoi la baleine ne mange que des animaux très petits. D'ailleurs les fanons ne laissent rien passer de gros. La baleine a de petits yeux, pas plus grands que ceux d'un bœuf et qui sont placés de chaque côté de la tête à une assez grande distance l'un de l'autre.

La baleine a la peau très lisse; sous cette peau est une épaisse couche de graisse, pour tenir chaud à l'animal. Sa chair est rouge et grossière et un peu comme celle du bœuf; les habitants du Groënland la mangent. Selon Buffon, un célèbre naturaliste français du dix-huitième siècle, la baleine pourrait exister mille ans; il est certain qu'elle peut vivre très longtemps.

C'est dans les mers boréales qu'on trouve la plupart des baleines, quoiqu'il y en ait aussi dans les mers australes. On les pêche pour en tirer de l'huile et aussi pour leurs fanons qui sont l'objet d'un grand commerce. D'une baleine adulte de grosseur moyenne on retire environ quinze tonnes d'huile et quinze quintaux de fanon de baleine. Les pêcheries les plus importantes sont au Spitzberg, au Groënland, dans la mer d'Hudson, dans le détroit de Davis et à la Nouvelle-Zemble.

10

La morue

La morue est un poisson assez grand. La partie supérieure de son corps ainsi que les côtés sont vert-foncé avec des taches jaunes, et la partie inférieure est blanche. Elle est couverte de petites écailles qui adhèrent à la peau et possède une barbe courte à la mâchoire inférieure. Sa langue est très large et ses dents forment plusieurs rangées comme celles du brochet.

Toutes ces dents sont très utiles à la morue, car c'est un animal extrêmement vorace. Elle mange des quantités de poissons plus petits qu'elle, des mollusques et des crustacés; la coquille des crustacés, avalée par la morue, est dissoute dans son estomac. C'est un poisson très destructeur. La demeure favorite de la morue est un banc de sable où l'eau est peu profonde.

C'est au mois de février que la morue dépose ses œufs, et cela en quantités immenses. Elle choisit des mers froides pour les y mettre et puis retourne dans des eaux plus chaudes afin de trouver de la nourriture. On a trouvé dans une seule morue environ huit millions d'œufs qui formaient la moitié du poids total de l'animal. Cependant, il n'y a qu'une petite partie de ces œufs qui éclosent; autrement les mers seraient infestées de morues.

La morue habite surtout les mers du nord et les mers tempérées de l'Europe, de l'Asie et de l'Amérique et en particulier les côtes de Terre-Neuve et de la Nouvelle-Ecosse, où il y a de grands bancs de sable. Quelques-unes vont au nord de l'Irlande, quoique cela soit rare. On les trouve dans l'Océan Atlantique jusqu'à la latitude de Gibraltar, mais elles n'entrent pas dans la Méditerranée.

Il y a de grandes pêcheries de morues, car c'est un article d'alimentation très recherché. Les pêcheries les plus célèbres sont celles de Terre-Neuve. La morue y est prise en grandes quantités; on la pêche à la linge avec un hameçon. Sa voracité lui fait saisir l'appât et ainsi on la prend assez facilement. On mange la morue fraîche ou bien séchée et salée. La pêche commence au mois de mai et la morue la plus longue peut avoir un mètre cinquante de long.

A part la chair de la morue, on utilise son foie, dont on extrait de l'huile connue sous le nom d'huile de foie de morue. Ce produit est un reconstituant et peut être un liquide transparent et jaunâtre ou foncé et d'une odeur très désagréable. Cela dépend de la façon dont on le retire du foie.

11

Le carbone

Le carbone est un corps simple qui existe dans la nature sous deux états, cristallisé ou amorphe. Sous ses formes variées, il est plus ou moins utile et précieux. Sa forme la plus pure est le diamant, qui est un cristal. C'est une pierre précieuse et aussi le plus dur et le plus brillant des minéraux. Aucun acide ne peut l'attaquer et rien ne peut le rayer.

Le diamant ne peut être taillé qu'avec sa propre poussière. Non seulement il est très recherché comme objet de parure, mais il est aussi fort utile pour couper le verre et polir d'autres pierres précieuses. Les mines les plus célèbres de diamants sont dans l'Afrique du Sud, les Indes anglaises et l'Australie. Quelques diamants sont devenus fameux, entre autres le Régent qui ornait la couronne royale de France, et le Koh-i-noor, la montagne de lumière, qui est le plus beau joyau de la couronne britannique.

On trouve du carbone cristallisé sous la forme de graphite. C'est un minéral très différent du diamant en apparence, bien que sa composition soit presque la même. Il est noir et très peu dur; on l'utilise pour écrire sur le papier et on en fait alors des crayons. Il n'est pas brillant et ne reflète pas la lumière comme son proche parent le diamant. Il est même assez difficile de croire que c'est vraiment du carbone presque pur, tant il est différent du diamant.

Parmi les formes amorphes du carbone nous avons en premier lieu le charbon, un des minéraux les plus importants que nous ayons. Le charbon est le produit qui résulte du bois brûlé à l'abri du contact de l'air.

Toutes nos grandes mines de houille sont d'origine végétale; dans des temps très éloignés des nôtres, probablement à une période antidéluvienne, d'immenses forêts furent enfouies sous terre.

Les arbres et les fougères gigantesques qui les composaient sont devenus, par le travail des siècles, ce charbon qui nous est si utile. On voit souvent, sur un morceau de charbon de terre, l'empreinte d'une feuille ou d'une tige qui nous rappelle son origine. L'Angleterre et l'Allemagne ont de grandes mines de charbon et en Angleterre elles vont même sous la mer en certains endroits. Les mines de la France, de la Belgique et des Etats-Unis sont aussi très abondantes.

12

Les coquillages des mollusques

Les mollusques sont des animaux à corps mou et sans vertèbres, ce sont donc des invertébrés. Quelques-uns sont terrestres, mais la plupart habitent les eaux. Ils possèdent une sorte de squelette, mais de squelette extérieur et non pas intérieur comme chez les animaux supérieurs. Cependant, quelques-uns font exception et ont une sorte de squelette intérieur, bien qu'il ne soit pas du tout semblable à celui des vertébrés.

La seiche possède une coquille interne connue sous le nom d'os de seiche. Cependant, les autres membres de sa famille ont des coquilles extérieures; d'ailleurs, il y en peu de vivants, ils sont presque tous fossiles. Certains mollusques qui en mangent d'autres sont munis d'un appareil très intéressant avec lequel ils perforent la coquille de leurs ennemis. D'autres peuvent même faire des trous dans le bois et les rocs et causent de grands dégâts.

Les mollusques possèdent un cœur, un appareil digestif, etc., et sont presque tous ovipares. Ils n'ont pas tous une tête et on peut dire que ceux dont la coquille est formée d'une seule partie ont une tête; ceux dont la coquille est composée de deux parties n'en ont pas. Cependant, il ne faut pas que cette règle soit considérée une vérité absolue, car, comme toute autre règle, elle a ses exceptions.

Un des mollusques les plus communs, la patelle, a la coquille la plus simple qui soit, c'est-à-dire une sorte de bouclier qui protège l'animal. Ces coquilles des mollusques peuvent avoir des formes variées, et être allongées, rondes, en forme de tubes et souvent ornementées. A part leur coquille, les mollusques ont un autre moyen de défense; ils ont quelquefois une membrane qui ferme l'entrée de la coquille.

Les escargots de terre dorment pendant l'hiver. Ils construisent une sorte de barrage formé de carbonate de chaux qui disparaît en été. La coquille des mollusques n'est pas toujours dure; au commencement de leur existence, elle est flexible et mince. Peu à peu, la chaux la rend plus dure et plus épaisse. Ce sont les mollusques vivant dans les eaux courantes contenant beaucoup de chaux qui ont les plus dures et les plus lourdes coquilles.

13

Le corps humain

Le corps de l'homme est l'une des choses les plus intéressantes qui soient. C'est une machine très compliquée et d'une grande perfection. Il est composé principalement d'un squelette, de muscles et d'organes variés. On le divise en trois parties, le tronc, la tête et les membres. L'homme, étant un vertébré, n'a que deux paires de membres, ce nombre étant le maximum dans cette classe.

Le squelette est l'ensemble de tous les os. Quelques-uns sont grands, comme par exemple le fémur, tandis que d'autres sont fort petits et même microscopiques, comme certains os de l'oreille moyenne. Chez les jeunes enfants le nombre des os est plus considérable que chez les grandes personnes, parce que quelques os qui sont soudés plus tard étaient d'abord séparés. C'est le cas de l'os frontal, qui est toujours divisé à la naissance et ne forme généralement qu'un seul os chez les adultes, quoiqu'il y ait des exceptions a cette règle.

La propriété fondamentale des muscles est la contractivité. Le muscle, c'est la chair; un muscle est formé par la réunion de faisceaux musculaires constitués eux-mêmes par des fibres musculaires. Au point de vue physiologique, on distingue deux espèces de muscles: les muscles striés ou muscles volontaires et les muscles lisses ou involontaires. Il y a une exception à cette règle: le cœur, quoiqu'il soit un muscle involontaire, est strié.

Le système nerveux sert à nous mettre en relation avec le monde extérieur; il règle les fonctions des divers organes, les coordonne et les harmonise. Il est divisé en deux parties: le système cérébro-spinal et le système du grand sympathique. Le cerveau fait partie du premier système; il est situé dans le crâne et est divisé en deux parties qu'on appelle les hémisphères. A leur surface, les hémisphères présentent des circonvolutions formées par la matière grise. Les circonvolutions de l'hémisphère gauche et de l'hémisphère droit ne sont pas symétriques.

Si l'on examine le corps humain et qu'on l'étudie, on trouve des quantités de choses intéressantes. Les différentes fonctions des organes offrent un vaste champ d'étude et il semble qu'on puisse toujours découvrir quelque chose de nouveau et digne d'attention.

14

La terre et ses mouvements

La terre est une planète et a la forme d'une sphère. Pendant longtemps on croyait que la terre était plate, jusqu'à ce qu'on eût prouvé qu'elle était ronde. Pour démontrer ce fait, on a les voyages de circumnavigation, la disparition graduelle d'un navire à l'horizon en mer, l'ombre toujours ronde de la terre sur la lune et l'élévation toujours plus grande de l'étoile polaire à mesure qu'on approche du pôle. La terre n'est pas une sphère parfaite; elle est aplatie aux deux pôles d'environ 21 kilomètres.

Comme toutes les planètes, la terre tourne autour du soleil. A ce sujet les anciens avaient une autre théorie, la théorie géocentrique. Ils étaient persuadés que tout tournait autour de la terre, mais maintenant on sait que ce n'est pas le cas. Ce mouvement de la terre autour du soleil est appelé mouvement de translation; il a été découvert par Copernic et prouvé par Képler.. Il faut que la terre avance pendant environ 365 jours pour achever le tour du soleil.

A part son mouvement de translation, la terre accomplit un autre mouvement, celui de rotation. La terre, en effet, tourne sur elle-même, et cela une fois par jour. Pendant longtemps personne ne voulait croire que la terre tournait. Galilée, un des plus célèbres astronomes que le monde ait produits, déclara qu'elle tournait. Il passa en jugement, fut condamné, et dut rétracter ce qu'il avait avancé.

Nous savons que la terre fait environ 464 mètres par seconde à l'équateur, ce qui est une vitesse considérable. On donne plusieurs preuves du mouvement de rotation. La première est l'aplatissement polaire. D'autres sont: la déviation des corps vers l'est dans la chute libre, la déviation des vents, des courants, etc.; il ne nous faut pas oublier de citer une preuve absurde, c'est-à-dire l'impossibilité que des milliers d'étoiles fassent en 24 heures le même tour autour de notre planète.

15

L'EAU

L'eau est un liquide transparent, inodore, incolore, insipide, très répandu dans la nature. C'est un corps composé, résultant de la combinaison de deux volumes d'hydrogène pour un volume d'oxygène, chose que les hommes ont eu de la peine à comprendre; il leur semblait étrange que deux gaz pussent former un liquide, quoique, à nous, cela paraisse assez naturel.

Prenez n'importe quelle eau naturelle et analysez-la. Vous y trouverez toujours des impuretés ou des minéraux en dissolution. En effet, l'eau pure ne se rencontre guère que dans le laboratoire du chimiste. A cause même de sa nature, l'eau peut dissoudre beaucoup de corps étrangers et absorber des gaz, de sorte que la combinaison d'oxygène et d'hydrogène sans additions est pour ainsi dire introuvable.

Mettez de l'eau sur le feu et attendez que la température monte à 100° C. Vous remarquerez alors qu'elle est changée en vapeur; on appelle cela le point d'ébullition. Plus l'eau est froide, plus elle est lourde; cependant le maximum de poids est à 4° C. De ce point jusqu'à 0° C. elle devient plus légère. La glace flotte sur l'eau. Nous connaissons l'eau dans la nature sous les trois formes—gazeuse, liquide et solide.

L'eau de mer est différente de l'eau de source ou de l'eau de rivière. Goûtez un peu d'eau de mer et vous lui trouverez un goût très salé et fort désagréable. En effet, elle renferme beaucoup de sel et par conséquent est inbuvable. Toutefois, il est possible d'en tirer de l'eau douce. Pour cela, prenez de l'eau de mer et mettez-la dans un vase fermé duquel sort un tuyau. Faites communiquer ce tuyau avec un autre vase également fermé placé à quelque distance.

Ces préparatifs achevés, allumez du feu sous le récipient contenant l'eau. Bientôt cette eau commencera à bouillir et la vapeur cherchera une issue; elle passera par le tuyau et ira dans le second vase. Ce vase n'étant pas exposé à l'action de la chaleur, la vapeur redeviendra de l'eau. Goûtez cette eau ; elle n'est plus salée, car le sel est resté au fond du premier vase, ne pouvant devenir vapeur. On appelle ce procédé distillation.

16

Les Alpes

Regardez la carte de l'Europe, si possible une bonne carte où le relief est bien marqué. Alors il est facile de voir les Alpes, cet énorme massif montagneux de l'Europe occidentale. Remarquez où la chaîne commence et finit et considérez-en la longueur. Les Alpes vont en effet du col de Cadibone près du golfe de Gênes jusque près de Vienne. Quand vous avez bien vu cela, regardez tout le massif et divisez-le en chaînes.

Vous en trouvez trois principales : les Alpes occidentales, les Alpes centrales qui possèdent les plus hauts sommets, et les Alpes orientales. Comparez maintenant les Alpes aux autres montagnes de l'Europe : il est facile de voir sur la carte que ce sont les plus hautes de ce continent. Le sommet le plus élevé est le Mont-Blanc, qui est situé dans les Alpes centrales entre la France, la Suisse et l'Italie et a 4810 mètres de hauteur.

Quittons maintenant la carte et allons dans une des régions alpines les plus frappantes. Qu'y voyez-vous ? D'immenses pics couverts de neige dont la cime semble toucher le ciel ; plus bas des pâturages où paissent les belles vaches suisses ; plus bas encore des forêts, et tout au pied la campagne cultivée. Allons sur un des sommets environnants et regardons en bas. Nous voyons des vallées plus ou moins grandes, des rivières et ces magnifiques lacs aux eaux bleues pour lesquels les Alpes sont renommées.

Essayons de franchir les Alpes. Il nous faut pour cela trouver un col, c'est-à-dire un passage entre deux sommets dont l'altitude peut varier. Les plus célèbres sont : le col du Grand-Saint-Bernard entre la Suisse et l'Italie à 2472 mètres d'altitude, où l'on remarque le fameux couvent et où, en hiver, le froid est terrible, le Simplon, le Saint-Gothard et beaucoup d'autres.

Pour faciliter les communications dans les régions des Alpes, on a percé des tunnels sous plusieurs massifs élevés. Le tunnel du Mont-Cenis unit Turin et Lyon; celui du Saint-Gothard, Bâle et Milan. Pendant longtemps le Saint-Gothard était le plus long tunnel en existence, mais maintenant on a percé le Simplon, qui le dépasse de beaucoup.

17

L'AIR

L'air est la substance qui entoure la terre et que nous respirons. Il est inodore, invisible et sans goût et c'est un gaz. Il se compose de plusieurs gaz et les contient dans les proportions suivantes: 21% d'oxygène, 78% d'azote qu'on appelle quelquefois nitrogène, 1% d'argon et un peu d'acide carbonique, environ 0·04%. Toutefois, la quantité d'acide carbonique varie selon le degré de pureté de l'air et la quantité d'oxygène varie en conséquence.

L'oxygène est un gaz très actif. C'est lui que nous respirons et qui purifie le sang veineux dans les poumons, le changeant en sang artériel. C'est aussi lui qui permet à n'importe quelle substance de brûler, car rien ne peut brûler sans son aide. L'azote est très inactif, il sert à neutraliser l'action de l'oxygène. Il ne peut pas brûler et aucune substance ne peut y brûler non plus.

L'argon est encore plus inactif que l'azote; il ne fut découvert qu'en 1894, quoique Cavendish l'eût réellement rencontré plus d'un siècle auparavant et eût pensé que c'était une impureté de l'air qu'il avait laissée de côté en faisant son expérience. L'acide carbonique se compose d'oxygène et de carbone et provient en grande partie de la respiration des animaux.

La masse d'air qui entoure la terre exerce sur tous les corps une pression qui s'appelle pression atmosphérique. Cette pression n'est pas toujours la même, mais elle est en moyenne de quinze livres par pouce carré (1·003 kilogramme par centimètre carré). L'épaisseur de l'atmosphère ne paraît pas dépasser une soixantaine de kilomètres, quoique nous ne soyons pas tout à fait sûrs de ce fait.

Ce furent Galilée et son disciple Torricelli qui établirent que l'air est pesant, et la vérité de cette assertion peut facilement se prouver par expérience. On pèse un certain vase plein d'air, puis on en retire l'air au moyen de la machine pneumatique. Ensuite on pèse le vase et la différence de poids s'observe, qui prouve que l'air est lourd. L'air a aussi des températures variées: plus on s'éloigne de la terre, plus il est froid. Cela peut fort bien se remarquer en montant sur une montagne élevée. Au sommet du Mont-Blanc par exemple, l'air est si froid qu'on n'y peut rester longtemps.

Les observations astronomiques ont démontré que d'autres planètes et leurs satellites, la lune exceptée, sont aussi entourés d'une atmosphère; quant à la nature de cette atmosphère, elle peut être la même que la nôtre ou non, nous n'en savons rien ou du moins pas grand'chose.

18

Les déserts

Les déserts proprement dits forment une sorte de ceinture autour du monde. En partant du nord de l'Afrique, nous rencontrons le Sahara, cet énorme désert qui pourrait couvrir trois Méditerranées, l'Arabie, l'Iran, le désert Indien, le désert du Turkestan, le Thibet, le Tarim et le désert de Gobi. Sur cette immense surface, les hommes mènent le même genre de vie: ils sont nomades et vont d'une oasis à l'autre montés sur leurs chameaux.

Le Sahara s'étend de l'Egypte à l'océan Atlantique et de la Méditerranée au Soudan. Il faut à une caravane trois mois pour le traverser du nord au sud. Il ne consiste pas seulement en une vaste étendue de sable; on y voit aussi des plateaux pierreux d'où émergent quelques massifs de montagnes et de rares oasis. Il est habité par plusieurs races de nomades, les Touaregs, les Maures et autres.

Tous les déserts ne sont pas semblables au Sahara, mais tous se caractérisent par l'absence plus ou moins complète d'eau. Partout où l'on peut trouver de l'eau, la végétation est assez belle. Dans les oasis se voient des dattiers dont la verdure attire les voyageurs et qui ne se trouvent qu'où il y a de l'eau. Certains déserts, le Sahara entre autres, renferment des lacs salés ou des dépôts de sel, et cela a donné naissance à la théorie que ce sont les lits d'anciennes mers desséchées.

A part les grands déserts mentionnés ci-dessus, il s'en trouve d'autres quelque peu différents. Il y a par exemple des déserts qui prennent la forme de grandes plaines herbeuses au moins pendant une certaine partie de l'année, comme c'est le cas dans l'Asie centrale et dans les steppes de Sibérie. Les steppes se rencontrent sur les bords des déserts proprement dits et leur font une sorte de cadre pas entièrement fertile, mais pas non plus tout à fait stérile.

Les savanes sont une espèce de steppes; elles sont couvertes d'herbe haute et possèdent des buissons et des fleurs de toutes couleurs, quoiqu'on n'y voie pas d'arbres. Les savanes couvrent certaines parties de l'Amérique, en particulier de la Guyane. Au nord de la Sibérie se rencontrent d'immenses régions gelées qui s'appellent toundras et dans lesquelles ne poussent guère que des mousses.

Si le désert ne nous offre en général qu'une très faible proportion de vie végétale, il possède en revanche une belle collection d'animaux sauvages. Le Sahara et les autres déserts composant la grande ceinture sont habités par des gazelles, des ânes sauvages, des autruches, des hyènes et des lions.

19

Le travail de la mer

La mer use et construit; elle possède une force extraordinaire. En général, c'est lorsqu'elle contient des pierres et des débris qu'elle use le plus. Le vent, soufflant dans la direction de la terre, soulève des vagues plus ou moins fortes qui peu à peu entament les côtes; les débris arrachés par la vague reculent avec elle, puis reviennent et attaquent la base de la falaise. Bientôt, il se forme une sorte de caverne, et par suite de l'infiltration, la partie supérieure de la falaise se détache et tombe dans la mer.

Au bout d'un certain temps, la terre diminue et c'est ainsi que la mer envahit peu à peu les terres. Les mers à marées désagrègent davantage que les mers sans marées. La mer tend à raser les caps et les îles. D'un autre côté, si un obstacle brise la vague, les débris se déposeront près de la plage et formeront une nouvelle plage empiétant sur la mer. Si par exemple il y a une petite île à l'entrée d'un golfe, les débris s'accumuleront derrière elle jusqu'à ce qu'elle devienne une presqu'île.

Dans beaucoup d'endroits, en effet, la mer recule, mais c'est aussi souvent le contraire qui a lieu. Certaines côtes perdent beaucoup de terrain, comme par exemple les côtes de France près de Bordeaux où la mer avance de vingt à vingt-cinq centimètres par année. L'île Helgoland a, en quatre ou cinq siècles, perdu la moitié de son territoire.

Une quantité d'archipels ne sont que des restes de continents. L'archipel indien est un reste de l'Asie et on appelle ces îles îles continentales. Les débris emportés aux côtes vont exhausser le fond de la mer; des calcaires en poudre se forme le squelette des animaux marins. Les îles madréporiques s'élèvent aux points où les débris s'entassent sur l'édifice madréporique et apparaissent alors au-dessus de l'eau.

La mer est aussi la cause de l'existence des dunes, excepté peut-être de celles du Sahara. Quand elle se retire à marée basse, elle laisse derrière elle une zone sableuse; sur cette zone il se produit une grande évaporation, les sables sèchent et le vent peut alors les transporter facilement. Supposons alors qu'un obstacle se présente en route: le sable s'y arrêtera et formera un amas; quand cet amas sera à la hauteur de l'obstacle, le sable s'écoulera de l'autre côté. Ainsi une dune se sera formée; la pent en sera douce du côté de la mer et abrupte de l'autre côté.

Ces dunes ne sont pas immobiles, elles avancent ou reculent plus ou moins vite. On en a vu qui faisaient cinq cents mètres par an; celles de Gascogne faisaient de vingt à vingt-cinq mètres et on a dû les « clouer » à l'aide de plantations de pins maritimes. Le Sahara possède deux sortes de dunes, maritimes et continentales; ces dernières proviennent de la région intérieure du Sahara.

20

LES SOURCES

Une grande partie de l'eau de pluie est reçue par la terre, où elle disparaît. Sous le sol, elle s'accumule jusqu'à ce qu'elle rencontre une couche imperméable et que la pression de la colonne descendante la force à se trouver une issue à la surface. L'endroit où cette eau quitte la terre est une source.

La température des sources nous renseigne sur la profondeur de laquelle l'eau arrive; cependant, cette information n'est pas toujours correcte, et on ne peut trop s'y fier. Cette température est très variable: certaines sources sont seulement à un degré au-dessus de zéro, tandis que d'autres ont la température de l'eau bouillante.

On peut dire que les sources très froides sont originaires de montagnes froides ou couvertes de neige. Mais elles sortent quelquefois de glacières situées dans des districts comparativement chauds et bas.

Les sources les plus chaudes se rencontrent dans les régions volcaniques, quoique, même à de grandes distances d'un volcan, on ait trouvé des sources thermales. C'est le cas des sources thermales de Bath, dont la température est de cent vingt degrés Fahrenheit et qui ne sont dans le voisinage d'aucun volcan. Elles sont probablement sorties d'une très grande profondeur.

Une source renferme toujours une certaine quantité de minéraux en dissolution chimique. Cette substance minérale est obtenue des rochers à travers lesquels l'eau a passé, soit en descendant de la surface à l'intérieur, soit aussi en remontant de l'intérieur à la surface. La qualité et la quantité des substances minérales dépend de la composition des rochers, des acides et des gaz présents dans l'eau, de la profondeur qui a été atteinte par l'eau et de la température à laquelle elle est montée.

Les sources ordinaires contiennent de l'air, du carbonate ou du sulfate de chaux, du sel et quelquefois des matières organiques. Ces substances en dissolution n'excèdent pas de 0·5 à 1 gramme par litre. On appelle eaux minérales celles dans lesquelles ont été dissous des minéraux en quantités plus grandes, de 1 à 300 grammes par litre. Les sources minérales sont arrangées en groupes selon leur composition.

Il y a les sources calcaires, qui renferment beaucoup de chaux. Si l'eau s'évapore, la chaux qui est restée forme une sorte de croûte blanche, parfois très épaisse quand une grande quantité de chaux est dissoute dans l'eau. Les eaux calcaires se trouvent dans les régions calcaires, comme on peut facilement le deviner.

Les eaux ferrugineuses sont celles qui sont très riches en fer. On les rencontre dans les districts où les rochers renferment du fer en combinaisons variées. Ces eaux ont le goût d'encre et déposent où elles passent une substance jaunâtre ou brunâtre. Elles sont beaucoup employées en médecine. Certaines eaux déposent du silex, on les appelle eaux siliceuses. Elles sont généralement chaudes, comme par exemple les geysers de l'Islande, de la Nouvelle-Zélande et de la Yellowstone dans les Etats-Unis.

On rencontre quelquefois des eaux salées dans les endroits où il y a des blocs de sel sous la terre, comme c'est le cas à Bex en Suisse et dans le comté de Chester. A part les eaux mentionnées ci-dessus, il y en a d'autres employées en médecine et à qui l'on a donné le nom un peu vague d'eaux médicinales. Ce sont les eaux alcalines comme celles de Vichy, les eaux sulfureuses comme celles d'Aix-la-Chapelle et de Harrogate.

Quelquefois l'eau contient de l'huile en petite quantité. Alors cette huile flotte à la surface en gouttes. Mais il peut arriver que l'huile sorte de la terre avec peu ou point d'eau, et on a alors une source d'huile minérale telle que celles qu'on rencontre au Canada.

21

La répartition et l'éruption des volcans

On peut dire qu'il existe des volcans sur toute la surface de la terre, mais il faut remarquer que l'intérieur des continents actuels ne renferme aucun volcan actif. C'est au bord de la mer qu'ils sont distribués, surtout autour de l'Océan Pacifique où ils forment une suite ininterrompue, le Cercle de Feu.

Dans le Pacifique lui-même, il se trouve quelques groupes isolés et une multitude de volcans sous-marins.

Une autre ligne volcanique va des Antilles aux îles Sandwich en passant par la Méditerranée, le Caucase, l'Océan Indien, la Polynésie. Il est important de remarquer que les volcans n'existent pas sur les côtes plates, ni près des mers peu profondes. Ils sont toujours sur les côtes bordées de ressauts brusques de l'écorce terrestre, ces ressauts coïncidant naturellement avec de grandes profondeurs océaniques.

Or, partout où ces grands mouvements de l'écorce se sont produits, des cassures se sont effectuées. Ces cassures établissent naturellement une communication entre la partie du globe au-dessous de l'écorce et la surface. C'est par ces fentes que s'écoulent les matières ignées. Chaque bouche volcanique marquerait ainsi le sommet d'une cassure.

Quant à la cause qui détermine l'éruption, les géologues ne sont pas d'accord. Les uns admettent l'intervention des eaux marines uniquement, d'autres ne la trouvent pas forcément nécessaire. Les gaz qui sont dissous dans le réservoir interne suffiraient à expliquer le phénomène.

L'éruption volcanique débute généralement par un panache de fumée qui couronne le cratère pendant un certain temps, puis par des craquements qui font souvent écrouler les parois du cratère. Ensuite, une immense colonne de vapeur d'eau (qui peut s'élever à plus de dix mille mètres) s'élance vers le ciel, entraînant avec elle des pierres et des cendres.

Après ce phénomène d'explosion arrive le courant de lave, c'est-à-dire de pierre fondue, qui s'échappe, soit par dessus les bords du cratère, soit par des fentes pratiquées le long des flancs du cône. La vitesse de la marche de la lave dépend de sa fluidité, de l'importance de sa masse et de la pente du terrain. La plus grande vitesse observée est de huit mètres per seconde.

La lave forme des coulées de constitution variable et de composition qui diffère. Mais toutes sont des silicates, c'est-à-dire des substances riches en silice. Leur fluidité varie; il y en a qui sont pâteuses, d'autres presque liquides.

La température de la lave doit être très élevée. On admet qu'elle dépasse mille degrés centigrade.

Les volcans modernes sont beaucoup moins puissants que ceux des temps passés. Il y en a aussi beaucoup moins, comme la présence de volcans éteints semble l'indiquer. C'est probablement le résultat du refroidissement graduel de la terre et selon toute probabilité les volcans deviendront de moins en moins nombreux et de plus en plus faibles.

22

La pêche en Islande

On a pêché la morue sur les côtes d'Islande depuis le onzième siècle. Pendant longtemps le seul port de départ des bateaux de pêche était Dunkerque, mais maintenant ils partent d'autres endroits, par exemple de la Bretagne et surtout de Paimpol et de Saint-Brieuc. Les navires se mettent en route au mois de février, emportant la provision de sel nécessaire à la préparation du poisson. La vie des pêcheurs islandais est très pénible, car il y a beaucoup de privations à endurer avec un tel métier. Il leur faut se contenter d'une nourriture grossière et peut-être insuffisante, travailler souvent au milieu de la brume et des orages et braver de grands dangers. Nombreux sont les bateaux qui ne reviennent pas, et l'on ne voit que trop l'inscription « Disparu en mer » dans les cimetières de Bretagne.

Tous ces pêcheurs sont fort pauvres: ils ne gagnent que de quatre à cinq cents francs par an. A part les pêcheurs venus du continent, il y a les Islandais eux-mêmes qui vivent du produit de leur pêche. Etant pauvres, ils vendent tout le poisson qu'ils ont pris et ne gardent que les têtes pour leur alimentation. C'est en juillet que les morues sont le plus abondantes et on travaille alors jour et nuit, quoique, à vrai dire, il n'y ait pas de nuit. On a vu quelquefois des bancs de morue de cent mètres de profondeur défiler pendant des heures entières. Ces morues se voient poursuivies non seulement des hommes, mais aussi des dauphins et des requins. Il faut ajouter que deux navires de guerre suivent la flottille de pêche. Ils sont chargés de faire observer les règlements et de régler les contestations. En outre ils doivent recueillir les malades et réparer les avaries. Pendant la saison de pêche ils vont d'un port à l'autre et tiennent les bateaux pêcheurs au courant de leurs mouvements.

La capitale de l'Islande, Reikiavik, n'est qu'une petite ville. Les habitants islandais en sont pauvres, comme ceux du reste de l'île; ils ont de petites maisons avec un jardin où croissent quelques légumes et quelquefois une sorte de cabane où ils font sécher leur poisson. Cependant, si l'Islandais n'est pas assez riche, il se passe de cette cabane et les morues doivent sécher en plein air.

Quand la saison de pêche est passée, l'île reste presque sans communications avec le reste de l'Europe, la navigation étant très dangereuse près des côtes à partir du mois d'octobre.

23

L'HYGIÈNE

« Hygiène » a pour étymologie un mot grec signifiant « santé. » L'hygiène est donc une science qui s'occupe de la santé des hommes et de toutes les choses s'y rapportant. Elle démontre comment l'on doit vivre pour éviter les maladies et les infirmités. En Angleterre, on ne s'occupa guère d'hygiène jusqu'à une période très rapprochée de la nôtre, et cela explique les épidémies sans nombre du moyen-âge et même des temps modernes.

Tout le monde a entendu parler de la grande peste qui sévit à Londres en 1665 et pendant laquelle cent mille personnes périrent. Nous savons que de telles catastrophes étaient occasionnées par le mauvais état des maisons et des rues, la saleté et l'absence de drainage.

L'hygiène s'occupe de tout cela et donne des règles précises concernant l'homme, son corps et son habitation. En obéissant à ces lois, on évite non seulement le danger de terribles épidémies, mais aussi celui de maux individuels.

Ce fut après le grand incendie de Londres, qui se chargea de désinfecter la capitale en 1666, que les gens commencèrent à réaliser l'importance de la propreté et du bon air. On bâtit alors des maisons d'une façon plus saine, on eut des rues plus larges, et la santé générale s'améliora beaucoup en conséquence.

Peu à peu, l'hygiène fit des progrès, mais c'est surtout pendant ces trente dernières années que des travailleurs sérieux et zélés se sont efforcés de découvrir la cause des maladies et d'être ainsi à même d'y remédier.

Toutefois, cela ne suffit pas; il faut aussi que chacun soit en état de comprendre la valeur des lois hygiéniques et de s'y conformer. Nous n'avons plus à redouter, il est vrai, les grands ravages engendrés par la peste, mais il y a encore nombre de maux qui pourraient et devraient être évités. Malheureusement, cela ne peut se faire en un jour et nous devons nous en remettre à l'avenir.

24

Histoire de la chimie

L'histoire des sciences naturelles ressemble à celle des nations en ce qu'elles ont toutes deux pour but d'étudier les causes qui produisent certains effets et d'en tirer des lois. Le point de départ des deux est le même: ni l'une ni l'autre ne proclame de lois pour commencer. Nous pouvons observer les efforts lents et persévérants de l'esprit humain qui s'évertue à remonter des effets aux causes avant de poser des lois.

La chimie ne doit pas l'existence à une seule nation ou à un développement soudain. On peut la comparer à un édifice dont les fondations ont été posées par les Orientaux, dont les murs ont été élevés au moyen-âge et continués par Lavoisier, Cavendish et d'autres. Les savants de notre époque construisent le toit. Toutefois, il est probable que les générations de l'avenir agrandiront et orneront de plus en plus l'édifice.

Les anciens réalisaient jusqu'à un certain point bon nombre de faits chimiques. Nous ignorons comment ils obtinrent leur information; ce fut peut-être par hasard, peut-être aussi à la suite d'expériences. Aristote reconnaissait quatre éléments dont il donne les propriétés physiques. En Orient on avait des idées semblables aux siennes et les Hindous admettaient cinq éléments: le feu, la terre, l'eau, l'air et l'éther.

Pendant le moyen-âge il y eut une grande école, celle des alchimistes. Du onzième au quinzième siècle, l'alchimie fut étudiée avec enthousiasme en Italie, en France, en Angleterre et en Allemagne. Les alchimistes avaient certainement des idées étranges et pensaient qu'un métal pouvait se changer en un autre métal. Cette croyance est d'origine grecque: les Grecs pensaient que l'or pouvait se fabriquer et que ce métal prenait différents aspects avant de devenir ce qu'il est. Cependant, les alchimistes connaissaient quelques-unes des propriétés des corps.

Le but suprême des alchimistes avait été de trouver la pierre philosophale qui leur eût donné le pouvoir de faire de l'or. Avec Paracelse une nouvelle phase commença, celle de mettre la chimie au service de la médecine. Ce savant déclara que l'utilité de la chimie consistait à préparer des remèdes et il eut beaucoup de partisans et de disciples.

Le nom de chimie avait fait son apparition au onzième siècle, mais ce mot signifiait alors « préparation de l'or et de l'argent.» Plus tard la chimie montra les relations et les limites des propriétés des corps. Au dix-huitième siècle il y eut grand progrès en chimie. Parmi d'autres noms célèbres nous pouvons citer ceux de Cavendish, qui découvrit la différence entre l'hydrogène et l'azote et démontra que l'oxygène et l'hydrogène sont les constituants de l'eau; Lavoisier, qui soutint que l'oxygène et l'hydrogène sont des éléments et infusa une nouvelle vie à la chimie; Faraday et beaucoup d'autres.

L'électro-chimie prit naissance vers le commencement du dix-neuvième siècle, époque à laquelle l'eau fut décomposée sous l'action de la pile de Volta. Depuis lors il y a eu des découvertes importantes et il s'en fait de nouvelles chaque jour. Les grands savants ont, par leurs travaux incessants, élevé la chimie à la place d'honneur qu'elle occupe aujourd'hui.

ABBREVIATIONS OF TITLES OF PERIODICALS

Ann. Fac. Sci. Marseille	*Annales de la Faculté de Sciences*, Marseilles.
Ann. Mines Carb. ..	*Annales des Mines et des Carburants*, Paris.
Ann. Sci. Nat.	*Annales des Sciences Naturelles*, Paris.
Ann. Soc. Linn. Lyon ..	*Annales de la Société Linnéenne de Lyon.*
Ann. Phys.	*Annales de Physique*, Paris.
Arch. Zool. Exp. ..	*Archives de Zoologie Expérimentale et Générale*, Paris.
Bull. Soc. Bot. Genève	*Bulletin de la Société Botanique de Genève.*
Bull. Soc. Chim. Fr. ..	*Bulletin de la Société Chimique de France*, Paris.
Bull. Soc. Géol. Fr. ..	*Bulletin de la Société Géologique de France*, Paris.
Bull. Soc. Zool. Fr. ..	*Bulletin de la Société Zoologique de France*, Paris.
Comm. Lab. Inst. Opt.	*Communications des Laboratoires de l'Institut d'Optique*, Paris.
Helv. Chim. Acta. ..	*Helvetica Chimica Acta*, Basle, Geneva.
Helv. Phys. Acta. ..	*Helvetica Physica Acta*, Basle.
J. Physique	*Journal de Physique et le Radium*, Paris.
Mémor. Sci. Math. ..	*Memorial des sciences mathématiques* (Gauthier-Villars), Paris.
Onde élect.	*Onde électrique*, Paris.
Rev. Étude Cal. ..	*Revue pour l'Étude des Calamités*, Geneva.
Rev. Gén. Bot... ..	*Revue Générale de Botanique*, Paris.
Rev. Gén. Élect. ..	*Revue Générale de l'Électricité*, Paris.
Vie Comm.	*Vie Communale et Départmentale*, Paris.

PART V

PHYSIQUE

1

Explication de la couleur bleue du ciel

La lumière solaire est diffusée par les molécules des gaz de l'atmosphère; quand il n'y a aucun nuage, aucune brume et que le lieu de l'observation est suffisamment élevé pour que les poussières, rélativement abondantes près du sol, ne gênent pas trop, on obtient une vérification satisfaisante des lois de Lord Rayleigh. La lumière venant de la région du ciel située dans le vertical du soleil et à 90° de cet astre renferme une forte proportion de lumière polarisée, vibrant normalement aux rayons incidents et diffusés. La proportion de lumière polarisée dépasse souvent 70 pour 100. Ce n'est pas à l'ozone qui existe dans la haute atmosphère qu'est due la couleur bleue du ciel, comme on l'a autrefois avancé, bien à tort. Une absorption par l'ozone n'expliquerait pas pourquoi toute la voûte du ciel paraît bleue, ni pourquoi elle renvoie de la lumière polarisée.

London B.Sc. (*Special*).

2

Courants dans un gaz ionisé

Si l'on place dans un gaz ionisé deux conducteurs entre lesquels on établit une différence de potentiel, les ions positifs sont attirés par le conducteur chargé negativement et les ions négatifs par le conducteur chargé positivement, il s'établit un courant entre les deux électrodes. Le caractère essentiel des courants dans les gaz ionisés est le fait que, lorsqu'on augmente la différence de potentiel, le courant atteint une intensité limite, pour les potentiels élevés, son intensité est independante de la valeur de la différence de potentiel appliquée. On dit alors qu'on a atteint la saturation et le courant limite s'appelle le courant de saturation.

London B.Sc. (*General*).

3

Nouvelle forme de thermomètre à gaz

Un nouveau type de thermomètre à gaz (hélium) est décrit. Il est destiné à une installation permanente faisant partie d'un équipement à basse température ayant pour but de relier la région comprise entre les températures de l'hydrogène liquide et de l'hélium liquide. Ce thermomètre se différencie radicalement du thermomètre à gaz habituel par le fait que le volume de gaz qui se trouve dans la chambre thermométrique est considérablement plus grand que dans celui du thermomètre habituel. La théorie montre que par ce moyen la sensibilité est grandement accrue par les basses températures. Il est ainsi possible de construire un thermomètre suffisamment sensible dans la région située sous 20° K pour donner des températures à 0°·1 près à l'aide d'un manomètre à mercure ordinaire et ayant une pression de l'ordre de une atmosphére dans la chambre thermométrique. Un essai a montré que tout se passe régulièrement au point de vue pratique.

BONNEMAY. *J. Physique*, 1939, **10**, 18D.

4

Interprétation des expériences de Michelson-Morley et de Kennedy-Thomdike

Des articles ayant été publiés par H. E. Ives sur la signification des expériences de Michelson-Morley et de Kennedy-Thomdike concluant à une négation de l'explication relativiste, le présent article a un double but: premièrement, exprimer les idées de la théorie spéciale de la relativité de manière à faire ressortir leur caractère physique fondamental et leur distinction absolue de l'idée de la contraction de Fitzgerald-Lorentz et, deuxièmement, montrer que les objections de Ives à la théorie et son interprétation des expériences sont inadmissibles. Expériences et leurs interprétations alternatives; dans l'expérience de Michelson les chemins des faisceaux sont égaux, tandis qu'ils sont inégaux dans celle de Kennedy. La contraction expliquerait la première, mais non la seconde. D'après Ives, les deux expériences s'expliquent, si l'on admet que les longueurs des corps sont altérées dans le rapport $[(1 - v^2/c^2)^{1/2}]^{n+1} : 1$ dans la direction du mouvement et $[(1 - v^2/c^2)^{1/2}]^n : 1$ dans la direction perpendiculaire, et que les battements des horloges sont altérés dans le rapport $[(1 - v^2/c^2)^{1/2}]^{1-n} : 1$. Une expérience de Ives et Stilwell montre que la seule valeur admissible de n est zéro. D'après l'explication relativiste, le mouvement absolu n'a aucune signification. On peut résumer la différence entre les deux hypothèses en disant: avec celle de la contraction, une tige ou une horloge subit un changement physique intrinsèque quand sa vitesse change, avec l'hypothèse relativiste, le changement de vitesse de l'instrument est un processus conceptuel n'impliquant aucun changement physique intrinsèque. On montre le caractère de l'explication relativiste et l'on expose les hypothèses alternatives par rapport à l'expérience, en utilisant celle de Kennedy impliquant seulement les mesures de deux longueurs et de la position des franges d'interférences. Cette expérience ne peut s'interpréter que par la théorie de la relativité. Réfutation de quelques objections de Ives sur la simultanéité absolue en des lieux éloignés, sur l'aberration de la lumière, sur l'expérience de Michelson et l'éther et sur le paradoxe des horloges.

ACKERMANN. *J. Physique*, 1939, **10**, 234D.

5

LES LAMPES À TROIS ÉLECTRODES

L'expérience montre que la caractéristique de grille débute cependant pour un potentiel un peu inférieur à celui de l'extrémité du filament dont le potentiel est le plus bas. Cela tient à ce que des charges électroniques sont répandues dans l'espace autour du filament et modifient le champ qui serait dû aux seules charges des électrodes. Le courant commence, non pas dès que les lignes de force issues de la grille rencontrent le filament, mais lorsqu'elles commencent à se perdre dans les charges négatives qui entourent celui-ci.

London B.Sc. (General).

6

Les vibrations du chevalet d'un violon

Les mouvements du chevalet d'un violon pendant que l'instrument est joué sur différents sons sont étudiés au moyen d'un petit miroir qui est fixé sur ce chevalet et qui réfléchit un rayon lumineux. Il est utilisé également l'observation microscopique directe. Les mouvements du chevalet sont très compliqués et diffèrent beaucoup pour chaque son. D'autre part, ces mouvements se reproduisent d'une manière très précise et ne dépendent pas du mouvement de la corde sur laquelle le son est joué. Il faut distinguer: 1° Une vibration dans le plan du chevalet; 2° Une oscillation avec courbure perpendiculairement à ce plan; 3° Une oscillation avec torsion qui est également perpendiculaire à ce plan. Pour des sons déterminés, il apparaît des maxima caractéristiques pour l'une ou l'autre de ces vibrations et dont on peut démontrer expérimentalement qu'ils appartiennent à un système d'oscillations déterminé. A cause des entaillements du chevalet, ce sont seulement les oscillations du type 1° qui sont communiquées au couvercle de l'instrument qui subit un mouvement de rotation.

van Itterbeek. *J. Physique*, 1938, **9**, 45D.

7

Les symétries en physique

La tendance vers plus de symétrie se manifeste dans un grand nombre de phénomènes irréversibles comme le passage de la chaleur, la diffusion, etc. Ainsi lorsqu'on met en présence deux sphères identiques, l'une chaude, et l'autre froide, il se produit entre elles un échange de chaleur conduisant à l'égalisation des températures; de même lorsque dans l'expérience de Joule on met en relation deux récipients identiques, l'un plein de gaz et l'autre vide, la détente irréversible qui se produit rend le système symétrique. Dans tous ces cas il y a accroissement de l'entropie. Les phenomènes qui feraient diminuer la symétrie feraient diminuer l'entropie.

London B.Sc. (*General*).

8

La pile thermoélectrique

La pile thermoélectrique ne permet jamais que la mesure de différences de rayonnement. Ceci est dû, d'une part à ce que la pile est un récepteur différentiel, d'autre part à l'impossibilité dans laquelle nous nous trouvons de supprimer purement et simplement un rayonnement: un rayonnement supprimé est toujours en fait remplacé par un autre ; par exemple nous supprimons le rayonnement A, en l'interceptant par un écran, mais alors cet écran émet un rayonnement thermique B, lequel vient remplacer le rayonnement A.

Proposons-nous d'établir l'étalonnage d'une pile dont les soudures chaudes sont nues, les soudures froides étant par construction cachées derrière un écran.

Nous plaçons devant la pile une série de diaphragmes, dont le rôle n'est pas de supprimer le rayonnement venant du dehors du champ utile, ce qui est impossible, mais de donner à ce rayonnement une valeur stable dans le temps. Ces diaphragmes devront être noir mat pour toutes les longueurs d'ondes et avoir une grande masse calorifique. Devant le premier diaphragme, nous plaçons un corps noir à température connue, par exemple constitué par une cavité profonde, à parois intérieures noircies au noir de fumée et pratiquée dans un récipient plein de vapeur d'eau à 100°. Le rayonnement de ce corps noir est de 0,112 $\frac{\text{watt}}{\text{cm}^2}$. Nous observons une certaine position du spot du galvanomètre. Ensuite nous remplaçons le récipient par un autre identique empli de glace fondante pilée. Le rayonnement de ce corps noir est 0,032 $\frac{\text{watt}}{\text{cm}^2}$. Nous notons une nouvelle position du spot du galvanomètre. La distance qui sépare sur l'échelle du galvanomètre les deux positions du spot, correspond à la différence des rayonnements des deux corps noirs reçus par la pile.

Bayle. *Comm. Lab. Inst. Opt.*, 1944, **2**, 36.

9

ETUDE SUR L'ÉLASTICITÉ DE FLEXION

Le travail que nous publions maintenant fait suite à celui paru sous le même titre dans les Helvetica Physica Acta et la méthode employée a été la même.

Nous rappelons que les oscillations d'un pendule de torsion, auquel était adjoint un ressort en forme de spirale plane du matériel à étudier, étaient observées par la méthode de Poggendorff, l'échelle étant à 1090 mm. du miroir. La correspondance des amplitudes était la suivante:

Amplitude double du spot en divisions de l'échelle	Amplitude des oscillations en degrés
200 mm.	2° 39′
100 mm.	1° 19,5′
50 mm.	40′
20 mm.	16′

Les valeurs de la période, utilisées dans les calculs, ont généralement été déterminées pour des amplitudes doubles du spot comprises entre 200 et 20 mm.; il s'agit donc de faibles amplitudes. Nous sommes parfois descendus jusqu'à quelques millimètres seulement; dans ces dernières conditions, le mouvement de l'équipage mobile était pratiquement invisible à l'œil nu.

Par extrapolation, nous avons obtenu le module à 0°; il nous a servi à calculer le coefficient thermoélastique moyen pour l'intervalle 0 et 80°, tel qu'il figure dans différents tableaux de ce mémoire. Afin de mettre en évidence l'incurvation très faible de la courbe représentative du module *e* en fonction de la température, nous avons calculé les différences entre les valeurs expérimentales et une droite passant par les points relatifs à 0 et 80°, exprimée au moyen du coefficient thermoélastique ci-dessus.

De nombreux graphiques permettent une vue rapide de l'ensemble des résultats.

JAQUEROD et MÜGELI. *Helv. Phys. Acta*, 1931, **4,** 3.

10

Rayonnement ionisant émis par les métaux

Les métaux ordinaires possèdent la propriété d'ioniser les gaz placés dans leur voisinage, d'impressionner à distance la plaque photographique et de produire, dans l'atmosphère qui les environne, des centres de condensation pour la vapeur d'eau faciles à mettre en évidence avec l'appareil de Wilson.

Il nous a paru intéressant de rechercher si les actions de ces métaux sur l'électromètre et sur la plaque sensible, sont indépendantes l'une de l'autre ou ne sont que des manifestations d'une même propriété dont l'origine peut être discutée, mais dont l'existence paraît incontestable, savoir: l'émission d'un rayonnement peu pénétrant par tous les métaux. Et cette recherche nous a paru d'autant plus désirable que nous avons à notre disposition, tant au point de vue pratique qu'au point de vue théorique, des moyens qui manquaient à nos prédécesseurs. Les renseignements que nous possédons sur le rayonnement ultrapénétrant cosmique permettent de comprendre l'émission d'un rayonnement par les métaux ordinaires et d'en expliquer les particularités; d'autre part, dans l'étude des actions sur la plaque, on peut, grâce à la photométrie photographique, effectuer des mesures là où on ne pouvait guère autrefois que donner des résultats qualitatifs.

En somme, nous nous proposons d'étudier successivement l'action ionisante des métaux ordinaires et l'impression qu'ils produisent sur la plaque sensible, nous discuterons l'origine ou le mécanisme de ces actions, nous en déduirons les renseignements quantitatifs que l'on peut en tirer et enfin nous essaierons de voir, en rapprochant les résultats obtenus si ces deux actions présentent ou non quelque corrélation.

Reboul. *Ann. Phys.*, 1939, **11,** 354.

11

Absorption du son dans le CO_2 et autres gaz

Les mesures d'absorption dans l'acide carbonique à la pression atmosphérique et à 22° C. ainsi que d'autres mesures confirment la théorie de l'absorption anomale de Einstein, Kneser, etc. Le coefficient d'absorption à des fréquences descendant jusqu'à 2000 cycles atteint un maximum de 0,317 par longueur d'onde à 77000 cycles. Ce maximum est plus élevé que les valeurs obtenues par certains expérimentateurs; il indique que la déformation ainsi que la vibration des valences symétriques participent aux échanges entre l'énergie de translation et l'énergie de vibration. De faibles impuretés telles que l'eau ou la vapeur d'alcool affectent grandement l'absorption; la bande d'absorption entière est décalée vers les hautes fréquences. Des mesures dans des mélanges CO_2, O_2 ou CO_2, N_2 montrent que ni O_2, ni N_2 n'est excité d'une manière appréciable par les collisions avec CO_2. Les mesures dans le CS_2 montrent une absorption analogue à celle du CO_2, mais l'absorption commence à 10000 cycles. Dans des mélanges CS_2, O_2 l'absorption observée pour des fréquences inférieures s'explique si on suppose que seule la vibration des molécules de O_2 est excitée par collision avec le CS_2; à des fréquences plus élevées les molécules de CS_2 sont excitées elles aussi.

GANS. *J. Physique*, 1939, **10,** 110D.

12

CONFRONTATION DE LA THÉORIE DE DEBYE AVEC L'EXPÉRIENCE

Un premier contrôle de la théorie consiste à chercher s'il existe, pour un corps déterminé, une température caractéristique θ qui permette de représenter l'évolution expérimentale de la chaleur spécifique avec la température par une fonction de Debye. Ce contrôle s'effectue, le plus commodément, en portant simultanément, dans l'equation, les valeurs des températures et celles des chaleurs atomiques correspondantes et en detérminant, pour chaque couple de valeurs, la température θ satisfaisant l'equation. Une vérification parfaite de la théorie se traduirait par une constance de θ.

London B.Sc. (Special).

13

Points de Curie des métaux du groupe yttrique

Récemment, Urbain, Weiss et Trombe, puis Klemm et Bommer, ont montré que les métaux du groupe yttrique étaient des paramagnétiques à champ moléculaire: le gadolinium, notamment, est ferromagnétique à la température ordinaire.

L'étude de la constante de Curie montre que ces métaux possèdent le même moment magnétique que l'ion trivalent correspondant; les théories actuellement admises permettent donc d'en rendre compte. Il reste à interpréter les valeurs expérimentales des points de Curie. Par analogie avec le groupe du fer, l'auteur suppose que le champ moléculaire provient d'actions d'échange entre les électrons: il n'y a d'interactions qu'entre moments de spin, il n'y en a pas entre moments orbitaux qui, donc, ne sont pas couplés directement. Mais, comme sur un même atome, les spin et les moments orbitaux sont couplés pour donner un moment magnétique résultant, tout se passe finalement comme s'il y avait entre les moments résultants un couplage direct que l'auteur s'est proposé de calculer.

Il s'agit d'abord d'évaluer l'interaction entre deux spin placés chacun sur la couche magnétique $4f$ de deux atomes voisins. L'étude des éléments des groupes de transition a permis à l'auteur de montrer, qu'en première approximation, elle ne dépendait que de la plus courte distance entre les couches magnétiques de deux atomes proches voisins. Cette distance, par un hasard heureux, est pratiquement la même pour tous les éléments envisagés. Donc l'interaction entre deux spin conserve la même valeur pour ces mêmes métaux.

Comme chaque atome possède plusieurs spin couplés parallèlement, l'interaction totale entre les moments résultants est proportionnelle au carré du nombre des spin couplés parallèlement sur chaque atome, avec le même facteur de proportionnalité pour tous les éléments du groupe, d'après ce qui a été dit ci-dessus.

Soit donc L, S, J, les nombres quantiques résultants de l'atome, donnés par les règles de Hund, pour les ions trivalents;

un calcul simple montre que les points de Curie des métaux yttriques sont simplement proportionnels à $S^2(J+1)/J$. Le facteur de proportionnalité a été déterminé au moyen du point de Curie du gadolinium. Les points de Curie Θ, en degrés absolus des autres éléments ont été ensuite calculés et comparés aux résultats expérimentaux de Klemm et de Bommer, dans le tableau suivant:

ÉLÉMENT	Tb	Dy	Ho	Er	Tm
Θ calc. ..	202°	136	86	50	22
Θ obs. ..	205°	150	—	40	10

NÉEL. *J. Physique*, 1938, **9**, 50S.

14

TRANSMISSION DES MONOCHROMATEURS

L'étude spectrale énergétique d'un rayonnement impose avant toute mesure sa décomposition en radiations simples, au moyen d'un monochromateur; celui-ci va non seulement isoler ces radiations, mais les absorber, les réfléchir, les diffuser, ou réduire l'étendue des faisceaux. On ne peut mesurer la composition d'un rayonnement directement, mais seulement après les modifications qu'il a subies dans l'appareil dispersif. Une étude préalable des monochromateurs est donc indispensable. Elle s'impose non seulement pour l'étude des rayonnements, mais toutes les fois qu'il faut établir une relation quantitative entre les grandeurs photométriques ou énergétiques qui caractérisent le rayonnement monochromatique isolé par l'appareil dispersif et les grandeurs relatives à ce rayonnement avant qu'il ait traversé l'appareil.

Dans quelques cas particuliers, cette relation est une simple proportionnalité, avec un coefficient déterminé pour chaque longueur d'onde et chaque état de polarisation, appelé facteur de transmission. Ce facteur sera défini plus loin; d'autres auteurs en ont donné le calcul, en faisant intervenir les pertes par réflexion sur les faces des lentilles et des prismes.

Mais en général, le monochromateur impose des modifications géométriques aux faisceaux; de plus, ses aberrations, souvent importantes devant la largeur des fentes, ne permettent plus d'utiliser sans discernement certaines relations photométriques simples coutumières. C'est pourquoi il a paru utile de reprendre d'une façon plus générale, et en insistant surtout sur les facteurs géométriques, l'étude des relations quantitatives entre les qualités photométriques du rayonnement qui entre dans un monochromateur et les qualités de celui qui en sort.

Les séparateurs de radiations sont très divers, aussi est-il difficile de donner des conclusions valables pour tous. Nous avons surtout pensé aux monochromateurs à prismes, simples ou doubles.

TERRIEN. *Comm. Lab. Inst. Opt.*, 1945, **6**, 125.

15

IRRADIATION D'URANIUM ET DE THORIUM PAR LES NEUTRONS

On sait que les noyaux d'uranium et de thorium irradiés par les neutrons peuvent subir une explosion qui les scinde en deux atomes radioactifs. Certains des corps formés ont les propriétés des terres rares. Il existe, en particulier, un radioélément de période 3,5 h environ formé dans le thorium, que l'on avait pris tout d'abord pour un isotope de l'actinium, mais qui a été récemment reconnu semblable au lanthane. Un radioélément de période analogue, également semblable au lanthane, a été extrait de l'uranium irradié.

On peut supposer que ces produits de fission sont identiques, toutefois, il n'est pas très facile de le prouver: on n'a pas pu jusqu'ici déterminer les périodes avec une très grande précision; au point de vue chimique, on peut aisément montrer que le radioélément $R_{3,5\,h}$ se distingue de l'actinium, car le coefficient de fractionnement des oxalates de La et Ac est grand, mais on ne peut affirmer que $R_{3,5\,h}$ est identique à l'une ou l'autre des terres cériques, celles-ci se séparant difficilement les unes des autres, même par fractionnement. Les expériences faites sur le mélange de La + $R_{3,5\,h}$

extrait soit de l'uranium, soit du thorium, indiquent tantôt un coefficient de fractionnement nul, tantôt une légère concentration de $R_{3,5\,h}$ en tête du fractionnement, mais ces différences sont dans les limites des erreurs expérimentales.

CURIE, I., et TSIEN. *J. Physique*, 1939, **10**, 496.

16

LES ABERRATIONS GÉOMÉTRIQUES ET LA LIMITE DE LORD RAYLEIGH

Lorsqu'un instrument est parfaitement stigmatique, son pouvoir séparateur n'est limité que par les phénomènes de diffraction. Un instrument aberrant pourra donner des images de même qualité si ses aberrations sont suffisamment faibles pour ne pas modifier sensiblement la répartition de l'énergie dans la tache de diffraction: on dira alors que l'instrument est parfaitement corrigé.

Lord Rayleigh a montré à l'aide de quelques exemples que ce résultat est atteint si la différence de marche géométrique ne varie pas de plus de $\frac{\lambda}{4}$ pour les différents points de la surface d'onde. Ceci revient à dire que la surface d'onde est entièrement comprise entre 2 sphères concentriques, distantes de $\frac{\lambda}{4}$.

De nombreux calculs effectués en particulier par Conrady ont montré que si l'on applique cette règle, la répartition des éclairements dans la tache de diffraction est toujours sensiblement la même que pour un instrument parfait: pour l'observation des objets à grand contraste, la correction des aberrations sera donc satisfaisante. Il n'en est pas de même si l'objet a un faible contraste: de petites modifications de la répartition des éclairements pourront faire baisser notablement le pouvoir séparateur de l'instrument; la règle de Lord Rayleigh n'est plus assez sévère si l'on construit un instrument destiné à observer des objets à faible contraste (fins détails sur les planètes par exemple).

MARECHAL. *Comm. Lab. Inst. Opt.*, 1944, **3**, 60.

17

Variation thermique de la constante diélectrique

Tandis que la constante diélectrique des gaz a sa variation thermique bien représentée par les formules de Clausius Mosotti et de Debye, il n'en est pas toujours de méme pour les liquides. Les écarts entre les valeurs observées et les valeurs calculées par ces formules sont dus à la fois à ce que le champ électrique réellement appliqué aux molécules n'est pas exactement le champ de Lorentz et à ce que les molécules ne sont ni réparties ni orientées absolument au hasard. Ces écarts sont plus élevés chez les liquides polaires où on les attribue souvent à des associations de dipoles. Cette interprétation n'est pas valable pour les liquides homopolaires où cependant des anomalies allant de la variation de la pente $\frac{d\epsilon}{dT}$ au changement du signe de cette pente avaient été signalées pour les températures peu supérieures au point de fusion. On pouvait penser que ces anomalies étaient dues à la structure microcristalline des liquides. Ceci amenait à se demander comment varie la constante diélectrique lorsque la structure cristalline apparaît dans toute la masse à la solidification. Les résultats de différents auteurs peu concordants ne semblaient pas permettre de dégager de loi générale. Par exemple pour le benzène ϵ est plus élevé à l'état solide qu'avant la congélation suivant Errera, mais plus petit suivant Isnardi. Pour les liquides organiques étudiés par Isnardi ϵ décroît après la solidification et la polarisation $P = \frac{\epsilon - 1}{\epsilon - 2}\frac{1}{d}$ diminue notablement. Il semblait en être de même pour l'azote et l'hydrogène, au contraire le cyclohexane et le soufre montrent une augmentation de ϵ, P étant à peu près constant au changement d'état. Un fait particulièrement curieux était qu'Isnardi avait observé que ϵ diminuait déjà avant la solidification (sulfure de carbone, éther) et que Mazur avait confirmé ce fait et l'avait interprété par l'existence de points λ analogues à celui de l'hélium.

Guillien. *J. Physique*, 1940, **1**, 29.

18

EFFICACITÉ DES COMPTEURS GEIGER-MÜLLER

Au cours de nos recherches sur les spectres β à l'aide d'un spectrographe magnétique, fonctionnant avec deux compteurs Geiger-Müller en coïncidences, nous avions remarqué une variation du quotient $\frac{n_c}{N'_2}$, représentant le rapport du nombre des coïncidences au nombre des impulsions simples dans le deuxième compteur, en fonction de l'énergie des particules incidentes. Comme l'absorption ne pouvait jouer de rôle, étant donné que les particules devaient pénétrer dans le deuxième compteur aussi bien pour le faire fonctionner seul que pour donner des coïncidences, la variation constatée ne pouvait s'expliquer que par une diminution de l'efficacité des compteurs pour les rayons de grande énergie. Le pouvoir ionisant des particules β variant dans le même sens, il était naturel d'attribuer la diminution du rapport $\frac{n_c}{N'_2}$, dans le cas des particules rapides, au fait qu'un certain nombre de particules pouvaient traverser le premier compteur sans produire une seule paire d'ions à l'intérieur de celui-ci.

Nous avons jugé intéressant d'étudier la variation de ce rapport. Nous montrerons plus loin qu'il représente l'efficacité du premier compteur, et que sa connaissance permet la détermination du nombre de paires d'ions primaires formées dans l'air par centimètre de parcours, en fonction de l'énergie des particules ionisantes.

Inversement, la connaissance de l'ionisation spécifique primaire rend possible le calcul de l'efficacité d'un compteur donné, en fonction de ses dimensions et de sa pression.

En dehors de la numération des gouttelettes observées dans la chambre de Wilson, la méthode des compteurs est même la seule, à notre connaissance, fournissant une mesure directe de l'ionisation spécifique primaire des rayons β.

GRAF. *J. Physique*, 1939, **10**, 513.

19

Discontinuités au ventre d'un tuyau sonore

Un tuyau sonore, fermé à l'une de ses extrémités, est entretenu à l'autre par un téléphone depolarisé, branché sur le secteur. On se propose d'étudier les modifications apportées au mouvement par différentes discontinuités, introduites au ventre unique de ce tuyau. On étudiera simultanément les variations de phases et d'amplitudes. Beaucoup de ces mesures sont faites de deux manières: au moyen de capsules manométriques et au moyen de poussières d'amidon, entraînées par l'air en vibration.

Une première discontinuité consiste à ouvrir le tuyau en ventre, et à le mettre en communication avec un volume: soit infini (l'atmosphère), le diamètre de l'ouverture étant variable: soit fini, constitué par un tuyau de même diamètre et de longueur variable.

En mettant en communication avec l'atmosphère, on constate qu'il n'y a plus opposition de phase entre les variations de pression aux deux extremités du tuyau; ceci est dû au fait que l'onde existant dans le tuyau n'est qu'une onde quasistationnaire.

À ces phénomènes nous rattacherons l'ouverture d'un petit trou dans la paroi terminant le tuyau, à l'opposé du téléphone. Tant que ce trou est petit, il équivaut à un raccourcissement du tuyau; il y a modification de la différence de phase et diminution d'amplitude.

Mettre le ventre du tuyau en communication avec un second tuyau revient à étudier une dérivation, qui peut être soit en T soit en Y, la branche inférieure contenant le téléphone. Lorsque l'une des deux branches a seule une longueur convenable pour la résonance, elle prend à peu près toute l'énergie et le mouvement est inexistant dans l'autre, même si celle-ci a une longueur voisine de celle de la résonance. La première agit donc comme filtre.

Guittard. *Ann. Phys.*, 1946, **1**, 142.

20

SCISSION DE NOYAUX « À GOUTTELETTES » DE L'URANIUM ET DU THORIUM

Bohr et Fermi avaient rapporté à une conférence de Physique les expériences de Hahn et Strassmann concernant la formation de Rd-Ba par U + *n* et l'interprétation de Frisch et Meitner comme base du modèle nucléaire « à gouttelette » de Bohr. Afin d'observer les particules d'énergie très élevée qui doivent être émises dans cette scission, les auteurs ont recouvert des disques en Cu d'oxyde d'U et ils les ont placés sur le plateau collecteur d'une chambre d'ionisation se trouvant en-dessous d'une source de neutrons. En présence de neutrons émis par Li + D, on a observé. à l'aide d'un amplificateur et d'un oscillographe cathodique, en dehors des impulsions dues aux rayons α d'U, 2 groupes de particules dont les unes sont 4 fois plus ionisantes que les α et les autres de 20 à 90 fois plus que les précédentes, correspondant à des énergies de 75 à 150 MéV. En présence de paraffine, le rendement est environ 30 impulsions par min. et par μA de deutons sous 1000kV. Le rendement avec Th est du même ordre de grandeur. Résultat négatif avec Bi, Pb, Tl, Hg, Au, Pt, W, Sn et Ag. Pas d'effet avec U ou Th et rayons γ de Li + H ou F + H. L'effet avec U peut être produit, soit par neutrons rapides (après filtration par Cd) ayant nne énergie supérieure à 0,5 MéV (C + D) et inférieure à 2,5 MéV (D + D). Avec Th, seuls les neutrons rapides sont efficaces.

HAÏSSINSKY. *J. Physique*, 1939, **10**, 246.

21

L'ABSORPTION DES RAYONS GAMMA PÉNÉTRANTS

L'origine des rayons cosmiques n'étant pas encore connue d'une façon certaine, toute hypothèse qui pourrait l'expliquer doit être contrôlée. On admet généralement que les rayons cosmiques accompagnent la formation des éléments; la preuve n'a cependant pas encore été fournie directement. Si l'on regarde de plus près, on se rend compte qu'on n'a même pas encore exclu d'une façon certaine que les rayons cosmiques soient tout simplement des radiations de la décomposition radioactive ordinaire. En effet, en laissant de côté toute considération théorique, on pourrait supposer que ces rayons accompagnent les rayons gamma ordinaires, mais en une proportion aussi faible qu'ils ont échappé jusqu'à présent à l'observation dans l'étude des corps radioactifs. Par contre, si ces rayons, provenant de fortes sources radioactives dans l'espace, traversent l'atmosphère terrestre (qui est équivalente à environ 85 cm. de plomb), les rayons gamma ordinaires y seraient absorbés beaucoup plus fortement (à leur 10^{-18} ème partie environ !) que les composantes pénétrantes, de sorte que pratiquement seul ces dernières peuvent arriver jusqu'à la surface de la terre.—Des mesures d'absorption des rayons gamma ordinaires avec de très grandes épaisseurs des écrans absorbants pourraient éclaircir cette question.

A part ces spéculations sur l'origine des rayons cosmiques ces mesures présentent encore d'autres aspects intéressants: Dans un travail récent Steadman prétend avoir trouvé dans le spectre des rayons gamma du RaC des lignes correspondant à de très petites longueurs d'onde, jusqu'à 0,0003 Å, donc 20 fois moins que ce qu'on admettait jusqu'à présent comme limite des rayons gamma. De si courtes longueurs d'onde devraient se manifester par un très grand pouvoir de pénétration des rayons. Il y a en outre des discordances entre des mesures de Soddy et de Hoffmann; le premier trouve que le coefficient d'absorption des rayons gamma du radium reste constant au-dessus des filtrations de 5 cm. de Pb, le second interprète ses mesures en admettant qu'il y a à côté des rayons du coefficient d'absorption normale de 0,53 cm^{-1} une composante plus dure avec $\mu_{Pb} = 0{,}33\ cm^{-1}$.

Il nous a été possible de reprendre ces questions dans de très bonnes conditions parce que pour ces recherches le « Radium Belge » a mis 5 grammes de radium-élément à notre disposition.

Piccard, Stahel et Dony. *Helv. Phys. Acta*, 1931, **4**, 371.

22

Absorption des verres d'optique

Appelons facteur de transparence K d'un échantillon à faces parallèles le rapport du flux lumineux qui atteint la face de sortie au flux qui a pénétré dans le verre. Si k est le facteur d'absorption relatif à 1 cm. et x l'épaisseur de l'échantillon, on a:

$$K = (1 - k)^x;$$

soit r le coefficient de réflexion d'une surface de verre, R et T les facteurs de réflexion globale et de transmission de l'échantillon, on a:

$$R = r\left[1 + \frac{K^2(1-r)^2}{1 - K^2r^2}\right] \qquad T = \frac{K(1-r)^2}{1 - K^2r^2}.$$

Plaçons maintenant derrière l'échantillon un miroir métallique de pouvoir réflecteur ρ, voisin de l'unité et évaluons le flux réfléchi par l'ensemble de l'échantillon et du miroir. Si l'on peut considérer R, $1 - K$, et $1 - \rho$ comme petits, les flux lumineux réfléchis sur les faces de l'échantillon sont presque intégralement renvoyés vers l'appareil récepteur, tandis que le faisceau principal est affaibli par l'absorption du verre pendant ses deux traversées de la lame et par sa réflexion sur le miroir.

Le rapport du flux lumineux Φ réfléchi par l'ensemble échantillon miroir au flux Φ', réfléchi par le miroir seul est donc voisin du facteur de transparence K^2 relatif à une épaisseur $2e$.

Marechal. *Comm. Lab. Inst. Opt.*, 1944, **5**, 120.

23

Les phénomènes électromagnétiques dans les métaux

Le présent ouvrage par Arkadiew est consacré à l'examen des phénomènes électromagnétiques dans les métaux et principalement dans les corps ferro-magnétiques, donc à un domaine des plus complexes de l'électricité et ayant, en même temps, préoccupé le moins les divers auteurs. On dispose actuellement de nombreux ouvrages sur la conductivité électrique, sur le magnétisme ou sur les corps magnétiques.

Les uns traitent les problèmes fort abstraits de la mécanique des atomes, d'autres examinent les questions technologiques, sans effleurer presque le développement mathématique des processus ayant lieu dans les corps magnétiques sous l'action des champs et des courants alternatifs.

Le premier chapitre du présent ouvrage est consacré aux problèmes de la structure des métaux et des alliages, ainsi que d'autres chapitres où l'on traite spécialement de leurs propriétés conductives et magnétiques.

L'examen des conceptions fondamentales de la théorie du potentiel est basé sur les phénomènes du courant électrique, sur les lois de la diffusion, de la transmission de chaleur et de la filtration des liquides, ce qui nous mène à l'élucidation des lois fondamentales du flux du vecteur et en particulier à celle de la loi de la proportionnalité inverse au carré de la distance.

En liaison avec l'exposé des problèmes concernant l'aimantation des corps séparés et de circuits magnétiques ouverts on introduit les conceptions de l'aimantation des corps de la substance et de la forme, appliquée par l'auteur dès 1913. Ces conceptions sont adoptées actuellement par d'autres auteurs et facilitent notamment l'exposé de la théorie moderne des aimants où la valeur $P_A \frac{l}{s}$ n'est autre chose que notre perméabilité de la forme *m*. Certains cas d'aimantation dans un champ non homogène ainsi que la théorie des images magnétiques sont donnés par nous d'après Hague. L'auteur a élaboré, d'après Evershede et Picou, la théorie des aimants permanents, sous une forme plus appropriée.

La seconde partie est consacrée aux champs alternatifs et à la propagation des phénomènes électromagnétiques dans les métaux. C'est ici qu'est introduite la conception de la dynamique magnétique. Le processus de la naissance du champ électromagnétique dans les métaux est comparé à celui de la conductibilité de la chaleur. On emploie un schéma commun pour tous les cas particuliers de l'effet superficiel, notamment en ce qui concerne la pénétration du champ extérieur à l'intérieur des métaux.

Ces phénomènes sont étudiés pour les corps ferromagnétiques à l'aide de la notion de la perméabilité complexe introduite par l'auteur et dont l'application est maintenant généralement admise. La théorie du comportement des fils cylindriques et de la tôle dans un champ magnétique ou électrique, élaborée dans tous ses détails, a permis d'établir des diagrammes commodes pour les calculs pratiques.

Dans la même partie de l'ouvrage est exposée la conception du spectre électromagnétique général de la substance et en particulier des spectres magnétiques des métaux et des corps ferro-magnétiques.

Il est impossible d'énumérer ici tout ce que l'ouvrage contient de nouveau ou d'original et que seulement son étude approfondie permettra d'apprécier.

J. Physique, 1937, **8**, 229D.

24

NOUVEL AMPLIFICATEUR POUR COURANTS CONTINUS

L'appareil a pour but de mesurer les faibles courants produits au moyen de cellules photoélectriques. Les auteurs ont recherché non pas tant une très grande sensibilité (laquelle est quasi illimitée avec les amplificateurs à lampes) qu'une grande stabilité.

L'expérience montre que le défaut de stabilité provient surtout de ce que le chauffage des filaments des lampes au moyen des batteries d'accumulateurs n'est pas constant.

Or une variation de température de la cathode entraîne une variation de la résistance interne ρ et par suite une variation de la pente de la caractéristique $i = f(u)$, u désignant la tension de la grille, le potentiel de la cathode étant pris pour origine. Alors même que le chauffage est constant, la pente S de la caractéristique n'est pas constante, elle dépend du courant anode-cathode i.

Le pouvoir amplificateur k est sensiblement constant. On doit donc rechercher d'une part un chauffage aussi constant que possible pour assurer une stabilité et d'autre part un montage tel que la résistance interne apparente de la lampe soit également constante de façon à ce que les variations de tension de grille entraînent des variations proportionnelles de courant anode-cathode.

On obtient un courant de chauffage pratiquement constant en utilisant deux lampes à chauffage indirect dont les filaments sont montés en série avec un régulateur fer-hydrogène. Le tout est alimenté par le secteur. Le courant se maintient très constant et égal à 0,3 A lorsque la tension du secteur varie entre 100 et 120 V. Pour diminuer l'influence des variations de ρ nous avons utilisé un montage à rétroaction de courant. Les montages à rétroaction soit de tension, soit de courant, ont fait leurs preuves dans le cas des amplificateurs de tensions variables; ils rendent les mêmes services dans les cas où l'on désire amplifier des faibles courants continus.

BEDEAU et HERMAN. *J. Physique*, 1938, **9**, 47S.

25

MATÉRIALISATION DE L'ÉNERGIE

Un phénomène tout à fait analogue à la matérialisation d'un photon peut avoir lieu avec une particule matérielle chargée, de grande énergie cinétique. Cette particule peut, par interaction avec un noyau, perdre une certaine quantité de son énergie cinétique, qui sera matérialisée en donnant naissance à une paire d'électrons de signes contraires. Pour que cela soit possible il faut évidemment que l'énergie cinétique de la particule initiale soit d'au moins $2m_0c^2$, soit environ 1,024 MeV, énergie correspondant à la masse propre des particules créées. Comme dans le cas de la matérialisation du photon la présence du noyau est indispensable pour qu'il puisse y avoir en même temps conservation de l'énergie et de l'impulsion. L'impulsion prise par le noyau est importante mais l'énergie est très faible de sorte qu'on peut la négliger dans le bilan énergétique du phénomène. Si l'on ajoute les énergies cinétiques des trois particules existantes après la matérialisation plus 1,024 MeV nécessaires pour la création des deux électrons provement de la matérialisation on doit obtenir l'énergie cinétique de la particule incidente.

Un tel phénomène a pu être observé dans une chambre de Wilson quand il prend naissance, soit dans le gaz même de la chambre, soit dans un écran mince placé à l'intérieur de cette chambre. Dans toutes les expériences la particule initiale était un rayon β. Le phénomène devait se présenter alors de la façon suivante: la trajectoire du rayon β en un certain point se diviserait en trois branches, deux correspondant à une charge négative et une correspondant à une charge positive. Si dans la chambre règne un champ magnétique uniforme, toutes ces trajectoires seront circulaires et le sens de la courbure indiquera immédiatement le signe de la charge.

On n'a jamais obtenu, à notre connaissance, des clichés correspondant exactement à l'aspect que nous venons de décrire; on a seulement observé la subdivision de la trajectoire initiale en deux trajectoires correspondant à des électrons de signes contraires. Cela s'explique si on admet que l'énergie finale de la particule incidente est nulle.

Le bilan énergétique du phénomène est assez difficile à établir puisque les rayons de courbure de la trajectoire initiale et des trajectoires résultantes sont d'un ordre de grandeur très différent en raison de la grande perte d'énergie cinétique subie par la particule initiale. Cela fait que si on choisit pour le champ magnétique une valeur convenable pour que le rayon de la trajectoire initiale soit mesurable avec précision, il sera trop fort pour permettre une mesure précise des autres rayons de courbure et inversement. On a pu une seule fois établir le bilan énergétique et on s'est trouvé en bon accord avec la théorie.

DA SILVA. *Ann. Phys.*, 1939, **11**, 513.

26

PRESSION DE VAPEUR SATURANTE DU LITHIUM

La pression de vapeur saturante du lithium a été déterminée par Hartmann et Schneider entre 931° et 1080°, en mesurant la température d'ébullition de ce métal dans une atmosphère d'argon de pression variable.

Lester C. Lewis a tracé la courbe de tension de vapeur du lithium au voisinage de 600°, en étudiant l'équilibre entre atomes et molécules dans les vapeurs de métaux alcalins par la méthode des jets atomiques. Il trouve des valeurs plus élevées que celles que l'on peut déduire par extrapolation des mesures de Hartmann et Schneider ; j'ai pensé que ces valeurs trop élevées pouvaient être dues à la présence de sodium dans le lithium utilisé et j'en ai conclu à la nécessité d'un contrôle de la pureté du lithium fait avant son emploi.

Bogros utilisant entre 459° et 572° la méthode des jets atomiques avec détection sur une plaque à condensation, a trouvé des nombres coïncidant à peu près aux basses températures avec ceux déduits par extrapolation des résultats de Hartmann et Schneider, mais qui leur sont inférieurs aux températures élevées dans un rapport croissant avec la température. Il pensait que cette divergence pouvait s'expliquer en partie par les inégalités de température existant entre le lithium fondu et la paroi supérieure du creuset et que de nouvelles expériences, où cette cause d'erreur serait éliminée, seraient nécessaires pour trancher la question.

MAUCHERAT. *J. Physique*, 1939, **10,** 441.

27

LE RADAR

Ce livre a pour but d'initier le grand public aux principes essentiels du radar, cela en 60 pages seulement, c'est dire combien est condensée cette étude, qui, examinée en ses détails, exigerait un volume d'importance considérable.

Après avoir montré les origines du radar, l'auteur indique comment se présente son installation, il ouvre ensuite une parenthèse pour décrire d'un façon sommaire le fonctionnement des tubes à rayons cathodiques puis montre à l'aide d'un schéma simplifié comment s'effectue l'envoi des signaux et leur réception sur l'écran de l'oscilloscope après réflexion.

Des considérations sur la triangulation et l'emploi des ondes ultracourtes complètent cette description. La base de temps, le balayage sont également examinés, puis, l'auteur après quelques indications sur la technique des courants de haute fréquence et sur l'emploi du magnétron termine son exposé en résumant les applications du radar au matériel d'armement et à ses développements au temps de paix. Des schémas très clairs et de curieuses photographies donnent à ce travail un caractère attrayant qui contribuera au succès de cet ouvrage auprès du grand public à qui il s'adresse plus particulièrement.

Rev. Gén. Élec., 1946, **40**, 290.

28

Collisions nucléaires

Les données expérimentales accumulées dans les dernières années sur les phénomènes de collision nucléaire ont permis à Bohr, Breit et Wigner de dégager plusieurs traits caractéristiques de ces phénomènes. On a reconnu ainsi le caractère essentiellement dispersif de ces collisions en opposition avec les processus de collision atomique. Dans certains cas limites les états intermédiaires intervenant dans la description de la collision se réduisent à un ou plusieurs états, donc à un nombre fini d'états, auquel cas ils correspondent à des niveaux d'activation réels du noyau formé dans le premier stade de ce processus dispersif limite. Le caractère de réalité de ces niveaux exige que les théorèmes de conservation jouent dans le stade du processus où ils sont créés. Plus le caractère de réalité d'un niveau intermédiaire devient prononcé, plus les relations de phase entre les éléments dynamiques entrant dans la description du phénomène de collision dispersive s'estompent. On peut donc, dans ce cas limite, se borner à l'étude de ces noyaux intermédiaires activés semi-stables produits dans la première étape du phénomène ainsi qu'à leur disparition conduisant à un état final déterminé. Mais le fait reconnu par Bohr que la plupart des niveaux d'énergie nucléaire doivent correspondre à des états d'activation simultanée de plusieurs constituants nucléaires conduit à adopter comme méthode de description de ces états une méthode où l'on peut tenir compte simultanément du mouvement d'un grand nombre de corpuscules. Alors qu'en principes les méthodes générales de la théorie des quanta paraissent être applicables à ce problème, les méthodes pratiquement utilisables, telle que la méthode du champ self-consistent de Hartree, ne lui seraient pas adaptées. On doit se contenter ici, actuellement, d'une méthode limite de statistique quantique.

Une première limitation importante inhérente aux méthodes courantes de statistiques quantiques consiste en ce que ces méthodes, tout en maniant *h*, le quantum d'action de Planck, laissent encore de côté la non-commutation des grandeurs mécaniques canoniquement conjuguées. Il en résulte alors qu'un modèle nucléaire statistique ne peut prétendre, précisément dans un domaine où *h* paraît jouer un rôle extrêmement important, qu'à une description assez grossière de la véritable situation.

GOLDSTEIN. *J. Physique*, 1938, **9**, 96.

29

ACTION DE LA CHALEUR SUR LES ISOLANTS

Les températures élevées constituent un facteur de destruction des isolants des câbles, et se trouvent ainsi être une cause de production de défauts sur les fileries secondaires.

Bien que généralement les fileries bien calculées ne travaillent pas à une forte densité de courant et ne sont pas portées à une température nocive pour les isolants, il est cependant certains cas où l'isolant peut atteindre des températures dangéreuses. Tel est le cas, par exemple, des conducteurs des moteurs de régulateurs des turbines qui sont portés, par la chaleur dégagée autour des machines, à des températures élevées.

Egalement, dans les coffrets étanches où la dissipation de chaleur est médiocre, des températures gênantes peuvent être atteintes. Enfin, dans les postes extérieures, certains conducteurs et les boîtes d'extrémité situées sous la couvercle même de transformateurs pouvant s'échauffer à 80° C. sont eux-mêmes portés à des températures du même ordre.

L'élément chaleur n'est donc pas à négliger, et dans la détermination de la nature de l'isolement des câbles, il sera nécessaire d'en tenir compte, et chaque cas particulier devra faire l'objet d'un examen pour permettre le choix de la solution optimum à adopter.

JOSSE. *Rev. Gén. Élec.*, 1946, **40**, 296.

30

SUR LE RAYONNEMENT γ DE FREINAGE INTERNE DE Ra*E*

Parmi les éléments naturels qui se décomposent avec émission de rayons β, seul le Ra*E* présente la particularité de ne pas émettre de rayonnement γ nucléaire dû à une désexcitation du noyau.

Cette particularité permet à l'expérimentateur d'étudier de cet élément le « rayonnement de freinage interne » excité par les rayons β au moment de leur expulsion du noyau. Dans les autres cas, en effet, le r.f.int., s'il existe, est absolument masqué par le rayonnement γ nucléaire, toujours beaucoup plus intense.

Stahel et Kipfer ont montré d'autre part que ce que l'on nomme généralement le rayonnement γ du Ra*E* se compose en grande partie de « rayonnement de freinage externe » excité par le rayonnement β dans la matière environnante. Ces auteurs ont en même temps indiqué une méthode permettant de déterminer la valeur du r.f.int. seul.

Si l'on absorbe entièrement le rayonnement β dans une faible quantité de matière formant « radiateur » et entourant la source, ce radiateur émet un r.f.ext. dont l'intensité dépend de la nature du corps dont il est constitué. La théorie montre que cette intensité tend vers zéro en même temps que le nombre atomique Z du radiateur. La méthode consiste à porter en graphique l'intensité du rayonnement γ observé pour différents radiateurs en fonction de leur Z, et à extrapoler la courbe obtenue jusque l'axe $Z = 0$. L'ordonnée du point de rencontre représente l'intensité du r.f.int. seul.

En répétant ces opérations pour différentes épaisseurs de filtres de plomb placés sur le trajet des rayons γ, il est possible de tracer la courbe d'absorption du r.f.int. dans le plomb.

STAHEL et GUILLISSEN. *J. Physique*, 1940, **1**, 12.

31

MAGNÉTOSTRICTION

Sous l'influence de champs magnétiques extérieurs, les milieux ferromagnétiques manifestent des déformations dont on a fait l'objet d'études expérimentales très nombreuses. Les résultats apparaissent très compliqués, surtout dans les champs de moyenne intensité. Les variations des dimensions ne sont des fonctions déterminées ni du champ inducteur H, ni de l'aimantation *observable* I. On constate couramment des renversements de signes, alors même que H et I varient dans un seul sens.

Je me propose de montrer dans ce travail que ces effets peuvent, dans leur ensemble, être reliés rationnellement par un petit nombre d'hypothèses simples et générales à la fois, et qui sont en harmonie avec les données et les notions acquises récemment sur la structure de la matière.

C'est un avis mainte fois exprimé déjà que les lois les plus nouvelles de la mécanique atomique ne sont pas en mesure de rendre compte de diverses propriétés, d'ailleurs très générales et marquées, des corps solides (par exemple les irréversibilités d'ordres divers). Cette assertion est justifiée si elle s'applique aux tentatives d'application *immédiate* des concepts et lois de la nouvelle mécanique aux faits de l'expérience brute. Pour relier les propriétés de l'atome aux lois de l'expérience directe, je crois indispensable de faire souvent usage de notions et lois intermédiaires, qui d'ailleurs, bien entendu, doivent autant que possible viser à la réalité elles aussi. On est déjà entré dans cette voie et avec fruit.

Les notions mises en œuvre ici participent de ce caractère; elles sont provisoirement hypothétiques en ce sens qu'elles ne sont pas directement exprimables par des mesures et servent à décrire des mécanismes invisibles. Mais si l'expérience les justifie, elles pourront être conservées quelle qu'en soit l'interprétation atomistique; en sens inverse elles cherchent à être immédiatement accessibles à cette interprétation.

J'esquisse encore dans cet article quelques raisonnements montrant comment on peut rattacher à la théorie proposée nombre de phénomènes et de lois ressortissant à des domaines différents, plus ou moins éloignés de l'objet précis du travail, tels les propriétés énigmatiques de la susceptibilité initiale, les effets irréversibles des traitements thermiques et mécaniques, etc. Beaucoup de leurs manifestations sont restées sans explication plausible jusqu'ici. Le lien logique que je crois trouver entre elles, puis les questions nouvelles posées à l'expérimentation témoigneront, j'espère, au moins de la fécondité de cette tentative.

Cet article est une première étude dans un cadre restreint pour l'exploration d'un domaine particulièrement vaste et extrêmement confus (irréversibilités !). Il doit se borner à établir des bases précises autant qu'elles peuvent l'être et à en faire voir l'usage; la discussion expérimentale est réduite aux phénomènes typiques. Des figures eussent été sans aucun doute éminemment désirables, mais il eût fallu en choisir d'emblée une notable quantité pour qu'elles soient utiles, et j'y ai renoncé complètement. On trouvera dans le texte les points de départs de plusieurs travaux détaillés que je n'ai pas la possibilité de rédiger immédiatement.

PERRIER. *Helv. Phys. Acta*, 1931, **4**, 213.

32

RÉFLEXION DU SON

Le coefficient de réflexion du son par un solide est déterminé par la proportion de l'énergie qui passe dans l'onde réfractée qui se propage dans le solide: pour un matériel dense et dur, cette proportion est très faible, aussi est-il surprenant de voir par exemple Curtiss R. V. mesurer 70 pour 100 de réflexion seulement pour un son aérien de 10^6 périodes par seconde tombant sur du laiton. L'auteur trouve, par une analyse assez difficile mais entièrement du type classique, que l'effet serait dû à la conductibilité thermique du métal, qui perturbe la distribution de température dans le gaz et introduit un certain manque d'adiabatisme dans la compression et dilatation, les variations de pression n'étant plus en phase avec les variations de densité. A 10^6 périodes par seconde, on explique par ce procédé quelques unités pour cent perdues dans la réflexion. La diffusion des molécules par la paroi, la diffraction des ondes sonores par les petites irrégularités de surface sont aussi étudiées, de même l'adsorption des molécules par la paroi. Toutes ces causes ne joueraient dans l'absorption du son qu'à des fréquences encore plus élevées (10^7).

ROCARD. *J. Physique*, 1939, **10**, 170D.

33

POLYMÉRISATION DES MOLÉCULES ET ÉQUATIONS D'ÉTAT

Les équations d'état simples, comme celle de Van der Waals, ne s'appliquent pas pour les fortes pressions et pour les températures au-dessous de la température critique. Pour accorder les formules avec les expériences, Van Laar admet que *b* est une fonction du volume. Bridgman obtient un accord en admettant six constantes dans sa formule. Toutes ces formules empiriques ne peuvent évidemment être acceptées sans justification théorique. On peut néanmoins justifier les écarts observés entre les formules du type Van der Waals et les expériences en tenant compte de la polymérisation des molécules, au moins pour les gaz simples. L'existence de ces molécules, même dans le cas des gaz pour lesquels on n'a pas jusqu'ici trouvé de molécules doubles d'après leurs spectres d'absorption ou d'après leurs propriétés chimiques, semble prouvée par la théorie des valences résiduelles. Comme nous venons de la voir, London ainsi que Slater et Kirwood ont montré que les molécules même non polaires peuvent s'attirer par suite de l'action mutuelle due aux mouvements des électrons. A cette force d'attraction se superpose la force de répulsion due aux noyaux. Cette dernière l'emporte pour les faibles distances. Il en résulte que la courbe de l'énergie potentielle en fonction de la distance entre les molécules présente un minimum. A ce minimum correspond une molécule stable. Cette stabilité est évidemment d'autant plus grande que l'énergie de dissociation est plus grande. Le nombre de molécules formées par seconde dépend d'autre part du nombre des chocs, et, par suite, de la pression. A chaque pression, il y a donc un mélange de gaz, formé par les molécules simples et les molécules doubles.

Il est aisé de calculer (très approximativement) la concentration de ces constituants en admettant la loi de Van der Waals pour les constituants.

Nous admettrons que les écarts entre l'expérience et les relations de Van der Waals ne sont pas dus aux approximations admises dans l'établissement de ces relations. On est conduit à une relation du type de Van der Waals si on admet l'existence d'une force d'attraction entre les molécules. London, en considérant cette attraction résiduelle, calcule les valeurs de *a* de la formule de Van der Waals et trouve des valeurs voisines de celles obtenues à partir des équations d'état ou à partir des données critiques.

Wohl a généralisé empiriquement pour les gaz simples, les résultats de London, Slater et Kirwood pour l'hydrogène et l'hélium.

La comparaison avec les données du « second coefficient de viriel » lui permettent de calculer, comme nous l'avons vu plus haut, la courbe d'énergie potentielle qui montre bien un minimum. L'énergie de dissociation de la plupart des molécules est de l'ordre de quelques centaines de calories par molécule-gramme. Les molécules doubles ainsi formées entraînent un écart notable entre les valeurs de *b* calculées pour les faibles et fortes pressions.

Le gaz étant considéré comme formé d'un mélange d'au moins deux constituants, on ne peut lui appliquer la formule simple de Van der Waals. Pour mieux se rendre compte des variations qui peuvent prendre naissance dans le gaz lorsqu'on le comprime, on peut calculer la constante *b* par exemple pour chaque état du gaz. S'il y a polymérisation, le co-volume ainsi calculé qu'on peut appeler co-volume apparent, présentera des anomalies dues à la condensation.

HERMAN. *Ann. Phys.*, 1939, **11**, 591.

34

LE DIAMAGNÉTISME CRISTALLIN DU THALLIUM

On fait une étude de la susceptibilité magnétique du thallium contenant de petites quantités d'un métal étranger cadmium, plomb, étain et bismuth. Tandis que le cadmium augmente l'anisotropie magnétique du thallium, les autres métaux la diminuent. Bien que l'étain et le plomb contiennent le même nombre d'électrons de valence, l'étain est deux fois et demi plus efficace que le plomb. Un graphique donne la variation de l'anisotropie magnétique en fonction de la concentration en métal étranger. Les résultats obtenus sont semblables à ceux obtenus avec le bismuth par Goetz et Focke et par Shœnberg et Uddin. Ils ont trouvé que tandis que l'étain augmentait l'anisotropie magnétique du bismuth, le tellure la diminuait. Cette observation a été attribuée au fait que tandis que les atomes de bismuth contiennent 5 électrons de valence, ceux d'étain en contiennent 4 et de tellure 6. Le thallium en a 3, le plomb ou l'étain 4 et le cadmium 2. On trouve que tandis que l'étain ou le plomb diminuent l'anisotropie, le cadmium l'augmente. On observera ainsi que bien que le plomb et l'étain contiennent le même nombre d'électrons de valence, ils affectent différemment l'anisotropie magnétique du thallium. Cela semble être vrai à toutes les concentrations entre 0 et 2 pour 100. Ce résultat est en accord avec ceux de Shœnberg et Uddin qui trouvent que avec du bismuth, l'étain est 4 fois plus efficace que le plomb. On a ainsi observé que l'influence du bismuth sur le thallium est presque la même que celle du plomb bien que le bismuth ait 5 électrons de valence par atome et le plomb 4 seulement. De là, on peut conclure que la décroissance de l'anisotropie du cadmium due à la présence d'une petite quantité de métal étranger n'est pas entièrement réglée par le nombre d'électrons de valence par atome du métal introduit. Considérant ensuite l'influence de métaux étrangers sur les susceptibilités principales, on observe que tandis que le plomb influe sur la valeur de χ_{11} plus que sur χ_1 l'étain altère les deux considérablement.

BONNEMAY. *J. Physique*, 1939, **10,** 114D.

35

TECHNIQUE DE LA COLORIMÉTRIE

Les premiers points à préciser concernaient les qualités visuelles des observateurs, l'élimination des conditions de mesures anormales, et, pour l'étude des corps colorés, la nature de leur mode d'éclairage.

La sensation colorée dépend à la fois de la lumière qui la provoque et de l'observateur qui la perçoit: deux lumières de même constitution physique paraissent à tous identiques, mais deux lumières nettement différentes peuvent être jugées « équivalentes » par certains yeux, et non par d'autres. C'est, en définitive, en fonction de leur appréciation par la plupart des observateurs qu'il convient de repérer les couleurs; on a donc été conduit à définir un « *œil moyen de référence* »,—par une convention basée sur une étude statistique, laissant de côté les vues franchement anormales. Les écarts individuels sont d'ailleurs notables; il faudrait pouvoir en réduire l'influence, mais les éléments nécessaires aux corrections correspondantes manquent encore bien souvent.

D'autre part, les résultats obtenus ne s'interprètent simplement que si l'on respecte certaines conditions expérimentales: adaptation de la vue à la lumière du jour—donc brillance suffisante des plages observées,—intervention d'une zone rétinienne fovéale assez petite, mais pas trop,—élimination des effets de fatigue et de contraste. . . . Ainsi se trouve délimitée une « *métrique simple des couleurs* » qui seule semble pouvoir servir de base à un repérage commode: nous ne nous en écarterons pas, malgré l'intérêt évident d'une étude plus générale.

Considérons enfin que la couleur des corps non lumineux par eux-mêmes, et qui ne font donc que nous renvoyer une partie des radiations qu'ils reçoivent, varie suivant la source avec laquelle on les éclaire. Sauf indication contraire, celle-ci doit être « blanche », mais la notion de blanc est elle-même à préciser, car ce qualificatif s'applique à la lumière du jour, qui est assez variable, et aussi à des lumières artificielles, qui ne s'en rapprochent que grossièrement. Il n'a pas paru possible de fixer conventionnellement une lumière unique pour l'éclairage des corps dont on étudie la couleur. Trois *sources étalons*, dites A, B, C, ont alors été codifiées; ce sont des lampes au tungstène, de régime électrique convenablement réglé, associées (en ce qui concerne B et C) à des filtres liquides déterminés. Leurs lumières sont comparables respectivement à celles des sources artificielles les plus usuelles, à celle du soleil, et à celle de l'éclairage diurne moyen.

FLEURY. *Comm. Lab. Inst. Opt.*, 1944, **4,** 76.

36

LIBÉRATION DES NEUTRONS DE L'URANIUM

Poursuivant l'étude de la libération des neutrons au cours de la rupture nucléaire de l'uranium, nous avons étudié la distribution de la densité des neutrons thermiques présents dans une masse à haute teneur d'uranium, irradiée par une source de photoneutrons.

Une sphère en cuivre (diamètre 50 cm.), immergée dans un réservoir plein d'eau, peut être remplie d'eau ou de 300 kg. de poudre de U^3O^8 (sèche ou additionnée d'une quantité variable d'eau). Une source de photoneutrons (1 g Ra +160 g Be) est placée au centre de la sphère. La densité des neutrons présents dans la sphère est mesurée par l'activité I induite dans des détecteurs en dysprosium placés à l'intérieur de la sphère ou dans l'eau environnante. La courbe $Ir^2 = f(r)$, r étant la distance entre la source et l'un des détecteurs, a été tracée pour les cas suivants:

— sphère pleine d'eau;

— sphère pleine d'oxyde sec d'uranium;

— sphère pleine d'oxyde mouillé de façon à contenir 1 H pour 1 U;

— sphère pleine d'oxyde mouillé à 2 H;

— sphère pleine d'oxyde mouillé à 3 H.

Nous nous sommes assurés que la distribution de l'eau dans la poudre mouillée était suffisamment uniforme.

La forme des courbes obtenues présente des particularités intéressantes, notamment dans le cas 2 H, où deux maxima distincts sont visibles (pour $r = 14$ cm. et $r = 28$ cm.). Elle montre qu'il est illusoire de tirer des conclusions quantitatives de la détermination de la densité neutronique *à une seule distance*: les conclusions qu'on tirerait ainsi changeraient avec la distance choisie. L'intérêt principal réside dans la valeur numérique de la surface des courbes obtenues. En effet, cette valeur étant l'intégrale de la densité neutronique étendue sur un volume, la surface de la courbe est proportionnelle au nombre des neutrons thermiques présents à chaque instant dans le système. Rappelons que ce nombre est égal au produit $Q\tau$, Q étant le nombre de neutrons de toute origine qui, par unité de temps, sont ralentis jusqu'à l'état thermique dans le milieu, et τ la vie moyenne d'un neutron thermique, c'est-à-dire le temps moyen pendant lequel il diffuse dans le milieu, en équilibre thermique, avant d'être absorbé. Le produit $Q\tau$ n'est réellement proportionnel à l'intégrale $\int Ir^2 dr$ qu'à condition d'employer un détecteur qui absorbe les neutrons thermiques suivant la loi $1/v$; les données expérimentales dont on dispose actuellement semblent montrer que le dysprosium suit cette loi d'assez près.

HALBAN *et al.* *J. Physique*, 1939, **10,** 428.

37

LA MÉCANIQUE ONDULATOIRE DES SYSTÈMES DE CORPUSCULES

Louis de Broglie publie ensuite deux fascicules sur une nouvelle conception de la lumière, où il développe sa très originale théorie des photons et des neutrinos (Hermann, 1934 et 1936). Reprenant alors la suite de ses exposés sur les théories quantiques, l'auteur consacre un livre à la question des interactions entre le matière et le rayonnement (Hermann, 1938) et présente de manière systématique la théorie de la dispersion de la lumière, l'effet Raman, l'effet Compton et l'effet photoélectrique.

Dans l'intervalle, et un peu indépendamment, la mécanique ondulatoire s'était développée du côté des applications aux systèmes atomiques et moléculaires, c'est-à-dire, à des ensembles de corpuscules groupés; c'est à ce type de problèmes qu'est consacré le dernier volume, qui vient de paraître chez G. Villars. Dès qu'on veut étudier à la fois plusieurs corpuscules, on se trouve contraint de renoncer aux méthodes relativistes. La théorie relativiste classique (non ondulatoire) n'est pas arrivée à fournir une formulation rigoureuse d'un problème de plusieurs corps, et l'on n'a pu traiter que le cas des faibles vitesses, où tout se réduit à la mécanique ancienne. En théorie ondulatoire, les difficultés rencontrées sont analogues, et l'on doit laisser de côté les effets de relativité, en se contentant de l'approximation de Schrödinger, valable pour les faibles vitesses, et qui constitue la transcription ondulatoire de la mécanique analytique classique. A signaler toutefois les très intéressantes tentatives de Destouches, qui arrive à une mise en équation relativiste ondulatoire de la mécanique d'un système de corpuscules; mais cet essai tout récent n'a guère fait l'objet d'applications pratiques, et il nécessite un appareil mathématique très abstrait; L. de Broglie n'a pas introduit, dans son traité, cette dernière théorie de Destouches; il fait par contre un large usage des recherches antérieures de Destouches et de ses démonstrations générales. Avec juste raison, il reprend ainsi les théorèmes principaux de la mécanique rationnelle et discute leurs conditions d'application en mécanique ondulatoire. Ayant défini la notion d'intégrale première d'un mouvement, en théorie ondulatoire, il discute la théorie du centre de gravité. C'est une grosse question, en général insuffisamment exposée; et l'on peut se rendre compte, à la lecture d'autres travaux, des graves malentendus qui subsistent à ce sujet. Dans ses exposés et dans de récentes discussions, Eddington a formulé des critiques sur la présentation de la mécanique ondulatoire, et proposé une modification de la théorie. Cette modification d'Eddington repose sur l'hypothèse de la validité des équations ondulatoires seulement dans un système de référence lié au centre de gravité du système; or L. de Broglie rappelle avec raison que le théorème du centre de gravité ne peut plus être énoncé tout à fait de la même façon en mécanique ondulatoire qu'en mécanique classique. La méthode d'Eddington n'est donc pas cohérente et ne peut être justifiée en mécanique ondulatoire d'une façon générale; ce n'est que dans un certain nombre de cas particuliers bien définis qu'on peut, en première approximation, traiter séparément le mouvement du centre de gravité, d'une part, et le mouvement des particules autour de ce centre, d'autre part.

BRILLOUIN. *Ann. Phys.*, 1939, **11**, 257.

38

HYPOTHÈSE CONCERNANT L'EXISTENCE D'UNE FORCE TENGENTIELLE COMPLÉMENTAIRE AUX FORCES TRANSVERSALES DE BIOT-SAVART ET LAPLACE

On sait qu'en appliquant les lois classiques à deux circuits fermés, on obtient toujours une force sur chaque circuit telle que les deux forces se fassent équilibre. Si l'on étudie par contre des segments de courant, les deux forces ne s'équilibrent généralement pas. On dit souvent qu'on n'a pas le droit d'appliquer la loi de Biot et Savart à des segments ouverts sans tenir compte des « courants de déplacement » de Maxwell. Mais comme l'ensemble d'un courant de déplacement produit par des sources électrostatiques donne en chaque point de l'espace un champ zéro et que d'autre part la déviation par un champ magnétique d'un projectile chargé est un fait observé, nous croyons que cette défense ne se justifie pas. Il faut donc compléter la théorie.

On peut éviter toutes les difficultés en introduisant une force longitudinale qui doit agir sur un élément de courant ($i\mathrm{d}l$) lorsque d'autres charges se rapprochent latéralement de lui. Si ces autres charges sont situées dans le plan de symétrie perpendiculaire à dl cette force doit être

$$P = \frac{\mathrm{d}V}{\mathrm{d}t}\frac{ev}{c^2} \text{ (V et e en unités statiques)}$$

où e est la charge qui constitue l'élément de courant, V sa vitesse (donc $ev = i\mathrm{d}l$), $\frac{\mathrm{d}V}{\mathrm{d}t}$ la variation de potentiel produite par l'approche des autres charges, et c la vitesse de la lumière. Si les autres charges sont distribuées symétriquement dans l'espace et y ont des directions quelconques, la force devient

$$P = \frac{2}{3}\frac{\mathrm{d}V}{\mathrm{d}t}\frac{ev}{c^2}.$$

Aussi longtemps que nous étudions deux segments de courant ouverts, cette force est toujours d'un ordre de grandeur bien inférieur à tout ce que l'on peut espérer rendre mesurable.

PICCARD. *J. Physique*, 1937, **8**, 87S.

39

Les causes et les effets de la trempe du verre

Le phénomène de la trempe du verre est presque toujours identifié avec celui des tensions engendrées par la viscosité pendant un refroidissement plus ou moins rapide. La question de la trempe du verre est bien plus complexe et les tensions n'en constituent qu'un des facteurs, le moins important. Le phénomène est dominé surtout par l'hétérogénéité physicochimique du verre, c'est-à-dire par l'existence de deux variétés allotropiques: α, stable à froid; β, stable à chaud; les deux variétés étant d'ailleurs caractérisées par des constantes physiques différentes (par exemple, l'indice de réfraction, le coefficient de dilatation, etc.). La proportion de α et β, en équilibre dans un verre donné est déterminée pour une pression et une température données, mais change lorsqu'un de ces facteurs varie. Les propriétés du verre changent en même temps: en particulier l'indice de réfraction change et la surface se déforme, parce que α et β ont des coefficients de dilatation différents. Si par un moyen quelconque on obtient un verre dans lequel la proportion de α et de β n'est pas celle qui correspond à l'équilibre (pour la température et la pression données) cette proportion varie spontanément, en tendant vers sa valeur d'équilibre. Pour obtenir un tel verre non stable il faut le refroidir brusquement. Ainsi s'expliquent l'éclatement spontané des verres trempés et la loi de variation avec le temps.

J. Physique, 1937, **8**, 209.

40

Le contact entre un métal et un isolant ou un semi-conducteur

Selon la mécanique des quanta, il existe dans n'importe quel cristal non métallique une bande de niveaux électroniques permis qui sont inoccupés quand le cristal est dans son état de moindre énergie. Nous appelons cette région, la région de conduction. Le cristal peut conduire l'électricité si les électrons arrivent au niveau de conduction à partir des niveaux inférieurs. Selon la théorie des semi-conducteurs de Wilson, il existe dans ces substances des imperfections du réseau pour lesquelles un électron peut exister à un état stationnaire sous la bande de conduction, les électrons peuvent atteindre à partir de ces niveaux la bande de conduction par l'agitation thermique des atomes voisins.

En appelant χ le travail nécessaire pour amener un électron depuis le niveau de conduction le plus bas à un point extérieur au cristal. Supposant que l'on place une surface métallique au voisinage ou au contact d'un isolant ou d'un semi-conducteur. Si les deux corps ne s'influencent pas, l'énergie nécessaire pour faire passer un électron du métal à la bande de conduction de l'isolant est ϕ-χ, où ϕ est la fonction travail du métal. Développant les calculs, appelant n_0 le nombre d'électrons à la surface de l'isolant tel que

$$n_0 = - N_1 e^{\frac{-U_0}{kT}},$$

où N_1 est le nombre d'atomes par unité de volume qui sont dans la bande de conduction $V(n)$ désignant le potentiel d'un électron à la distance x de la surface. La solution pour un nombre n d'électrons par unité de surface est

$$n = \frac{n_0 l^2}{(x + l)^2} \quad \text{où} \quad l = \frac{k\mathrm{T}}{2\pi n_0 e^2}.$$

BONNEMAY. *J. Physique*, 1939, **10**, 173D.

41

MESURE DES HAUTES TENSIONS CONTINUES PAR DIFFRACTION D'ÉLECTRONS

On sait que la mécanique ondulatoire associe à un faisceau d'électrons monocinétiques de vitesse v une longueur d'onde donnée par la formule

$$\lambda = \frac{h}{mv}$$

dans laquelle h est la constante de Planck, m la masse de l'electron et v sa vitesse. Ce résultat prévu par Louis de Broglie a donné lieu aux confirmations expérimentales les plus précises; c'est ainsi qu'en faisant traverser par un faisceau électronique de vitesse v un film suffisamment mince d'une substance cristalline, ce faisceau est dévié en des directions qui sont exactement celles que l'on obtiendrait si l'on envoyait à travers ce film un pinceau de rayons X de longueur d'onde λ. La diffraction des électrons s'est considérablement développée depuis les premières expériences. Elle constitue maintenant une méthode d'investigation de la structure fine de la matière, qui complète celle basée sur la diffraction des rayons X.

D'une façon accessoire, on peut remarquer que la diffraction électronique peut être indirectement une méthode de mesure des hautes tensions continues et être utilisée ainsi pour l'étalonnage des appareils de mesure de hautes tensions tels que les voltmètres électrostatiques.

TRILLAT. *Rev. Gén. Élect.*, 1946, **40,** 307.

42

RADIOACTIVITÉ ARTIFICIELLE

Les phénomènes de radioactivité artificielle produits dans l'uranium par choc de neutrons ont été l'objet de nombreux travaux expérimentaux, sans avoir été éclaircis, à cause des très grandes difficultés techniques que l'on rencontre. Il était cependant généralement admis qu'un groupe (A) de ces radio-éléments (c'est-à-dire celui que l'on précipite avec les sulfures) devait être attribué aux éléments transuraniens; en particulier, l'on acceptait le schéma d'analyse en diverses familles d'isomères proposé par Meitner, Hahn et Strassmann. Un autre groupe (B) était attribué à des éléments de numéro atomique inférieur à celui de l'uranium, c'est-à-dire à des isotopes du thorium ou même de l'actinium et du radium; la production de ces corps aurait été due à des processus nucléaires du type (n, α) ou même $(n, 2\alpha)$. L'étude du groupe (B) présente les difficultés les plus grandes et n'avait pu jusqu'ici avancer beaucoup. Curie et Savitch ont reconnu dans ce groupe un élément de vie moyenne de 3,5 h, homologue chimique des terres rares, qui devait d'après cela raisonnablement être un isotope de l'actinium; il pouvait cependant être séparé de l'actinium en montrant une plus grande analogie avec les terres rares. Tout récemment, Hahn et Strassmann ont analysé plus à fond la fraction (B) et proposé un schéma d'analyse en familles, qui tireraient leur origine de différents isomères d'un élément alcalino-terreux, à partir duquel dériveraient les éléments du groupe des terres rares, qui sont encore radioactifs. Ils ont trouvé que ces éléments sont bien plus semblables au baryum et aux terres rares, qu'à leurs homologues supérieurs Ra et Ac. D'après cette propriété, ils ont émis l'hypothèse hardie que la radioactivité du groupe (B) doit être attribuée réellement à des isotopes du Ba et des terres rares, qui se produiraient à la suite de la décomposition du noyau de ${}^{238}_{92}U + {}^{1}_{0}n$ en deux parties de masse beaucoup plus petite. Le reste de la masse de l'U pourrait d'après eux se retrouver dans un isotope du ${}_{43}Ma$ (${}^{238}_{92}U + {}^{1}_{0}n$, ${}^{138}_{56}Ba + {}^{101}_{43}Ma$); à cet élément, d'autre part, on pourrait peut-être attribuer l'origine de la fraction (A), puisque Hahn et Strassmann déclarent à présent que les expériences sur les « transuraniens » prouvent que les éléments « 93 » sont vraiment des homologues du ${}_{75}Re$, et non du Re lui-même, mais n'établissent pas s'ils sont des homologues supérieurs (93) ou inférieurs (43).

Nous nous trouvons donc en présence d'une idée physique qu'on n'avait pas jusqu'ici envisagée sérieusement: les noyaux lourds pourraient se décomposer en parties de masses à peu près égales (« fracture » des noyaux), tandis que jusqu'ici on ne connaissait que des décompositions en parties très différentes (émission de noyaux légers par un noyau lourd). Cette idée doit être aujourd'hui acceptée avec la plus grande réserve. D'autre part, il paraît certain qu'on ne peut faire rentrer les phénomènes de radioactivité artificielle de l'uranium dans aucun des schémas connus de la physique nucléaire, de sorte que l'on doit être prêt à envisager des possibilités nouvelles. Il peut pourtant être bon d'examiner dès à présent si l'hypothèse émise par Hahn et Strassmann peut être considérée comme vraisemblable d'après les connaissances théoriques actuelles sur la physique du noyau.

Il est bien connu que la décomposition d'un noyau lourd en deux noyaux de poids moyen—réaction que nous allons appeler « fracture »—serait non seulement énergétiquement possible, mais même fortement exothermique: d'après les données sur les masses des noyaux la fracture de l'$^{238}_{92}U$ en deux noyaux stables pourrait libérer au moins 220 MV d'énergie.

Ce gain d'énergie est dû au fait que la fracture serait accompagnée d'une réduction ΔE_{Coul} énorme de l'énergie électrostatique du système, tandis que la réduction ΔE_{Nucl} de l'énergie totale de liaison des particules est beaucoup plus petite; la différence $\Delta E_{Coul} - \Delta E_{Nucl}$ qui constitue le bilan énergétique de la fracture est donc positive et très grande: sa valeur expérimentale approchée a été donnée plus haut.

La stabilité expérimentale de l'$^{238}_{92}U$ vis-à-vis de la fracture est d'habitude expliquée de la manière suivante: on observe que, au moment où le noyau lourd est déjà décomposé, mais que les noyaux résultants n'ont pas encore été éloignés l'un de l'autre, ΔE_{Nucl} a déjà atteint sa valeur finale, tandis que le gain d'énergie électrostatique ΔE_{Coul} est encore très petit et beaucoup inférieur à sa valeur finale. Si donc il y avait une phase intermédiaire de la réaction, au cours de laquelle $\Delta E_{Coul} - \Delta E_{Nucl}$ serait négatif, la réaction même serait impossible du point de vue de la mécanique classique. Le cas de l'émission α, qui est à ce propos tout à fait analogue, montre qu'à la suite d'un effet quantique cette impossibilité n'est pas rigoureuse (perméabilité de la barrière de potentiel de Gamow). Cet argument quantique serait, d'autre part, beaucoup moins important dans notre cas, car les noyaux de poids moyen, à cause de leur grande masse, ont une pénétration beaucoup plus petite dans la barrière de potentiel de Gamow.

Puisqu'on n'avait jusqu'ici pas douté de la stabilité, même des noyaux les plus lourds, vis-à-vis de la fracture, on n'a pas pris grand soin de vérifier quelles sont les limites de validité des arguments, que nous venons d'exposer en sa faveur. Pour atteindre notre but, nous allons justement analyser ces limites, pour vérifier si l'$^{238}_{92}U$ ne se trouvait pas tout près d'elles; nous devrions encore tâcher de nous expliquer comment la capture d'un neutron dans un noyau d'U peut être suffisante pour faire disparaître sa stabilité.

FANO. *J. Physique*, 1939, **10**, 229.

43

UNE THÉORIE DE L'UNIVERS EN EXPANSION

Il semble que la méthode permettant de donner des théories valables dans un certain domaine puisse être basée sur les considérations suivantes : à l'échelle humaine, dans le domaine où les théories classiques sont valables, l'univers physique peut être regardé comme constitué par un ensemble de systèmes matériels caractérisés par un faisceau de propriétés intrinsèques, caractéristiques et immuables : leur charge et leur masse, sources des champs dont les lois de distribution sont fixées une fois pour toutes, indépendamment de toute opération de mesure, leur configuration géométrique spatio-temporelle, leur fonction de Hamilton qui détermine leur structure dynamique (oscillateur harmonique, système planétaire, etc.); les transformations qui décrivent l'évolution de tels systèmes ainsi que les changements de variables les plus généraux doivent être compatibles avec ces exigences, et appartiennent ainsi à une famille bien définie; c'est ce qui se passe en relativité générale où les transformations canoniques admettent la fonction caractéristique invariante fonctionnellement et numériquement :

$$H = \sqrt{g^{ik}(q)\, p_i\, p_k} = m^\circ c \; (i, k = 1, 2, 3, 4)$$

Mais lorsque l'on s'écarte du domaine humain, ces restrictions ne peuvent plus être justifiées, car l'existence de telles propriétés intrinsèques caractérisant les systèmes matériels comme des objets existant « en soi » ne peut être justifiée au delà de ce domaine; en particulier, la notion d'objet ou de substance en soi doit subir un échec sur deux points principaux : le premier se présente à l'échelle atomique, où l'emploi des transformations de contact n'est plus justifié par le fait de l'apparition des phénomènes de diffraction; le second a lieu dans la théorie cosmologique de l'univers où il s'agit de décrire la représentation que pourrait avoir de cet univers un super-observateur pour lequel les nébuleuses spirales se réduiraient à des points; comme cet univers constitue un système clos, il réaliserait l'image parfaite de l'objet isolé, et chacune de ses propriétés serait une propriété « en soi »; il aurait ainsi une certaine courbure, une certaine géométrie, se composerait d'un certain nombre de particules, aurait pris naissance de telle ou telle manière; cette idée est inexacte, si l'on admet que les notions d'objet et d'objectivité sont douées de significations seulement à l'échelle humaine; ce n'est que parce qu'à cette échelle, nous étudions de petits systèmes relativement éloignés les uns des autres et d'interactions assez faibles que la croyance en l'existence des objets en soi a pu être soutenue; mais aucune extrapolation de cette notion n'est permise; c'est parce qu'en définitive, toutes nos expériences aboutissent à des résultats constatables à l'échelle humaine qu'une physique théorique objective applicable aux domaines éloignés de l'échelle humaine peut se constituer.

L'interprétation correcte de ce que l'on a coutume d'appeler le « principe de Mach » d'après lequel l'inertie ne peut être regardée comme une propriété « en soi » des corps matériels ne consiste pas, comme l'ont cru beaucoup de physiciens relativistes, et, en particulier Eddington, à admettre que les propriétés dynamiques des corps sont conditionnées par l'ensemble des corps de l'univers ce qui conduit à des calculs inextricables, mais qu'elles sont relatives, au même titre que la notion de vitesse ou celle de contraction de Lorentz, c'est-à-dire qu'elles ne peuvent s'exprimer que lorsqu'on dispose d'un instrument de mesure et d'un système observé; elles caractérisent les opérations du groupe géométrique que l'on effectue au moyen de l'instrument de mesure sur le système observé.

MARIANI. *J. Physique*, 1939, **10,** 35.

44

L'ÉMISSION ÉLECTRONIQUE DES CATHODES À OXYDES

L'émission électronique des cathodes à oxydes est un phénomène dont la complexité théorique est due au grand nombre de variables entrant en jeu, variables dont l'importance vis-à-vis du phénomène principal à étudier, l'émission électronique, est souvent difficile à déterminer par suite de la réaction des variables entre elles.

Pour fixer les idées, citons les différences considérables que peuvent présenter les cathodes d'une même série de tubes au moment de leur formation. Après transformation des carbonates en oxydes, si l'on applique une tension positive sur les électrodes du tube, on constate pour certaines lampes de la série une émission électronique très bonne alors que d'autres lampes ont une émission électronique beaucoup plus faible s'améliorant en quelques heures.

Signalons aussi les différences considérables dans la durée de vie des cathodes, 2.000 ou 50.000 heures suivant la manière dont a été effectuée l'activation.

L'étude systématique des différentes variables conditionnant l'émission électronique n'est donc pas chose aisée et depuis près de vingt ans de nombreux chercheurs ont, en plus de 200 publications et articles divers, tenté d'apporter une théorie entièrement satisfaisante de l'émission électronique des cathodes a oxydes et des facteurs qui la conditionnent.

De cette somme d'efforts, certaines lois générales se dégagent:

1) L'émission électronique est un phénomène de surface.

2) La surface émissive est constituée par du baryum créé, au sein de l'oxyde par transformation chimique de ce dernier.

3) Cette transformation chimique peut avoir son origine soit dans une réduction d'origine thermique de l'oxyde à l'aide d'un réducteur énergique qui serait introduit dans la masse même de l'oxyde, ou dans le métal support, soit par un phénomène d'électrolyse activant par suite du passage du courant dans la masse de l'oxyde, une faible réduction d'origine thermique servant à amorcer le phénomène.

4) Le baryum libéré diffuse dans la masse de l'oxyde, d'une part vers l'extérieur où il constitue la couche adsorbée nécessaire à l'émission, d'autre part vers le métal support dans lequel il peut pénétrer.

5) Le courant électronique traverse la couche de l'oxyde qui est semi-conductrice par suite de la présence de baryum libre.

Cette théorie d'ensemble est séduisante mais n'est pas complète car elle n'explique pas le déclin de l'émission.

BIGUENET et MANO. *Le Vide*, 1946, **1**, 137.

45

LE RÔLE DE LA TECHNIQUE DU VIDE EN PHYSIQUE NUCLÉAIRE

La technique du vide a joué un rôle essentiel dans la réalisation des expériences grâce auxquelles on a pu isoler, étudier et appliquer la plupart des phénomènes élémentaires de la physique atomique. En effet, pour réaliser ces expériences dans les meilleures conditions possibles, il faut généralement que la molécule ou l'atome qui doit être le siège de ces phénomènes soit très éloignée de tout autre molécule ou atome susceptible d'exercer sur elle une influence parasite; de plus, le phénomène élémentaire que l'on désire étudier est généralement produit par le choc d'une particule incidente, par exemple d'un électron, et son énergie cinétique doit être définie avec le maximum de précision possible; il ne peut en être ainsi que si cette particule se déplace dans un vide suffisant pour éviter sur son parcours tous les chocs parasites susceptibles de donner naissance à d'autres phénomènes et de modifier la vitesse que l'on a voulu lui donner.

Les phénomènes élémentaires de la physique nucléaire sont produits, dans des conditions similaires, par le choc de particules incidentes sur des noyaux atomiques. Mais les dimensions des noyaux sont de l'ordre de 10^{-13} cm.; les dimensions des édifices atomiques sont de l'ordre de 10^{-8} cm.; il en résulte que, même à l'intérieur d'une cible à l'état solide, les noyaux atomiques doivent être considérés comme complètement isolés les uns des autres. En physique nucléaire, la technique du vide n'intervient donc que pour l'obtention de faisceaux convenables de particules incidentes ou pour la réalisation de certains appareils de mesure spéciaux dont nous ne nous occuperons pas ici.

De nombreuses expériences de physique atomique sont réalisées avec de très faibles puissances, dans des appareils d'un volume très réduit, souvent de l'ordre du cm^3. Si l'on prend des précautions convenables, il y a donc peu de gaz à évacuer. Tout l'art de la technique du vide consiste alors à obtenir, dans un temps raisonnable, le meilleur vide possible et à le conserver pendant le fonctionnement des appareils, ce qui peut généralement être fait avec une pompe moléculaire ou avec de petites pompes à condensation de vapeurs de mercure ou d'huile dont les modèles sont maintenant classiques. Dans les installations très puissantes qu'utilise maintenant la physique nucléaire, ce n'est plus seulement la qualité du vide qui est en cause; la rapidité d'évacuation de l'air ou des gaz, soit pendant la période de préparation des appareils, soit pendant leur fonctionnement, pose, en réalité, les problèmes essentiels. Par exemple, les dimensions de ces installations sont devenues telles qu'il faut, lors de leur mise en route, évacuer, à partir de la pression atmosphérique, des enceintes d'un volume de plusieurs mètres cubes. Et il faut y réaliser, le plus rapidement possible, une pression de l'ordre de 10^{-6} mm. de mercure. Pour cela, on a dû construire des pompes primaires, à palettes, dont le débit est considérable; la vitesse d'aspiration de ces pompes est encore si grande, pour une pression de 10^{-3} mm. de mercure, qu'elle permet d'amorcer directement une grosse pompe à condensation d'huile. On peut ainsi supprimer toute pompe intermédiaire à vapeur de mercure et éviter, de la manière la plus radicale, toute contamination des pièces métalliques par les vapeurs de mercure, dont l'effet n'est jamais bon et qui sont toujours difficiles à éliminer. La construction de ces pompes à palettes à très grand débit doit donc être considérée comme un progrès très appréciable; d'ailleurs, dans le schéma d'installations qui deviennent de plus en plus compliquées, il y a intérêt à supprimer tout organe qui n'est pas absolument indispensable.

MORAND. *Le Vide*, 1946, **1**, 105.

46

DIFFERENTES CATÉGORIES DE CONTRÔLEURS UNIVERSELS

Les appareils classiques simplifiés sont de prix relativement réduits, fabriqués en série, généralement sous une forme réduite et portative. Ce sont des appareils robustes, dont la précision n'a pas besoin d'être « poussée », mais dont la sensibilité est généralement désormais de l'ordre de 1.000 ohms par volt, et nous reviendrons sur ce point. Ces appareils simples permettent la mesure des tensions et des intensités en continu, et, généralement, la mesure des tensions en alternatif; ils peuvent servir de sonnettes, et comportent des échelles de lecture directe pour la vérification des résistances et des condensateurs, tout au moins sur certaines gammes. Certains permettent même de vérifier le niveau de sortie, en fonctionnant en outputmeters.

Quelques rares modèles comportent des galvanomètres rudimentaires à fer mobile, mais, la plupart sont équipés avec des milliampèremètres à cadre mobile. Les perfectionnements ont surtout porté sur la disposition mécanique et électrique, et le montage des pièces détachées, de façon à rendre l'emploi plus pratique et plus rapide; les échelles de lecture sont plus développées et plus distinctes, ce qui rend les mesures plus faciles. La disposition des organes de commande est plus claire et plus ordonnée: d'ailleurs, le nombre de ces organes, ainsi que des bornes de connexion, est généralement réduit.

Certains constructeurs français utilisent encore *la méthode de connexion par douilles et fiches*, mais l'emploi des *rotamètres*, c'est-à-dire des répartiteurs rotatifs à plots, adoptés à l'origine par les fabricants américains, se répand de plus en plus.

Il devient ainsi possible de placer essentiellement deux combinateurs sur le tableau de commande: un *combinateur d'utilisation*, permettant de disposer les circuits en vue de l'opération à effectuer, et un *répartiteur de sensibilités*, permettant le choix de la gamme de mesures à considérer suivant cette utilisation. Un *bouton de tarage* assure, en outre, la remise à zéro pour certaines mesures, en particulier, des résistances et des capacités.

Les manœuvres à effectuer peuvent encore être rendues plus rapides par l'adoption d'un système mécanique ou électro-mécanique, rendant les connexions automatiques ou semi-automatiques, suivant l'utilisation considérée, et la gamme de mesures désirée.

Notons encore certains montages électriques permettant de simplifier les échelles de lecture, en faisant coïncider l'échelle en continu avec l'échelle en alternatif.

A côté de cette catégorie d'appareils pratiques simplifiés, mais de plus en plus perfectionnés, et pour lesquels certains constructeurs prévoient, d'ailleurs, des *dispositifs adaptateurs* permettant d'étendre le champ des applications, il existe désormais d'assez nombreux modèles de *contrôleurs universels de précision,* destinés aux ateliers ou aux laboratoires. Ces appareils de qualité se distinguent par leur précision et leur très faible consommation correspondant à une résistance de 5.000 à 20.000 ohms par volt en courant continu; ce sont, en outre, des appareils à multiples usages, souvent plus complets que les modèles similaires étrangers, même américains. Ils permettent, en particulier, la mesure des intensités en alternatif, l'emploi en décibelmètres et wattmètres dans des conditions rationnelles.

Certains constructeurs sont même allés plus loin; ils ont adjoint au contrôleur universel proprement dit, formé, en réalité par la combinaison d'un milliampèremètre avec des séries de résistances convenablement disposées en dérivation ou en série, d'autres appareils de mesure comportant des organes communs, tels qu'un voltmètre électronique, ou un pont de mesure. On réalise ainsi un ensemble de contrôle et de mesure très complet, dont il existe évidemment d'assez nombreuses variantes en France et à l'Etranger, mais nécessairement réalisés sous une forme complexe, plus ou moins encombrante, et relativement coûteuse.

HÉMARDINQUER. *T.S.F. pour Tous,** 1947, **23**, 35.

47

CONTRÔLE D'AMPLITUDE DE L'ONDE PORTEUSE ET CONTRÔLE D'ACCORD VISUEL

Dans un récepteur, il est toujours intéressant de pouvoir se rendre compte de l'amplitude de l'onde porteuse de la station que l'on désire entendre.

* *La T.S.F. pour Tous,* revue mensuelle, Editions Etienne Chiron. 40 rue de Seine, Paris VIème.

On a, d'abord, la possibilité de vérifier très exactement l'accord et, au besoin, de parfaire l'alignement du récepteur. En cas d'anomalie d'une courbe de résonance, on s'en aperçoit immédiatement en faisant lentement et régulièrement défiler le condensateur variable de part et d'autre de la résonance.

On est renseigné immédiatement sur l'importance et la fréquence des phénomènes d'evanouissements, ce qui permet rapidement de faire le choix entre deux émetteurs qui transmettent le même programme, par exemple.

Enfin, on peut comparer deux collecteurs d'ondes différents. L'indication fournie est, en effet, fonction de la tension transmise par le collecteur d'onde au circuit d'accord.

L'œil magique, ou le trèfle cathodique, permet de trouver facilement l'accord exact. Mais il manque totalement de précision. On ne pouvait graduer les secteur lumineux en microvolts, par exemple. . . .

Or, c'est précisément de cela qu'il s'agit, A défaut de « microvolts », nous désirerions avoir tout au moins sous les yeux des graduations arbitraires et savoir que telle graduation correspond toujours à telle amplitude de l'onde porteuse. . . .

CHRÉTIEN. *T.S.F. pour Tous*, 1947, **23**, 44.

48

LES CRITÈRES DE QUALITÉ

Quels sont les éléments principaux d'appréciation de la qualité des communications téléphoniques ?

Ce sont ceux qui expriment l'intelligibilité et la discrétion des communications, leur sécurité de fonctionnement et la facilité plus ou moins grande d'exploitation et d'entretien.

On peut les énumérer:

a) Fidélité de reproduction { bande passante / distorsion harmonique.

b) Rapport signal/bruit.

c) Rapport signal/diaphonie.

d) Constance de l'équivalent de transmission.

e) Servitudes d'exploitation et d'entretien.

f) Sécurité des liaisons; fraction du temps de dérangement.

D'après ces critères de qualité, nous serons à même d'apprécier les mérites de tel ou tel système.

Au préalable, il nous paraît utile d'analyser d'un peu plus près la signification que donnent les téléphonistes à certains de ces critères.

Pour l'intelligibilité, les conditions de fidélité admises dans chaque voie prise individuellement correspondent à 3 à 4 Kc. pour la bande passante, 5 à 10% pour la distorsion harmonique.

Les parasites du type bruit (contrairement aux produits de distorsion harmonique) ont un niveau indépendant du niveau actuel des signaux à percevoir. Du point de vue de l'intelligibilité, ils sont à comparer aux valeurs les plus faibles possibles de ces signaux. Les niveaux réels, en valeur instantanée comme en valeur moyenne, des signaux transmis sur une ligne téléphonique étant très variables, les téléphonistes, compte tenu des limites pratiques de ces variations, ont trouvé commode d'utiliser comme référence de signal utile une tonalité d'un niveau défini, en un point particulier du réseau; à savoir une tonalité d'une puissance d'un milliwatt à la table interurbaine.

Par convention, le rapport signal/bruit des téléphonistes exprime par comparaison à 1 milliwatt, le niveau du bruit mesurable à la table interurbaine sur la voie téléphonique examinée. Une valeur usuelle pour ce rapport signal/bruit est de 55 db.

Les parasites produits par la diaphonie sont ceux dont la présence dans une voie est due aux paroles transportées par une autre voie. La diaphonie étant éventuellement intelligible, on demande, pour des raisons de discrétion, que les produits de diaphonie soient couverts par les bruits, soit par exemple 10 db au-dessous de ceux-ci. De nouveau, le rapport signal/diaphonie se réfère au niveau de comparaison, d'un milliwatt pour une tonalité d'essai à la table interurbaine. La diaphonie susceptible d'être perçue dans une voie est due principalement aux crêtes de niveau dans les autres voies. Nous avons donc à nous intéresser, non plus aux minima de ces niveaux, mais à leurs maxima. Cette étude est pour les systèmes multiplex d'un intérêt tout particulier. En effet, dans tous les systèmes multiplex connus avec transmission par fil ou radio, existe un compromis entre le rapport signal/bruit et le rapport signal/diaphonie. Cela tient à la présence d'organes amplificateurs, modulateurs, démodulateurs, ayant, du point de vue des niveaux des voies à transmettre, une capacité de charge définie au delà de laquelle se produisent des distorsions des signaux transmis, introduisant des diaphonies entre voies.

ALTOVSKY. *L'Onde Électrique,** 1946, **26,** 401.

49

TÉLÉCOMMUNICATION RADIO-MULTIPLEX EN HYPERFRÉQUENCES

L'appareil n° 10 de l'armée anglaise est un multiplex à impulsions, probablement le premier en date en exploitation. Le premier modèle est de 1942. Il a été utilisé peu après le « D day » par l'armée Montgomery.

Ses principales caractéristiques:

— 8 voies téléphoniques.

— Liaison avec relais intermédiaires, intervalle normal 20 miles (32 km.) exceptionnellement 50 miles (80 km.) utilisation jusqu'à 10 intervalles (Luxembourg-Bruxelles 200 km.).

— Longueur d'onde 5,85 et 6 cm.

— Modulation par impulsions de durée variable; multiplex à division de temps.

— Fréquence de récurrence 9 kcs;

par période, 8 impulsions de voie de durée $3{,}5 \pm 2{,}33 \mu s$ plus une impulsion de synchronisation de 20 μs.

Une dent de scie générale à 9 kcs donne naissance à des dents de scie individuelles de voie décalées dans le temps, se transformant elles-mêmes en combinaison avec la modulation BF, en impulsions de durée variable. A la réception, la séparation a lieu au moyen de dents de scies individuelles recréées au moyen de l'impulsion de synchronisation. Le décalage dans le temps des dents de scies individuelles est

* *L'Onde Électrique,* revue mensuelle pour les ingénieurs. Editions Etienne Chiron, 40 rue de Seine, Paris VIème.

produit à partir de la dent de scie générale au moyen d'une série de lampes de polarisation croissante.

— Les impulsions sont appliquées à un magnétron qui fournit une puissance moyenne de 300 mw.

L'aérien est un miroir parabolique de 1 m. 50 de diamètre excité par un guide d'onde.

Il est monté sur le toit d'une remorque à 4 roues et un pylone démontable permet de le hisser à 20 m. de hauteur.

Une station-relais complète comprend 4 camions.

Le rapport signal/bruit est de l'ordre de 40 à 45 db.

Le principal reproche que l'on fait à l'équipement est le manque de stabilité du système de distribution et de séparation des impulsions dans le temps. Les réglages doivent être contrôlés et retouchés tous les jours, ce qui oblige à disposer d'une équipe de spécialistes auprès de chaque relais.

L'équipement a été très apprécié par les militaires pendant la campagne d'Allemagne pour sa grande flexibilité, sa sécurité d'emploi par tous les temps, et surtout sa discrétion.

ALTOVSKY. *L'Onde Électrique*, 1946, **26,** 415.

50

LIAISONS MULTIPLEX PAR CABLES HERTZIENS

Les liaisons multiplex sur câbles ont été instituées afin de réaliser une économie d'installation en utilisant une seule ligne au lieu de plusieurs, grâce à un supplément d'équipements terminaux et intermédiaires moins onéreux que la pose des câbles autrement nécessaires.

Dans une liaison par câble hertzien, aucune ligne matérielle ne réunit les deux correspondants. Cependant le mot de câble ne doit pas être considéré comme une simple métaphore, et on ne doit pas confondre une liaison par câble hertzien avec une liaison radio-électrique ordinaire: celle-ci occupe une région spatiale si étendue qu'elle ne peut ignorer les autres liaisons analogues, tandis qu'une liaison par câble hertzien occupe un faisceau spatial si étroit qu'en principe elle est absolument maître de choisir ses fréquences tout comme il en est pour une liaison par câble ordinaire.

Pour que cette condition-ci soit satisfaite, il est clair qu'il est nécessaire d'utiliser une fréquence très élevée de l'ordre du millier ou de quelques milliers de mégacycles. Dès lors, la partie matérielle du câble hertzien nous apparaît : c'est tout l'appareillage nécessaire pour transformer un courant à fréquence audible en une onde radio-électrique canalisée dans un étroit faisceau par des aériens fortement directifs et pour inversement retrouver, à partir de cette onde, le courant audible initial. Il est évident que la première économie à rechercher est celle réalisable sur cet appareillage même. Ceci conduit tout de suite, pour réduire le volume et le coût des aériens, à envisager des liaisons sur des longueurs d'onde les plus courtes possibles, sans être gêné par des phénomènes de propagation liés à la météorologie. On peut admettre, d'après les résultats de propagation aujourd'hui connus, qu'on ne dépassera guère la fréquence de 10.000 mégacycles ($\lambda = 3$ cm).

MAILLARD. *L'Onde Électrique*, 1946, **26,** 418.

51

ADAPTATION À L'ANTENNE DES ÉMETTEURS

Le problème de l'adaptation à l'antenne des émetteurs est l'un des principaux parmi ceux qui se posent à l'ingénieur radioélectricien. Ce problème peut d'ailleurs présenter deux aspects différents.

a) Etant donné un émetteur déjà construit, déterminer l'ensemble de ses possibilités d'adaptation pour une fréquence donnée. L'antenne étant caractérisée par son terme résistif (R) et son terme réactif (jX) ceci revient à déterminer l'ensemble des couples (R, X) assurant « l'adaptation » de l'antenne, c'est-à-dire, le fonctionnement de l'émetteur dans les conditions de puissance et de rendement imposées. C'est cet ensemble des valeurs (R, X) ainsi définies que nous appelons « plage d'adaptation » de l'émetteur pour la fréquence donnée. Un raisonnement très simple permet de montrer que, dans le cas ou l'ensemble du circuit oscillant de puissance et du système de couplage à l'antenne ne présente qu'un seul « degré de liberté » (un seul élément tel que self, capacité, etc... pouvant varier), la plage d'adaptation est une ligne dans le système d'axes (X, R). Si le nombre de degrés de liberté est égal ou supérieur à 2, la plage d'adaptation est une surface. Si maintenant nous voulons donner au problème envisagé toute la généralité dont il est susceptible, en faisant varier la fréquence du signal, il nous faudra envisager le système d'axes (*f*, R, X) dans l'espace. Il est facile de voir que dans le cas d'un seul degré de liberté la plage d'adaptation sera une surface gauche (dans l'espace), dans celui de 2 degrés de liberté—un volume. Le but de l'étude ci-dessous est d'indiquer une méthode permettant de construire ce volume d'adaptation dans le cas des couplages inductifs à l'antenne ce qui résout, en même temps, le premier aspect du problème posé.

b) Etant donné un émetteur à construire, déterminer rapidement la constitution et les caracteristiques de l'ensemble des circuits d'anode de puissance et de couplage à l'antenne de façon à satisfaire, de la manière la plus rationnelle et la plus économique, aux exigences posées par l'aérien à employer, compte tenu de la gamme d'ondes à couvrir.

Il est facile de voir que, si nous savons résoudre le problème décrit en *a*), il nous sera possible d'exécuter l'opération inverse, c'est-à-dire, en connaissant la forme du volume d'adaptation (se déduisant des caractéristiques de l'aérien à employer), déterminer les circuits de puissance et de couplage.

L'on voit immédiatement à quoi correspondent, dans la pratique de l'ingénieur radioélectricien, les deux aspects du problème dont nous venons de nous occuper:

a) Le premier intéresse surtout l'ingénieur exploitant et correspond au cas, assez fréquent, où l'on impose, à un émetteur déjà en fonctionnement, un aérien nouveau, différent de celui pour lequel il a été conçu. La connaissance de la plage d'adaptation de l'émetteur permet alors de déterminer immédiatement si le nouveau mode d'exploitation est possible et dans la négative, d'indiquer les moyens additionnels d'adaptation nécessaires. Il est évident que ces considérations s'appliquent surtout aux émetteurs transportables de moyenne et faible puissance devant souvent fonctionner avec des antennes quelconques dont les dimensions et les caractéristiques varient avec les lieux d'exploitation.

b) Le deuxième aspect du problème intéresse l'ingénieur d'études, chargé de la construction d'un émetteur. La notion des « plages d'adaptation » constitue une voie d'attaque nouvelle et souvent très commode.

GLAZER et FAMILIER. *L'Onde Électrique*, 1946, **26**, 430.

52

DÉPÔT DE COUCHES MINCES PAR ÉVAPORATION SOUS VIDE

Les surfaces à traiter et la substance à vaporiser sont placées dans une enceinte où règne un bon vide. En général, les surfaces sont réparties sur une sphère dont la source occupe le centre. Pour le travail en série, il est avantageux de placer la source au-dessous des lentilles à traiter, celles-ci occupant des logements découpés à la cote voulue dans une calotte sphérique. Lorsqu'on doit traiter simultanément des objets différents, il est plus commode de placer la source au-dessus des surfaces, celles-ci reposant dans un support permettant de régler individuellement leur distance au filament.

Les substances à déposer doivent être portées à la température de vaporisation. Pour cela on les dispose sur un filament ou à l'intérieur d'une hélice (surfaces au-dessous de la source), ou dans un creuset (surfaces au-dessus de la source). On utilise pour cela des fils ou des rubans, façonnés à la forme voulue, de métaux réfractaires, tungstène, molybdène ou tantale; les substances utilisées pour les couches antiréfléchissantes doivent, comme nous l'avons vu, posséder un indice de réfraction très faible. On est ainsi amené à utiliser les fluorures, dont l'indice varie de 1,32 à 1,38. Le fluorure de magnésium et la cryolithe, fluorure double de sodium et d'aluminium, sont, à l'heure actuelle, utilisés presqu'exclusivement. Leur vaporisation est facile à une température qui n'excède pas 1.000 degrés et ils n'attaquent pas les métaux réfractaires. Le réglage de l'épaisseur du dépôt est effectué, soit en observant la teinte de la lumière réfléchie, soit en évaporant une masse bien déterminée de substance fixée par des expériences préalables.

Pour les couches augmentant la réflexion, on utilise: l'oxyde de titane qui demande un chauffage très énergique, et une recuisson ultérieure d'une heure, à une température de 400°, destinée à oxyder des traces de titane métallique absorbant, provenant de la décomposition de l'oxyde lors du chauffage.

Le sulfure d'antimoine qui se vaporise à basse température.

L'oxyde de titane permet d'obtenir un facteur de réflexion maximum de 35% avec une absorption nulle. Le sulfure d'antimoine permet d'obtenir des surfaces de réflexion et de transmission identiques et égales à 47% avec seulement 6% d'absorption.

Dans tous les cas la qualité des couches obtenues, c'est-à-dire leur adhérence, leur dureté, leur résistance à l'eau ou aux agents atmosphériques, leur uniformité dépendent essentiellement de deux conditions qui doivent être réalisées d'une manière impérative: les surfaces doivent être propres et il faut opérer dans un bon vide.

ARNULF. *Le Vide*, 1946, **1,** 129.

CHIMIE

1

DÉTERMINATION DES CHALEURS DE COMBUSTION DE DÉRIVÉS NITRÉS DE LA SÉRIE BENZÉNIQUE

Dans une bombe à revêtement intérieur de platine et antérieurement étalonnée à l'acide benzoïque, on a brûlé les dérivés dinitrés et trinitrés du benzène, du toluène et du *m*-xylène qui possèdent les groupements NO_2 en méta les uns par rapport aux autres et en ortho ou en para par rapport aux CH_3, ce sont: le *m*-dinitrobenzène, 4131,7 Cal; le 2.4-dinitrotoluène, 4656,4 Cal; le 2.6-dinitrotoluène, 4683,5 Cal; le 2.4 dinitro *m*-xylène, 5132,4 Cal; le 4.6-dinitro *m*-xylène, 5112,1 Cal; le 1.3.5-trinitrobenzène, 3096,3 Cal; le 2.4.6-trinitrotoluène, 3593,7 Cal; et le 2.4.6 trinitro *m*-xylène, 4019,5 Cal. Les résultats sont exprimés en Cal 15° et correspondent à la chaleur de combustion isothermique, à vol Cte et à 17°, de 1 g de substance (vide).

A partir de ces résultats on a calculé qu'il existait une différence de 4,5 Cal, entre les chaleurs de formation des isomères dinitrés et on a vérifié que la chaleur de combustion de CH_2 restait voisine de 154 K Cal.

J. Physique, 1939, **10,** P.C. 195.

2

APPLICATIONS DE L'INFLUENCE DU RAYONNEMENT SUR LES CORPS

En chimie, l'étude de l'influence du rayonnement sur les corps a eu une grande importance; l'établissement de la formule de constitution chimique des corps, la distribution des atomes dans les molécules, la distribution des atomes et des molécules dans les corps cristallisés, la nature et les forces de liaison entre les atomes et les molécules, enfin l'analyse du mécanisme des réactions chimiques et la mesure de l'affinité chimique des corps sont des questions que l'on a pu faire avancer en grande partie grâce à l'étude de l'action du rayonnement sur le corps.

Enfin, en biologie, le fait de l'emmagasinement de l'énergie solaire par les plantes, ainsi que la grande quantité de réactions de synthèse que l'on peut provoquer par la lumière, permettent d'affirmer que l'origine de la vie doit être cherchée dans les actions photochimiques produites par le rayonnement sur des corps plus ou moins complexes.

London B.Sc. (*General*).

3

Méthode de mesures des produits actinifères

Les produits de lanthane actinifère ne peuvent être mesurés directement par l'activité propre de l'actinium, cette activité étant trop faible par rapport à celle des dérivés de l'actinium qui se forment immédiatement. D'autre part il est difficile d'éliminer les dernières traces des dérivés de l'actinium.

Les mesures se font sur des produits d'actinium en équilibre avec ses dérivés. Cet équilibre est obtenu à 1 pour 100 près en 5 mois. L'évolution du dépôt actif, dont on mesure le rayonnement β, peut être traité comme un problème de trois substances consécutives: l'actinium, le radioactinium et l'actinium X, ce dernier accompagné de ses dérivés responsables de la presque totalité du rayonnement.

Le courant d'ionisation produit par le rayonnement de l'actinium et de ses dérivés est mesuré au moyen d'un électromètre à quadrants et compensé par un quartz piézo-électrique. Pour les mesures des rayons β, on place le produit à mesurer à 5 cm d'une chambre d'ionisation fermée par une feuille d'aluminium de $2/100^{e}$ de millimètre d'épaisseur. L'ionisation due aux rayons α est mesurée dans une chambre ronde dans laquelle le rayonnement α est complètement utilisé.

Voulant suivre l'évolution de l'activité de produits de lanthane actinifère dès les premières heures qui suivent la dernière purification radioactive, j'ai constaté certaines anomalies dans l'évolution du rayonnement β de l'actinium privé de ses dérivés.

Perey. *J. Physique*, 1939, **10,** 435.

4

Synthèses asymétriques

Les méthodes de synthèses ordinaires, qui ne peuvent pas comporter un seul processus qui soit énantiomorphe, ne donnent pas directement des corps actifs: mais, si un des procédés dont il s'agit est lui-même énantiomorphe, ce résultat n'est plus nécessairement obtenu. On s'est donc efforcé de préparer des composés énantiomorphes dans des conditions qui seraient elles-mêmes énantiomorphes, espérant effectuer ainsi ce qu'on peut nommer une synthèse asymétrique. Par exemple, si on crée un nouveau carbone asymétrique dans un composé déjà actif, on peut espérer que la présence du premier élément de dissymétrie exercerait une action directrice sur le signe de l'énantiomorphisme du deuxième élément dissymétrique qui se forme; on a réussi, dans quelques cas, en partant d'un composé actif *d*-A, à produire par une méthode purement chimique, les deux dérivés *d*-A, *d*-B et *d*-A, *l*-B, en quantités inégales; on a ainsi demontré que la configuration du nouvel atome asymétrique, B, se laisse influencer par celle de l'élément dissymétrique, A, déjà présent.

London B.Sc. (Special).

5

La définition des acides et des bases

Les concepts classiques reposent sur les rôles antagonistes tenus par les acides et par les bases dans la génération des sels, le caractère acide correspondant à la production de l'anion et le caractère basique à la production du cation. Il arrive que les réactions procèdent apparemment par échange soit d'hydrogène, soit d'oxygène, soit de soufre, etc., mais le plus souvent elles ne peuvent être ramenées à un tel échange à sens unique. Le théorie classique a synthétisé l'ensemble des faits par la mise en valeur du rôle fondamental imparti pratiquement à l'eau. La richesse des faits est moindre dans les autres solvants, mais le résultat subsiste que les deux ions du solvant soient susceptibles selon les circonstances, de jouer des rôles équivalents, des phénomènes symétriques pouvant se retrouver pour chacun d'eux.

La conception de Brönsted supprime ce dualisme foncier. Elle implique, en vérité, deux idées distinctes, l'une qui généralise la notion d'acide et de base en l'étendant des composés aux particules, l'autre qui la restreint en ramenant toujours le phénomène acide-base à un échange d'hydrogène. La généralisation peut apparaître d'ordre formel par suite des différences intrinsèques de propriétés entre particules chargées électriquement ou non, mais elle maintient l'equivalence des rôles susceptibles d'être tenus par les deux ions de tout solvant ou, plus généralement, de tout composé. Il n'est possible de tout ramener à un échange d'ion hydrogène qu'en faisant un choix arbitraire parmi les phénomènes, cause de grande confusion, et en négligeant les phénomènes acide-base dans les systèmes non hydrogénés ou exempts d'ions hydrogène.

EMSCHWILLER. *Bull. Soc. Chim. France*, 1946, 200.

6

TEMPÉRATURES D'INFLAMMATION

Les déterminations de températures d'inflammation faites à divers pressions sur les mêmes mélanges par Townend et Mandlekar avec un dispositif expérimental assez différent (méthode statique) confirme les observations précédentes, mais l'apparition d'un domaine d'inflammation limité par deux températures se trouve très légèrement décalé, en raison du changement de méthode expérimentale, vers les températures et les pressions plus élevées (300° au lieu de 260°, 1,65 atm au lieu de la pression ordinaire). Cela s'explique facilement par le fait que la méthode dynamique met les gaz frais en contact avec des produits d'oxydation partielle, dont l'influence sensibilisatrice facilite l'apparition du phénomène explosif à pression et température moins élevées. La même explication justifie l'écart analogue entre les températures d'inflammation obtenues par la méthode statique selon que le chauffage est lent, ou très rapide.

London B.Sc. (Special).

7

La composition des solutions diluées de savons

En examinant au microscope des solutions contenant des cristaux obtenus en partant de solutions concentrées d'acides gras par dilutions successives on a constaté l'existence de trois sortes de cristaux: ceux des savons neutres aux fortes concentrations; ceux des savons acides aux concentrations moyennes; ceux des acides gras aux concentrations les plus faibles; et cette identification a été vérifiée par la détermination des points du fusion et par l'analyse élémentaire. Par exemple le myristate de Na, donne pour M: 500, des cristaux neutres; autour de M: 1000 des cristaux de savons acides et au-dessous de M: 1000 des cristaux d'acide libre. Lorsqu'on chauffe ces solutions contenant des cristaux on obtient la transformation des cristaux en globules et en même temps pour les sels des savons acides et neutres leur décomposition en liquide anisotrope. Dans les solutions saturées de savons neutres l'auteur admet l'existence d'un « Assoziations kolloïd » qui serait plutôt une phase qu'une micelle véritable; par contre dans les solutions plus ou moins diluées il y aurait équilibre entre la matière dispersée à l'état moléculaire et celle dispersée à l'état colloïdal. Mais l'auteur doute de la parfaite réversibilité de ces équilibres et de ce que l'état colloïdal puisse être essentiellement différent des solutions véritables.

Malfitano. *J. Physique*, 1939, **10,** C.P. 288.

8

Constitution de l'amidon

Il y a environ 10 ans, à la suite de travaux sur diverses autres substances hautement polymerisées, l'un de nous a apporté en collaboration avec H. Hopff et H. Mark une contribution à l'étude de la constitution de l'amidon. Irvine et Macdonald ainsi que Haworth avec ses collaborateurs avaient montré dans des travaux fondamentaux que la molécule d'amidon contenait probablement une chaîne formée de radicaux α-1,4-glucosiques; nous avions pu appuyer cette interprétation par l'analyse de l'activité optique de l'amidon au moyen de la règle de Hudson ainsi que par une étude de la cinétique de la dégradation par les acides. Des travaux très fouillés de Freudenberg ont modifié sur quelques points de détail notre argumentation, tout en confirmant sur tous les points essentiels nos conclusions, d'après lesquelles les liaisons sont formées d'une manière tout à fait prépondérante par des liaisons $\alpha 1,4$ (liaisons maltosiques) et les liaisons β sont absentes ou en tout cas très peu nombreuses. Mais l'étude de la viscosité de solutions d'amidon ainsi que que celle des propriétés mécaniques des films obtenus avec des dérivés de ce corps a montré qu'une formule à chaîne non ramifiée n'est pas satisfaisante. Dans notre monographe consacrée aux principes immédiats hautement polymérisés, nous sommes alors arrivés à la conclusion: « que nous n'avons pas uniquement de longues chaînes maltosiques filiformes, mais que ces chaînes doivent présenter en certains endroits des ramifications ou des réticulations. »

MEYER, BRENTANO et BERNFELD. *Helv. Chim. Acta*, 1940, **23**, 845.

9

DÉSHYDROGÉNATION DE LA β-VÉTIVONE SUR CHARBON PALLADIÉ

Le charbon palladié à 30% de palladium a été préparé selon la méthode de Zelinsky *et al*, améliorée par Linstead *et al.*

16 g de chlorure de palladium ont été dissous dans 50 cm^3 d'eau. Cette solution a été additionnée de 3 cm^3 d'acide chlorhydrique concentré, et elle a ensuite reçu en suspension 24 g de norit. Cette suspension bien agitée a été additionnée de 48 cm^3 de la solution commerciale d'aldéhyde formique et ensuite, peu à peu, en l'espace de 3 heures, de 96 cm^3 de potasse à 50% exempte de carbonate. Le mélange a encore été agité durant 3 heures au cours desquelles le liquide s'est décoloré. La partie solide a été essorée, lavée à l'acide acétique dilué et à l'eau, puis séchée à 140°. Le catalyseur ainsi préparé renfermait 30,6% de palladium.

20 g de β-vétivone additionnés de 10,2 g de charbon palladié ont été chauffés au bain de métal, en atmosphère de gaz

carbonique, pendant 22 heures, la température interne étant maintenue à 270-285°. La quantité de gaz recueillie sur lessive de potasse a été de 1650 cm^3 (20°, 730 mm) dont 1175 cm^3 au cours des 5 premières heures, 300 cm^3 au cours des 8 heures suivantes, soit approximativement 0,7 mol par rapport à la quantité de vétivone mise en œuvre. Le gaz avait une forte odeur de propylène.

Le résidu a été extrait par le benzène bouillant. L'extrait benzénique a été distillé sous 2 mm de pression. On a obtenu 7,2 g d'une huile violette passant de 92-161° et 5 g de résidu indistillable. Le distillat a été traité comme indiqué plus loin.

PFAN et PLATTNER. *Helv. Chim. Acta*, 1940, **23**, 781.

10

LA FLUORESCENCE DES COMPOSÉS ORGANIQUES EN SOLUTION

La raisons de la fluorescence de certaines molécules organiques sont mal connues. L'auteur rappelle la variation du pouvoir fluorescent dans la série uranine-éosine-étrythrosine-rose Bengale où le pouvoir fluorescent diminue quand croît le nombre des halogènes substitués. Il n'admet pas la théorie de Lewschin d'après laquelle cette diminution serait due à une « extinction interne » analogue à celle qui se produit par addition d'ions halogène à l'uranine. Il semble probable que les couleurs halogénées ont un caractère acide qui accroît leur interaction avec les molécules d'eau environnantes; ce sont celles-ci qui prendraient l'énergie d'excitation. D'ailleurs dans d'autres solvants que l'eau (acétone) le rendement de la fluorescence est bien meilleur. Franck a imaginé que les colorants sont souvent fluorescents parce qu'ils ont des bandes d'absorption dans des longueurs d'onde assez grandes pour que le quantum correspondant ne puisse déclancher un processus d'extinction. En contradiction avec cette hypothèse, on peut citer que les mêmes colorants deviennent fluorescents avec bon rendement dans leurs bandes ultraviolettes: de plus le nombre des colorants non fluorescents est très grand; des carbures aromatiques simples sont très fluorescents en lumière U.-V.; enfin il est rare que les composés inorganiques colorés (exceptions terres rares et uranyle) soient fluorescents. L'auteur pense que, si la substance est fluorescente, c'est que la transition électronique d'absorption est bien protégée contre les perturbations. Cette supposition est d'accord avec la finesse des bandes d'absorption des terres rares; elle est d'autre part en contradiction avec le fait que les sels de Pr et Nd à bandes fines ne sont pas fluorescents. On discute les publications où on a cherché si l'ion ou la molécule sont le siège de la fluorescence. Si on opère correctement avec le sulfate de quinine, on montre que le meilleur rendement en solution acide est seulement apparent; cela tient à ce que le sulfate acide absorbe 50 fois plus la lumière active que la solution neutre. Il peut y avoir une différence entre l'ion et la molécule dans d'autres cas; des exemples cités montrent qu'on ne peut dire son sens; il est également impossible d'affirmer que les colorants sont fluorescents uniquement à l'état dissous. Presque toutes les substances fluorescentes en solution perdent cette propriété quand on augmente la concentration. On discute les deux hypothèses émises à ce sujet (choc de 2e espèce, formation d'associations non fluorescentes). Le transfert d'énergie par résonance avec de grands rayons d'action (F. Perrin) semble improbable. On donne la courbe des pouvoirs fluorescents pour l'anthracène dissous dans le benzène; il semble dans ce cas que l'illumination déplace l'équilibre avec le dimère en faveur de celui-ci. Il est donc possible que la molécule excitée, qui est fluorescente à l'état isolé, s'unisse avec une autre quand la concentration augmente. Cette explication vaudrait pour tous les cas où l'association a été démontrée.

DARMOIS. *J. Physique*, 1939, **10**, 264.

11

PRÉPARATION DE L'ACÉTATE FERRIQUE NORMAL

De nombreux procédés, en vue de l'obtention de l'acétate de fer normal, ont été donnés jusqu'à ce jour. Toutes ces préparations réalisées en milieu aqueux, il en résulte que par suite de l'hydrolyse inévitable elles conduisent infailliblement à des sels basiques de formules plus ou moins déterminées.

Nous sommes arrivés, en opérant en milieu rigoureusement anhydre, à obtenir l'acétate de fer normal.

Nous partons du propylate de fer qui est attaqué par l'acide acétique cristallisable.

Le propylate de fer est obtenu par action du propylate de sodium sur le chlorure ferrique sublimé en solution alcoolique.

L'acétate ainsi obtenu est débarrassé de l'acide acétique qui l'imprègne en le soumettant à un acide très poussé. Nous obtenons un corps qui se présente sous la forme de petits cristaux rouge brique solubles dans l'eau, l'alcool et l'acide acétique.

Le dosage nous donne le rapport $Fe/CH_3COO = 1/2{,}97$.

Une solution alcoolique de notre produit soumise à l'essai de migration, révèle la présence de fer uniquement à l'état de cathion; l'anion étant constitué exclusivement par le radical acétique.

La solution fraîchement preparée donne une solution non colloïdale contrairement à celle preparée à partir des acétates obtenus avant nous.

Tous ces renseignements légitiment la représentation de notre corps par la formula $(CH_3COO)_3Fe$.

SCHWARTZ. *Bull. Soc. Chim. France*, 1946, 22.

12

LA BROMURATION DES HUILES

La façon dont se comportent les huiles vis à vis du Brome donne des indications très intéressantes qui permettent de les classer en divers groupes. Sur toutes les huiles le Brome réagit en donnant des produits d'addition sur les liaisons éthyléniques: l'acide oléique est transformé en acide dibromostéarique, l'acide linoléique en tetra et l'acide linolénique en hexabromostéarique.

La bromuration effectuée en solution éthérée donne naissance quand l'acide linolénique ou d'autres acides encore moins saturés sont présents à un précipité de glycérides hexa ou décabromés insolubles. Cette réaction peut être réalisée sous forme quantitative; j'ai utilisé la technique de Hehner et Mitchell.

1 à 2 grammes d'huile sont pesés dans une fiole et dissous dans 40 cm^3 d'éther additionnés d'acide acétique; la solution est refroidie à 0° et traitée par le Brome goutte à goutte jusqu'à coloration brune persistante. Après 3 heures de repos à 0°, le précipité, s'il y en a un, est recueilli sur un creuset de Gooch taré et lavé avec 40 cm^3 d'éther glacé, puis séché et pesé.

Certaines huiles n'ont donné dans ces conditions aucun précipité et sont demeurées parfaitement limpides plusieurs heures après l'addition de Brome. Ce sont celles dont les indices d'iode sont peu élevés et d'une façon générale inférieurs à 100. D'autres huiles ont fourni un précipité plus ou moins abondant. Je donne plus loin avec l'étude spéciale de chacun de ces produits les divers chiffres obtenus. L'acide linolénique a pu être caractérisé par le point de fusion et la teneur en brome de son dérivé hexabromé.

TIMON-DAVID. *Ann. Fac. Sci. Marseille*, 1930, **4**, 87.

13

PASSIVATION CHIMIQUE D'UNE SURFACE MÉTALLIQUE PRODUIT PAR LA BOMBARDEMENT

Les régions ayant subi le bombardement électronique font preuve par la suite d'une grande passivité chimique. Dans le cas de l'argent, par exemple, le métal peut être immergé pendant plusieurs heures dans une solution de sulfure d'ammonium sans qu'il y ait attaque de la zone bombardée, alors que le reste de la surface du métal se recouvre d'une couche brune ou noire de sulfure. La trace du faisceau apparaît alors, sur un fond sombre, sous forme d'une bande brillante.

Pour obtenir cette « passivation » il n'est pas nécessaire de prolonger le bombardement jusqu'à obtenir une trace directement visible à l'œil. Quelques minutes d'exposition suffisent pour que la sulfuration révèle le trajet du faisceau. En général, on peut dire que, dans le cas d'une surface donnant un diagramme électronique, il suffit d'une exposition juste suffisante pour faire disparaître le diagramme, pour qu'il y ait passivation.

D'autres expériences, effectuées avec divers métaux, confirment ce phénomène. C'est ainsi que les métaux sont protégés un certain temps contre l'attaque par des acides, et que l'or, exposé aux vapeurs de mercure, blanchit partout, sauf dans la région de l'impact électronique.

Il est certain qu'en prolongeant suffisamment le temps de pose, on arriverait à enregistrer ainsi un diagramme de diffraction sur une plaque métallique (Ag poli, par exemple) remplaçant la plaque photographique; le développement s'effectuerait simplement par immersion de l'argent dans du sulfure d'ammonium. Nous avons essayé d'obtenir ce résultat, en utilisant comme substance diffractante un film de platine, d'oxyde de zinc ou de mica disposé normalement aux électrons (méthode de transmission); mais si la trace du faisceau *direct* non diffracté s'obtient rapidement, la trace des anneaux ou taches *de diffraction* n'apparaît pas après quatorze heures de pose consécutives (ZnO). Il est d'ailleurs possible qu'elle apparaisse pour des temps de pose de l'ordre de plusieurs jours, mais l'expérience est difficilement réalisable au point de vue pratique, la préparation diffractante donnant un diagramme qui s'affaiblit progressivement pour disparaître après quelques heures.

TRILLAT et MÉRIGOUX. *J. Physique*, 1939, **10,** 247.

14

SÉPARATION DES CONSTITUANTS ACIDES DES CORPS GRAS

Une méthode tout à fait générale et qui revient au jour périodiquement, est la méthode de précipitation fractionnée des sels alcalins par les acétates ou chlorures de calcium, baryum, magnésium, lithium, thallium, etc., en milieu aqueux, alcoolique, acétonique ou benzénique: dans son principe, la méthode consiste à faire une solution, exactement neutralisée, de savon alcalin et à ajouter à cette solution successivement le quart (ou le cinquième) de la quantité théorique de chlorure de calcium, par exemple, nécessaire pour précipiter tous les savons à l'état de savon de calcium.

On obtient ainsi un premier précipité que l'on sépare et d'où on libère les acides.

Au liquide filtré on ajoute une nouvelle quantité, égale à la première, de chlorure de calcium: on a un nouveau précipité que l'on recueille et ainsi de suite.

On a ainsi quatre portions que l'on étudie séparément.

Nous avons essayé ces précipitations fractionnées: elles présentent le gros inconvénient qu'elles donnent lieu, souvent, à des entraînements considérables; pour obtenir quelque chose de relativement correct, il faut faire les additions de sels de calcium très lentement, en agitant très vivement, et laisser ensuite le précipité en contact avec la solution assez longtemps pour que les sels les plus insolubles soient vraiment ceux qui se précipitent d'abord.

Mais là comme ailleurs les solubilités réciproques jouent un rôle considérable, et si, dans quelques cas simples, comme celui de l'allanblackia, on arrive assez rapidement à des résultats corrects, il est des cas—tels les pépins de raisin—où l'on n'aboutit à rien de net; il semble que l'on ait, au contraire, multiplié la difficulté par quatre.

Au fond, la plupart de ces méthodes ont été étudiées sur des mélanges synthétiques simples et se trouvent en défaut en pratique, par suite de la présence de « l'impureté » qui change les taux de solubilité des divers constituants.

MARGAILLAN. *Ann. Fac. Sci. Marseille*, 1933, **5**, 22.

15

DÉVELOPPEMENT PHOTOGRAPHIQUE

Notre but étant d'accentuer les impressions vigoureuses, autrement dit, d'obtenir des clichés « durs » ou « contrastés », la technique photographique nous enseigne qu'il faut utiliser un révélateur fort, agissant en peu de temps. Après plusieurs essais, j'ai choisi un révélateur concentré, à la pyrocatéchine, composé comme suit:

Solution A: 250 g eau distillée,
25 g pyrocatéchine,
25 g sulfite de soude.

Solution B: 250 g eau distillée,
25 g potasse caustique,
45 g phosphate disodique.

Le révélateur est préparé au moment de l'emploi en mélangeant:

20 cm^3 de solution A,
10 cm^3 de solution B,
10 gouttes d'une solution saturée de bromure de potassium.

L'observation microscopique des plaques après des développements de durée croissante montre qu'au cours des 35 premières secondes, les grains fortement impressionnés (régions exposées aux rayons α) subissent seuls un commencement de développement. Après 35 sec, leur réduction se poursuit et leur noircissement s'accentue; mais en même temps, dans les autres régions de la plaque, les grains plus faiblement impressionnés (parasites) sont développés à leur tour.

J'ai été amené à arrêter le développement au bout de 35 sec et à l'achever par un séjour de 2 min dans un révélateur complémentaire, le *Vigutol Léonar*, qui renforce le noircissement des grains déjà partiellement réduits, sans avoir d'action sur les autres. Le Vigutol est généralement employé pour les tirages sur papier. Je l'ai utilisé pour les plaques avec la même concentration.

Les plaques doivent être lavées à l'eau distillée, une minute entre les deux révélateurs, et une minute après le Vigutol. Après quoi, on opère le fixage dans un bain contenant de l'hyposulfite et du bisulfite en quantités égales, avec un peu d'alun de chrome. Dans ces conditions, les noircissements parasites sont très peu nombreux et l'action spécifique des rayons α s'observe très facilement.

HALLWACHS. *J. Physique*, 1938, **9**, 126.

16

PRÉPARATION DE 4-MÉTHYL-3-CYANO-BENZOPHÉNONE

On dissout à chaud 40 g de 4-méthyl-3-amino-benzophénone pure, de p. de f. 109-110°, dans un mélange de 500 cm^3 d'eau et de 60 cm^3 d'acide chlorhydrique concentré, refroidit rapidement et diazote, entre 0° et 10°, avec une solution de 16 g de nitrite de sodium dans 84 cm^3 d'eau.

On traite par ailleurs sur le bain-marie, la solution de 50 g de sulfate de cuivre cristallisé dans 230 cm^3 d'eau par la solution de 55 g de cyanure de potassium dans 100 cm^3 d'eau. Puis on introduit dans le mélange ainsi préparé la solution du sel de diazonium par petites portions tout en agitant vigoureusement. La réaction terminée, on chauffe encore une demi-heure sur le bain-mairie et laisse refroidir. Le produit de réaction, encore très impur et fortement souillé de composés de cuivre, se dépose sous la forme d'un gâteau brunâtre que l'on pulvérise et sèche à 70°. On en tire le nitrile par distillation dans le vide dans un ballon de *Claisen-Anschütz*; le produit passe à 220° sous 14 mm. Le rendement est de 27,5 g, soit 65,5% de la théorie.

Pour l'usage ulterieur, la 4-méthyl-3-cyano-benzophénone ainsi obtenue est suffisamment pure. Pour l'analyse, on la cristallise encore dans le méthanol et dans le benzène. Feuillets incolores brillants fondant à 128°.

0,1693 g subst. ont donné 0,5040 g CO_2 et 0,0792 g H_2O
0,5080 g subst. ont donné 29,80 cm^3 N_2 (19°, 699,5 mm).

$C_{15}H_{11}ON$ (221,25) Calculé C 81,42 H 5,01 N 6,33%.
Trouvé „ 81,24 „ 5,23 „ 6,31%.

CHARDONNENS et SCHLAPBACH. *Helv. Chim. Acta*, 1946, **39**, 1418.

17

SUR UN MASTIC POUR JOINTS ÉTANCHES

Les résines vinyliques, associées à un plastifiant, peuvent servir de mastic pour joints étanches. Nous avons fait choix de l'acétate de polyvinyle plus connu sous le nom commercial de « rhodopas ».

L'acétate de polyvinyle résulte de l'enchaînement répété plusieurs centaines de fois, d'une petite molécule, celle de l'acétate de vinyle:

```
CH2 = CH
      |
      O—CO—CH3
```

D'où la formule:

```
... CH2 - CH - CH2  -  CH - CH2  -  CH - CH2 - CH...
          |            |            |
          O - CO - CH3 O - CO - CH3 O - CO - CH3
```

Les molécules de l'acétate de polyvinyle, très grandes et filiformes, sont remarquablement tenaces dans le sens de leur longueur et cependant facilement séparables les unes des autres, soit par agitation thermique soit par insertion entre elles de molécules de solvant ou de plastifiant.

En raison de son poids moléculaire élevé, l'acétate de polyvinyle présente une tension de vapeur très faible, ce qui autorise son emploi en technique du vide. D'autre part, il adhère bien aux métaux, au verre, aux produits céramiques. Il est insoluble dans l'eau, soluble dans un grand nombre de solvants courants tels que les alcools (éthylique, méthylique, amylique), la benzine, l'acétate d'éthyle.

L'acétate de polyvinyle est tenace à la température ordinaire; il est même cassant d'où la nécessité de lui ajouter un plastifiant pour former un mastic. Nous avons employé un mélange de celludol, qui est un dérivé de la paratoluènesulfamide et de triphénylphosphate. Ce sont de bons plastifiants mais qui ont l'inconvénient d'introduire une tension de vapeur non négligeable. (Nous avons en cours des essais avec d'autres plastifiants à tension de vapeur très faible.)

Le Vide, 1946, **1**, 176.

18

DISTILLATION DE LA CELLULOSE

Nous avons procédé comme suit: La cellulose (coton purifié et desséché) est introduite dans une cornue en verre peu fusible, qui est reliée à une série de récipients. Le premier de ces récipients plonge dans de l'eau maintenue à la température de 50°, les autres sont refroidis par un mélange de glace et de sel. On chauffe graduellement, dans un bain de sable, après avoir fait dans les appareils un vide de 12—15 mm. L'opération doit être conduite rapidement et ne pas durer plus d'une heure.

La distillation commence vers 210°. Il passe d'abord un liquide aqueux, qui se condense dans les récipients refroidis, puis une huile de couleur jaune; celle-ci se prend dans le premier récipient en une masse pâteuse, qui ne tarde pas à cristalliser en partie. Il reste dans la cornue un peu de charbon.

Les proportions pondérales moyennes de ces divers produits, par rapport à la cellulose employée, sont les suivantes:

Masse pâteuse	45%
Liquide aqueux	32 „
Gaz et pertes	13 „
Résidu	10 „

Le liquide aqueux ne contient que peu de matière organique en dissolution. Nous y avons constaté la présence d'une petite quantité de furfurol et d'acides de la série grasse; mais nous ne nous sommes pas attardés à son étude, notre intérêt allant surtout au produit principal de l'opération, qui est la masse pâteuse et semicristalline. Nous avons cherché à la purifier de différentes manières. La plus rapide consiste à la dissoudre dans l'eau, où elle est extrêmement soluble, à faire bouillir la solution avec du noir animal, puis à la concentrer jusqu'à cristallisation et à faire sécher le dépôt sur de la porcelaine poreuse. Une nouvelle cristallisation dans une petite quantité d'eau chaude, ou mieux encore dans l'alcool amylique bouillant, fournit enfin la substance à l'état de pureté. Son poids représente les 70% de la masse pâteuse, ce qui correspond à environ 30% de la cellulose. Ce rendement pourra sans aucun doute être notablement élevé lorsqu'on opérera sur une plus grande échelle et que l'on tirera parti des eaux-mères.

Le composé ainsi obtenu forme des tables ou de larges aiguilles plates, parfaitement incolores, stables à l'air humide, et fusibles à 179,5—180°. Chauffé plus haut, il se décompose vers 270°; il passe d'abord un liquide aqueux, puis, à plus haute température, un peu de goudron. Le liquide aqueux tient en dissolution une forte quantité d'acides formique et acétique, et des traces d'acétone; le goudron est constitué presque entièrement par des phénols. Ce sont là les produits principaux de la distillation sèche de la cellulose à la pression atmosphérique. On doit donc considérer, ici aussi, le produit de la distillation sous pression réduite comme représentant un stade intermédiaire de la décomposition pyrogénée de la cellulose.

PICTET et SARASIN. *Helv. Chim. Acta*, 1918, **1**, 88.

19

SÉPARATION DE L'ANTIMOINE EN PRÉSENCE ON NON D'ÉTAIN

Nous avons déjà indiqué que si l'on attaque par l'acide nitrique un alliage contenant de l'étain et de l'antimoine, il y a formation d'un composé complètement insoluble dans NO_3H, si la quantité d'étain est suffisante.

Pour atteindre ce résultat, l'expérience a montré que la quantité d'étain en présence doit être au moins égale a 5 fois celle de l'antimoine.

Dans le cas d'un métal antifriction pouvant contenir 20 à 25% de Sb et peu ou pas d'étain, si l'on opère sur une prise d'essai de 2,5 g il faut introduire dans l'analyse 2,5 à 3,125 g d'étain, au minimum, pour insolubiliser complètement l'antimoine.

Ce n'est pas sans appréhension qu'un chimiste introduira une telle quantité d'étain, même si l'étain est d'une grande pureté.

Il est possible de tourner cette difficulté si l'on opère de la façon suivante:

Attaquer l'alliage par l'acide nitrique étendu, insolubiliser en présence de nitrate d'ammonium, filtrer sur filtre à grains serrés, laver à l'eau nitrique et faire passer tout le précipité sur le filtre.

La solution filtrée est concentrée puis additionnée de 0,2 g d'étain. Chauffer pour dissoudre l'étain et après dissolution de celui-ci porter la solution à l'ébullition.

La petite quantité d'antimoine qui s'est dissoute au cours de la première opération se trouve insolubilisée et entraînée avec l'acide métastannique.

Après filtration et lavage à l'eau nitrique faire passer le précipité sur le filtre.

Les deux précipités contiennent tout l'antimoine et l'étain (y compris l'étain ajouté). Dans cette sorte de séparation, les oxydes d'étain et d'antimoine entraînent toujours une petite quantité des métaux en présence (Cu—Pb—Fe, etc. . . .) et le Bi en totalité s'il s'agit d'une impureté.

BERTIAUX. *Bull. Soc. Chim. France*, 1946, 102.

20

Sur la structure de l'acide nitrique et du nitrate de méthyle

Avant d'aborder la question de la structure de la molécule de nitrate de méthyle, nous avons repris celle de la molécule d'acide nitrique. L'acide nitrique tant dilué (ion NO_3^-) qu'aux concentrations élevées (molécule NO_2-OH) a été abondamment étudié par la spectrographie Raman. L'ion NO_3^- a été reconnu plan et possédant la symétrie D_{3h}; à notre connaissance le spectre et la structure de l'acide nitrique non dissocié n'ont pas été complètement expliqués.

Notre but est de montrer que l'examen des spectres et les considérations de symétrie permettent de passer du spectre et de la structure de l'ion NO_3^- à ceux de l'acide non dissocié NO_2-OH et à ceux du nitrate de méthyle. Autant qu'il se pouvait, on a cherché à confirmer les conclusions obtenues en utilisant les formules théoriques qui permettent en principe de calculer le spectre Raman lorsque sont connues la structure et les forces interatomiques de la molécule.

CHEDIN. *J. Physique*, 1939, **10**, 445.

21

La 3,6-diméthyl-fluorénone

Nous avons fait la synthèse de la 3,6-diméthyl-fluorénone, composé qui, à notre connaissance, n'a pas encore été décrit jusqu'ici, à partir de l'acide 4-méthyl-2-amino-benzoïque.

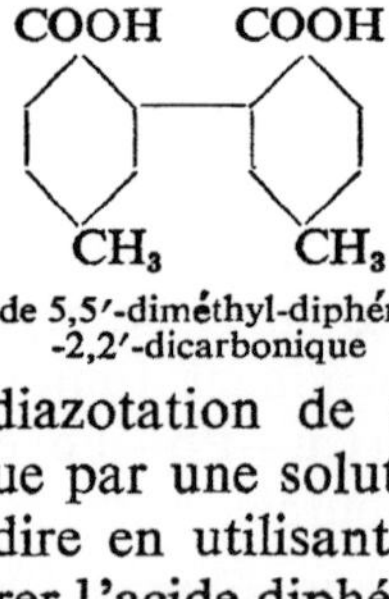

Acide 5,5'-diméthyl-diphényle-2,2'-dicarbonique

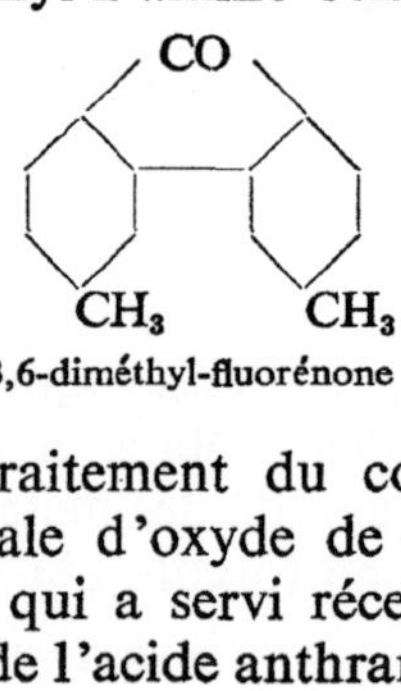

3,6-diméthyl-fluorénone

Par diazotation de cet acide et traitement du composé diazoïque par une solution ammoniacale d'oxyde de cuivre, c'est-à-dire en utilisant une méthode qui a servi récemment à préparer l'acide diphénique à partir de l'acide anthranilique, on obtient avec un bon rendement l'acide 5,5'-diméthyl-diphényle-2,2'-dicarbonique. Ce dernier, chauffé à 350° avec de l'oxyde de calcium, donne la 3,6-diméthyl-fluorénone cherchée. Le produit de la réaction est toujours accompagné d'une quantité notable d'un composé huileux de couleur rouge dont l'élimination, fort malaisée, ne réussit que par voie chromatographique. Le rendement en produit pur n'est guère que de 23 % de la théorie.

La méthode suivante, bien qu'un peu plus longue, conduit à de meilleurs résultats. On fait, à partir de l'acide 4-méthyl-2-amino-benzoïque, l'acide 4-méthyl-2-tosylamino-benzoïque, condense le chlorure d'acide de ce dernier avec le toluène en présence de chlorure d'aluminium et soumet le produit brut de la réaction à la saponification au moyen d'acide sulfurique concentré. On obtient ainsi la 4,4'-diméthyl-2-amino-benzophénone, qui, diazotée en solution sulfurique, donne, à côté d'un peu de 4,4'-diméthyl-2-oxybenzophénone, la 3,6-dimethyl-fluorénone.

CHARDONNENS et WÜRMLI. *Helv. Chim. Acta*, 1946, **29**, 922.

22

LES MOMENTS ÉLECTRIQUES DE LA *p*-BENZOQUINONE ET DE COMPOSÉS VOISINS

La *p*-benzoquinone est ordinairement considérée comme suffisamment symétrique pour être non polaire, cependant elle apparaît être légèrement mais certainement polaire. On montre que non seulement ce composé, mais aussi ses dérivés diméthylé et dichloré ainsi que la 2-2-4-4-tétraméthylcyclo-butane-1 : 3-dione présentent tous des différences équivalentes entre leur polarisation totale et électronique. Un modèle mécanique de la structure classique de la *p*-benzoquinone fléchit très facilement de sorte que les deux groupes carbonyle sont en dehors du plan du noyau, et il semble possible que les polarisations puissent prendre naissance lors de la flexion lente due à l'énergie thermique, qui permettrait à la molécule de s'orienter comme un dipole pendant la flexion. On conclut qu'une telle flexion ne peut pas être importante car si elle l'était, les polarisations des autres dicétones seraient sensiblement différentes ce qui n'est pas le cas, et la rigidité du noyau est attribuée à la résonance. De plus un calcul de la fréquence de cette sorte de flexion n'est pas en accord avec la fréquence d'absorption infra rouge. Les deux fréquences sont de loin trop élevées pour que l'orientation ait lieu, aussi la théorie de la flexion thermique est exclue et les différences de polarisation sont considérées soit comme des polarisations atomiques élevées, ceci en accord avec les mesures dans l'infra rouge lointain, soit comme des interactions soluté-solvant.

GROSS. *J. Physique*, 1939, **10,** 16.

23

LA BIOCHIMIE DES PHOSPHATASES

Un grand intérêt s'attache à l'étude des phosphatases, tant en raison de la multiplicité des combinaisons phosphorées naturelles qu'elles synthétisent et dégradent dans les cellules, que de leur participation aux processus physiologiques les plus divers, depuis le métabolisme intermédiare des glucides jusqu'à la régulation rénale de l'equilibre acide-base et à l'ossification. On connaît depuis près de quarante ans l'existence d'enzymes hydrolysant les esters orthophosphoriques, mais leur étude n'a pris un développement important qu'après la démonstration en 1923 du rôle joué par une phosphatase dans la calcification des os et, ultérieurement, avec l'identification de nombreuses réactions spécifiques de phosphorylation et de déphosphorylation dans le métabolisme cellulaire. L'ensemble de la biochimie des phosphatases ne saurait plus être présenté utilement que dans des mises au point assez étendues.

ROCHE. *Helv. Chim. Acta,* 1946, **29,** 1253.

24

FORMATION D'ANHYDRIDE IODIQUE PAR PHOTOXYDATION D'IODURES ORGANIQUES

L'oxydation photochimique de divers iodures organiques produit de l'anhydride iodique; les rendements sont les plus élevés dans le cas de l'iodure de méthylène.

Les quantités d'anhydride iodique vont croissant au cours de la photoxydation de l'iodure de méthylène en système clos, pour diminuer dès qu'elle est terminée; il y a en effet consommation d'anhydride iodique par d'autres produits de l'oxydation, le formaldéhyde en particulier. La production d'anhydride iodique est donc liée essentiellement à la présence de l'iodure de méthylène. Mais l'anhydride iodique n'est pas un produit primaire de la photoxydation de l'iodure de méthylène, car sa production est beaucoup plus faible lorsqu'on opère par irradiation d'un courant d'oxygène entraînant de l'iodure de méthylène. Le rapprochement d'expériences effectuées avec ou sans circulation d'oxygène permet aussi de reconnaître que la photoxydation de l'iodure de méthylène s'accompagne de phénomènes de catalyse oxydante. Diverses interprétations de tous ces faits sont proposées et discutées. L'hypothèse d'une union directe de l'oxygène et d'iode électroniquement excité est envisagée, puis deux mécanismes où la production d'anhydride iodique est reliée à la formation de composés organiques oxyiodés, la question se posant de savoir si ces composés oxyiodés sont issus directement de l'oxydation de l'iodure organique ou s'ils résultent de l'oxydation secondaire du formaldéhyde produit par photoxydation de l'iodure. Le rôle catalytique pourrait être tenu soit par des peroxydes, soit par des dérivés oxyiodés. La production d'anhydride iodique par photoxydation des iodures organiques est un phénomène général lié à leur oxydabilité; on l'observe sur des dérivés monoiodés (iodures de méthyle et d'éthyle) diiodés (iodures de méthylène et d'éthylidène), triiodé (iodoforme), tétraiodé (tétraiodure de carbone), mais non sur des dérivés comme les iodures d'éthylène et d'acétylène, dont la photolyse est insensible à l'oxygène.

La photoxydation de l'iodoforme ne produit pas d'anhydride iodique s'il est en solution, mais s'il est à l'état solide. Les rendements sont faibles, car l'iodoforme et l'anhydride iodique réagissent aisément l'un sur l'autre. Ces faits sont de nature à projeter une lumière nouvelle sur l'oxydation des solutions d'iodoforme, où des dérivés oxyiodés peuvent intervenir comme catalyseurs. Il est montré, à ce propos, que l'influence du solvant sur la vitesse de photoxydation de l'iodoforme est liée à sa nature chimique et non à son moment électrique, les résultats étant comparables dans les chlorures d'acétylène cis et trans.

La production d'anhydride iodique par photoxydation du tétraiodure de carbone solide oblige à rejeter l'hypothèse suivant laquelle la production d'anhydride iodique serait liée à la photoxydation secondaire de composés aldéhydiques. Il est confirmé non seulement que l'anhydride iodique n'est pas un produit primaire mais aussi que la réaction secondaire qui lui donne naissance est favorisée par la lumière. L'oxydation du tétraiodure de carbone dans l'obscurité ne donne pas d'anhydride iodique.

L'ensemble des résultats est interprété en supposant la formation, directe ou indirecte, de produits d'addition entre l'oxygène et l'iodure organique, assimilables aux composés iodylés de la série aromatique.

J. Physique, 1939, **10,** C.P. 196.

25

Chimie minérale—Recherches de Berg

Berg a étudié les chromoiodates et les chromotellurates. Les recherches très approfondies sur ces premières substances ont fait l'objet de sa thèse de pharmacie présentée en 1896 devant la Faculté mixte de Médecine et de Pharmacie de Toulouse.

Remarquant la facilité avec laquelle l'acide chromique donne des sels condensés tels que les bi, les tri et les tétrachromates, il avait tenté dès 1887 d'obtenir des sels complexes de cet acide dont on ne connaît qu'un petit nombre.

Il a pu obtenir des chromoiodates de potassium, sodium ammonium, lithium, calcium, magnésium, cobalt, nickel cuivre et argent. Il a décrit en outre l'acide libre lui-même

L'existence des sels d'argent et d'ammonium anhydres et du sel de potassium tantôt anhydre, tantôt hydraté, a permis à Berg de réfuter l'opinion de Blomstrand qui contestait la formule de constitution qu'il avait proposée pour ces composés.

La tendance de l'acide chromique à donner des sels complexes a été encore démontrée par l'obtention de chromotellurates dont il a préparé les sels de potassium et d'ammonium.

On sait que l'on connaît peu de composés dans lesquels le bioxyde de manganèse joue le rôle de base, et ils sont peu stables. Il en est ainsi de l'iodate de manganèse que Berg a pu préparer; ce sel est fortement hydrolysé en solution. Mais il a obtenu des combinaisons cristallisées bien définies et stables de ce corps avec d'autres iodates métalliques. Par exemple les iodates doubles de manganèse tétravalent avec les iodates de potassium, d'ammonium, de baryum et même de manganèse divalent, ont été préparés par Berg. Ces corps résultent de l'union d'une molécule d'iodate de bioxyde de manganèse avec deux molécules d'iodate d'un métal monovalent ou une molécule d'iodate de métal divalent; ils correspondent donc comme structure aux chloroplombates.

TIAN. *Ann. Fac. Sci. Marseille*, 1935, **8**, 184.

26

SIGNIFICATION PHYSIOLOGIQUE DES CORPS GRAS INSAPONIFIABLES

Les substances désignées en technologie des corps gras sous le nom d'insaponifiable représentent des produits très différents au point de vue chimique et qui ne peuvent manquer d'avoir une signification physiologique multiple. Ces produits sont:

1° des hydrocarbures (saturés et non saturés);
2° des alcools monoatomiques de la série grasse;
3° des alcools cycliques (stérols).

Une première distinction s'impose entre les hydrocarbures d'une part et les alcools (qu'ils soient acycliques ou aromatiques) d'autre part; les premiers représentent un produit en quelque sorte juxtaposé aux corps gras; ils n'ont avec eux que des relations de contiguïté; ils ne possèdent aucune liaison chimique avec les acides gras. La proportion de ces hydrocarbures est extrêmement variable dans les graisses de la série animale; on connait les masses prodigieuses d'hydrocarbures trouvées dans les huiles de foies de Sélaciens (Squalène, Spinacène). Ils ne font pas défaut non plus, bien qu'en quantités moins considérables dans les huiles d'insectes; j'ai déjà parlé du carbure saturé (Undécane) trouvé par C. Schall (1893) dans l'huile de fourmis rouges; plus tard, Menozzi et Moreschi ont signalé dans l'huile de chrysalides de vers à soie un carbure $C^{28}H^{58}$ (Octoeicosane) et ont indiqué qu'il doit exister à côté un autre carbure. J'ai trouvé dans l'huile de chenilles processionnaires du pin (*Thaumetopœa pityocampa* Sch.) 8,5 p. 100 d'insaponifiable dont une partie importante doit consister en hydrocarbures.

Quelle est la signification physiologique de ces produits? elle a paru évidente dans le cas des carbures fortement non-saturés des huiles de Sélaciens: on a considéré le Squalène comme une réserve prodigieuse d'énergie, douée d'un potentiel thermochimique très élevé. Dans un travail récent, André et Canal (1928) ont souligné les relations qui paraissent exister entre le squalène, les acides gras et le cholestérol. Se basant sur l'étude de l'huile de foie de *Cetorhinus maximus* Gunn. à divers âges, ces auteurs croient pouvoir considérer le squalène comme le terme d'une série de réactions dont le point de départ serait constitué par les acides gras fortement non-saturés et l'intermédiaire par le Cholestérol.

Le problème est encore plus obscur en ce qui concerne les carbures saturés tels que ceux que renferment les huiles d'insectes: dans ce cas, la chaleur de combustion ne présente pas une valeur très élevée; d'ailleurs on les trouve en quantité si minime qu'ils ne pourraient constituer qu'une réserve de peu d'importance.

Les alcools monoatomiques à poids moléculaire élevé (al. cerylique, mélissique, psyllostéarylique, etc.) se trouvent surtout dans les cires où ils remplacent la glycérine; le rôle de ces substances est souvent protecteur: les éthers de ces alcools constituent dans beaucoup de cas des enduits de revêtement imperméables et isolants (coccides, etc.).

L'importance du Cholestérol (et il en est probablement de même du Bombystérol et de ses isomères) est capitale. On sait depuis les travaux célèbres de Mayer et Schaeffer que cet alcool est un constituant fondamental de la cellule où il existe toujours en proportion déterminée par rapport aux acides gras totaux; son rôle dans l'intensité de l'imbibition des tissus est de toute première importance; il remplit peut-être aussi des fonctions de premier plan dans les phénomènes de la nymphose.

TIMON-DAVID. *Ann. Fac. Sci. Marseille*, 1930, **4**, 60.

27

LES DEUX SORTES D'IONS AMPHOTÈRES: INFLUENCE DE LA MÉSOMÉRIE

Si les deux pôles d'un ion amphotère sont séparés par une chaîne d'atomes coordinativement saturés, comme dans la série des acides amino-alcane sulphoniques $H_3N^+-(CH_2)_n - SO_3^-$, l'influence de l'ionisation d'une des fonctions sur la constante de dissociation de la seconde, d'origine inductive électrostatique décroît très rapidement en fonction du nombre d'atomes intercalaires.

Si, au contraire, la structure du dipôle se prête à une mésomérie, à laquelle ne peuvent donner lieu ni la molécule non chargée, ni les deux ions monovalents positif ou négatif, la second ionisation est favorisée par un effet supplémentaire, souvent considérable, qui résulte de cette mésomérie. Nous avons vu que c'est ce phénomène fondamental qui détermine les remarquables propriétés électrochimiques des dérivées hydroxylés du triphenylcarbinol. Il en est ainsi lorsque simultanément (*a*) le pôle positif est constitué par un atome coordinativement désaturé; (*b*) les deux atomes chargés sont unis directement, ou séparés par une série continue de liaisons multiples conjugées. Le cas le plus simple est la formation d'une double liaison:

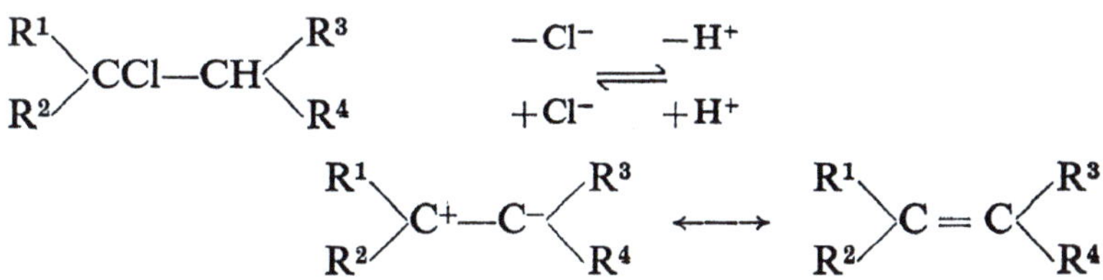

On observe alors un réarrangement électronique qui diminue la dissymétrie électrique de la molécule, en fournissant un état intermédiaire entre les structures limite, polaire et non chargée. Quand cette répartition stable des densités électroniques, autour des noyaux des atomes constitutifs, présenté une énergie interne beaucoup plus faible que celle qui correspondrait à la structure ion amphotère, toute modification électrochimique du système exige l'apport d'une grande énergie compensatrice. On doit opérer en milieu fortement acide ou alcalin pour obtenir un pourcentage élevé d'ions monovalents à partir de la molécule neutre qui se manifeste, dans les autres conditions, sous l'aspect caractéristique d'un véritable complexe. Mais il importe de faire remarquer à nouveau qu'une forte concentration en ions monovalents n'est pas nécessaire pour que la molécule puisse évoluer chimiquement à la faveur d'une réaction lente, ce qui justifie le point de vue de Lowry, Ebert, Euler. . . .

RUMPF. *Bull. Soc. Chim. France*, 1946, 439.

28

NATURE DE LA PARTIE HYDROPHILE DES ÉMULSIONNANTS

1) *Groupe hydrophile ionogène.*

a) *Anion actif.* — Les savons proprement dits constituent un type important des substances à anion actif. Si le cation fixé au groupe hydrophile–COO–, est alcalin les émulsions obtenues (du moins jusqu'à des concentrations moyennes en émulsionnant) sont du type *huile dans l'eau*; au contraire, si le cation est polyvalent le savon est insoluble et, d'après ce que nous avons dit précédemment, c'est l'émulsion *eau dans l'huile* qui correspond à l'équilibre le plus stable. Le cation peut être organique: une des meilleures bases que l'on puisse utiliser est la triéthylolamine.

Un autre groupe solubilisant est la fonction $-OSO_3H$ (ester sulfurique d'alcoyle, cétylsulfate de Na, par exemple). Ces corps qui sont des acides forts peuvent donner des émulsions même en milieu acide. Avec le groupe-SO_3H (dérivé sulfoné du carbure, cétyl-sulfonate par exemple) la solubilité, pour une même longueur de chaîne, est plus accentuée qu'avec le groupe précédent. De ce fait, à concentration égale, l'adsorption se fait moins rapidement à la surface des gouttelettes au moment de la dispersion, de sorte que les émulsions sont moins stables. SCHULMAN et COCKBAIN ont obtenu d'excellentes émulsions en associant de l'alcool cétylique ou du cholestérol au cétyle-sulfonate. Cela revient à réduire la solubilité de l'ensemble de la couche superficielle par l'adjonction du groupe OH de l'alcool ou du cholestérol lequel tout en étant hydrophile ne présente pas une aussi grande affinité pour l'eau que la fonction SO_3H et contribue à atténuer l'action solubilisante de celle-ci. Par ailleurs les parties liposolubles des deux constituants sont compatibles entre elles, ce qui favorise l'association en couche mixte.

b) *Cation actif.* — Le groupe ionogène peut être au contraire un cation (produits à cation actif). Peu de produits nettement émulsionnants appartenant à ce groupe ont été utilisés. Il est probable que le caractère émulsionnant doit apparaître avec des chaînes de l'importance de celle de l'acide palmitique ou stéarique.

c) *Groupe amphotère.* — Des corps actifs amphotères seraient susceptibles de donner des émulsions stables en milieu acide et en milieu alcalin, de part et d'autre du point iso-électrique. LOISELEUR a signalé la possibilité d'utiliser des amino-acides comme émulsionnants. Il a montré que seuls les amino-acides à chaîne apolaire avaient la propriété d'émulsionner; les meilleurs étant évidemment ceux qui ont une chaîne assez longue pour que le caractère lipophile soit suffisamment accentué.

2) *Groupe hydrophile non ionogène.*

Depuis une dizaine d'année est apparue une classe d'émulsionnants dont les fonctions solubilisantes ne sont pas de nature ionogène. De tels groupes hydrophiles sont obtenus par condensation d'une chaîne grasse (acide ou alcool) soit avec des molécules possédant un grand nombre de groupes hydroxyles, soit avec un certain nombre de molécules d'oxyde d'éthylène. A cette classe d'émulsionnants appartiennent les *Emulphores*. On constate qu'avec ces émulsionnants la dose nécessaire pour émulsionner une même quantité d'huile est relativement plus faible qu'avec les autres, et que les émulsions obtenues sont stables même en milieu très dilué. Cette particularité est probablement due au fait que la partie hydrosoluble est ici très encombrante, de sorte que l'aire occupée par chaque molécule à l'interface eau-huile étant beaucoup plus grande que dans le cas des savons, il peut se former une couche cohérente avec une quantité plus faible d'émulsionnant.

LACHAMPT et DERVICHIAN. *Bull. Soc. Chim. France*, 1946, 494.

29

SUR LES RÉACTIONS D'ESTÉRIFICATIONS DES ACIDES GRAS SUPÉRIEURS

H. GAULT et A. CHABLAY ont montré que les réactions d'estérification des acides gras supérieurs par l'alcool méthylique suivent les lois des équilibres chimiques de BERTHELOT:

$$RCOOH + CH_3OH \rightleftharpoons RCOOCH_3 + H_2O$$

dans la mesure où, naturellement, le milieu reste homogène, ce qui oblige les auteurs à opérer en présence d'un tiers-solvant car l'eau formée provoque la séparation des phases.

Si l'on veut déplacer cet équilibre dans le sens de l'estérification, il est tout à fait indiqué de profiter de la démixtion qui se produit lors de la réaction conduite sans tiers-solvant, en dehors des autres conditions habituelles, telles que l'emploi d'un excès de l'un des constituants, pratiquement l'alcool; la solubilité réciproque des phases est d'autant plus faible que la température est plus basse, mais la nécessité d'accélérer la vitesse de la réaction obligeant à faire agir la chaleur, on opère rarement au laboratoire comme dans l'industrie, des estérifications à température ordinaire. Toutefois, cette manière de procéder peut conduire dans le cas où l'on dispose d'alcool à bon marché et où l'on sépare les esters formés par simple décantation, à une remarquable simplification de l'appareillage et à une dépense insignifiante puisqu'on n'a pas besoin de fournir d'énergie extérieure.

Nous avons étudié comparativement l'estérification à température ordinaire des acides gras supérieurs, constituants principaux des huiles de palme et du beurre de palmiste (acides palmitique, oléique, laurique) par les alcools méthylique et éthylique absolus en excès (de 2 à 15 fois la quantité théorique) en présence de divers catalyseurs (acides chlorhydrique, sulfurique, benzène sulfonique, paratoluène sulfonique). Nous avons constaté, dans tous les cas, une nette supériorité du méthanol sur l'éthanol, en ce qui concerne les vitesses de la réaction.

L'alcool méthylique réunit des avantages appréciables, d'abord en dissolvant une moins forte quantité des esters, ensuite, par sa plus grande réactivité qu'on peut expliquer, soit par une moindre stabilité de la liaison R-OH, soit en faisant intervenir des phénomènes d'encombrement stérique qui jouent d'autant plus que le groupement R est plus volumineux.

Ainsi nous avons pu fabriquer industriellement les esters méthyliques des acides gras de l'huile de palme, avec un appareil rudimentaire. Rappelons que ces esters ont été proposés comme carburant colonial et ont donné d'excellents résultats sur les moteurs Diesel à grande vitesse de rotation, tels que ceux des véhicules de transports routiers.

LOURY et MELLIER. *Bull. Soc. Chim. France*, 1946, 461.

30

MACROMOLÉCULES EN ÉTAT D'EXTENSION

L'étude des clichés de diffraction de rayons X et le calcul des périodes de fibre, montrent que, dans les polyamides telles que la fibroïne ou le Nylon, les chaînes macromoléculaires présentent, à l'état normal, leur extension maximum. Les atomes constituant le squelette macromoléculaire dessinent le zigzag régulier imposé par les angles valenciels normaux et les distances interatomiques covalentes habituelles. Les chaînes macromoléculaires parallèles sont réunies à leur tour par des liaisons hydrogène et forment, de ce fait, un réseau supermacromoléculaire bidimensionnel dont les nappes ne sont plus assemblées que par des attractions très lâches du type Van der Waals.

Les observations s'accordent parfaitement avec ces particularités structurales. Les phénomènes d'élasticité, comparables à ceux d'un fil d'acier ondulé, doivent être attribués à une déformation temporaire et réversible des angles valenciels des chaînes macromoléculaires, alors que les déformations permanentes résulteraient du glissement relatif des nappes supermacromoléculaires et, éventuellement, des macromolécules individuelles dans ces nappes.

Le cas des chaînes cellulosiques de ramie est tout à fait analogue bien qu'il se complique du fait que, dans les fibres naturelles, les fibrilles cellulosiques constituées par les chaînes macromoléculaires, ne sont pas rigoureusement parallèles à l'axe de la fibre, mais décrivent une spirale à grand pas. On doit donc observer sous l'effort de traction, une rectification de la spirale, puis une déformation des angles valenciels, principalement des ponts oxydiques, des chaînes macromoléculaires parvenues à l'état de complète extension comme les chaînes polyamidiques; c'est ce que souligne l'analogie avec un fil métallique doublement ondulé. Dans ce cas encore, la déformation permanente accompagnant les déformations élastiques provient du glissement des nappes supermacromoléculaires ou des macromolécules les unes par rapport aux autres.

CHEVENARD et CHAMPETIER. *Bull. Soc. Chim. France*, 1946, 473.

31

TRANSFORMATION DE LA DIBROMO-1-2-TÉTRALINE EN DIHYDRO-3-4-NAPHTALÈNE

Action de la poudre de magnésium ou de zinc sur la dibromo-1-2-tétraline en solution alcoolique, application de la réaction classique de GLADSTONE et TRIBE de débromuration des dibromures.

a. Avec la *poudre de magnésium.*

Dans une fiole conique munie d'un réfrigérant à reflux, on met 100 gr de dibromotétraline et 150 cc d'alcool absolu, puis on ajoute 20 gr de magnésium en poudre, peu à peu. La réaction se déclanche. On laisse réagir en refroidissant par un bain d'eau, car elle se fait trop violemment. On chauffe ensuite une demi-heure lorsque la réaction est calmée. L'alcool est chassé au bain-marie, puis on décompose par l'acide chlorhydrique et la glace avec précaution. On épuise à l'éther, lave à l'eau deux fois, sèche sur chlorure de calcium, évapore l'éther et distille:

1 fraction entre 92° et 100° sous 17 mm: 22 gr: c'est le Δ_1-dihydro-naphtalène.

Il reste dans le ballon une assez grande quantité de résine colorée en brun vert: 20 gr environ.

b. Avec la *poudre de zinc*.

Dans un Erlenmeyer muni d'un réfrigérant à reflux, on met 100 gr de dibromotétraline avec 150 cc d'alcool absolu, puis on ajoute, par petites, portions, 28 gr de zinc en poudre. La réaction part violemment, aussi on est obligé de refroidir dans un bain d'eau glacée. Quand la réaction est calmée, on chauffe un quart d'heure au bain-marie, chasse l'alcool, puis décompose par l'acide chlorhydrique et la glace après refroidissement. On reprend ensuite par l'éther, en extrayant plusieurs fois les eaux-mères, lave à l'eau, sèche sur du sulfate de soude anhydre. Après évaporation de l'éther, on rectifie le produit qui passe:

entre 95° et 100° sous 20 mm (17 gr).

Il reste pas mal de résines dans le ballon (viscosité et couleur analogues aux précédentes).

c. Par la méthode au *couple zinc-cuivre*.

C'est celle qui m'a donné le meilleur rendement: 60% environ.

On prépare le couple zinc-cuivre en agitant 100 gr de zinc en poudre avec une solution de sulfate de cuivre, de façon qu'il y ait 1 atome de cuivre pour 1 atome de zinc. On filtre et lave à l'eau, puis à l'alcool et sèche ensuite à l'air.

Dans un Erlenmeyer contenant 30 cc d'alcool et 200 cc d'éther, on met le couple zinc-cuivre préparé ci-dessus. On ajoute ensuite 62 gr de dibromotétraline et on laisse la réaction se faire pendant quelques heures.

Après un chauffage d'environ trois quarts d'heure à reflux, puis refroidissement, le produit est versé dans l'eau, puis filtré en gardant le zinc cuivré dans le bécher, où il est lavé à l'éther plusieurs fois. La filtration est très longue. On décante la couche aqueuse, lave une fois la couche éthérée à l'eau, chasse l'éther, rectifie le produit qui passe entre 90° et 95° sous 18 mm. Poids: 16 gr.

COULOMB. *Ann. Mines*, 1946, 578.

MATHEMATIQUES

1

La notion de courbe

La notion de courbe, qui paraît au premier abord si simple quand on se donne x et y en fonction d'un paramètre t, est en réalité complexe. M. Peano a le premier appelé l'attention sur une circonstance extrêmement curieuse: Il est possible de former des fonctions continues $f(t)$ et $\phi(t)$ (de $t = 0$ à $t = 1$) telles que si l'on pose $x = f(t)$, $y = \phi(t)$, on puisse trouver une valeur de t, entre les limites indiquées, pour laquelle le point (x, y) coïncide avec un point arbitrairement choisi dans le carré de côté un construit sur OX et OY.

London B.Sc. (*Special*).

2

Les déformations infiniment petites

Il est important de signaler que le problème de la déformation infiniment petite dépend, linéairement, pour une surface donnée, de deux fonctions arbitraires d'une variable. De la sorte, pour une surface réglée, la déformation *finie* de la surface réglée encore, dépendant explicitement d'une fonction arbitraire, on en déduit, par un calcul simple, une solution explicite avec une fonction arbitraire de l'équation résolvante de la déformation infiniment petite; si donc il s'agit d'une quadrique, même si l'on ne sait pas déformer complètement la surface, on sait intégrer complètement le problème de la déformation infiniment petite, puisqu'il suffit de réunir les deux séries de déformation offertes par les deux systèmes de génératrices; cette simple remarque, due à Darboux, permet de prévoir, *a priori*, pourquoi les recherches de M. Bianchi sur les quadriques ne pouvaient manquer d'aboutir à des résultats élégants; il reste, dans cet ordre d'idées, à étudier toutes les surfaces pour lesquelles on sait résoudre explicitement le problème de la déformation infiniment petite; c'est alors qu'il est indispensable de suivre la méthode de Darboux et l'on est ramené au problème complètement traité par Darboux au Tome II de la *Théorie des surfaces:* trouver toutes les formes de $k(\alpha, \beta)$ pour lesquelles l'équation $\frac{\partial^2 \theta}{\partial\alpha\, \partial\beta} = k\theta$ s'intègre explicitement; Darboux a même montré que cela permet un premier progrès dans la détermination des surfaces que l'on sait complètement déformer.

GAMBIER. *Mém. Sci. Math.*, 1928, **31,** 41.

3

TRAVAIL DU POIDS

Lorsqu'un point pesant se déplace, il est soumis à son poids qui est une force $F = mg$, constante en grandeur, direction et sens. Quand le point passe d'une position M_0 à une position M_1, en suivant un chemin quelconque, le travail du poids est égal au produit de mg par le valeur algébrique de la projection du déplacement M_0M_1 sur la direction du poids, c'est-à-dire sur la verticale descendante; on peut dire aussi que c'est la produit de mg par la hauteur, dont descend le point en passant de sa position initiale à sa position finale.

London B.Sc. (*Special*).

4

LA THÉORIE DES CORPS PLASTIQUES ISOTROPES

Il va sans dire que pour les corps réels la distinction entre les deux domaines d'élasticité et de plasticité n'est pas précise. Par contre on pourrait même supposer que toute déformation élastique, quelque petite qu'elle soit, soit déjà accompagnée d'une petite déformation permanente négligeable. En tout cas il se trouve un certain domaine de transition entre l'état élastique et l'état plastique du corps.—Souvent on a affaire à des problèmes où la limite élastique n'est atteinte que dans *une partie* du corps tandis que le reste du matériau se comporte encore comme un corps élastique.—Pour le problème d'efforts plus généraux nous donnerons plus tard la définition complète de limite plastique.

L'ensemble des phénomènes au delà de la limite élastique est très étendu et l'on ne dispose pas encore d'une théorie satisfaisante assez générale et comprenant la totalité des faits qui se présentent, théorie en même temps assez précise et concrète pour donner lieu à des développements mathématiques. On ne connaît aujourd'hui qu'une seule théorie précise qui soit arrivée à établir des équations différentielles et à en déduire des conséquences mathématiques de même façon qu'on le fait par exemple dans la théorie d'élasticité. Certainement cette théorie de plasticité néglige différents détails des phénomènes observables. D'autre part il existe plusieurs conceptions plus générales embrassant les autres aspects des phénomènes.

GEIRINGER. *Mém. Sci. Math.*, 1937, **86,** 3.

5

DÉCOMPOSITION D'UNE FONCTION

La formule de décomposition d'une fraction ou fonction rationnelle R (x), en fractions ou éléments simples, a un double intérêt: d'une part, elle met en évidence les pôles de la fonction avec leurs parties principales; d'autre part, elle permet l'intégration de la fonction en la ramenant à l'intégration des éléments simples.

Ce double point de vue a été étendu par le génie d'Hermite à la plupart des fonctions connues de son temps, par exemple aux fonctions elliptiques, aux fonctions doublement périodiques de seconde espèce et à leurs dégénérescences, aux fonctions rationnelles de sin x et cos x, R (sin x, cos x), aux fonctions de la forme $e^{\omega x}$R (x), $e^{\omega x}$R (sin x, cos x), où ω est une constante.

Dans ses recherches Hermite a pris généralement pour élément simple une fonction d'une variable $f(x)$ devenant infinie au point $x=0$, en y remplaçant x par $x-a$, comme il arrive pour les fonctions rationnelles R (x) où l'élément simple se déduit de $\frac{1}{x}$ en y remplaçant x par $x-a$.

Mais dans certains cas, comme on le verra dans l'exemple des fonctions doublement périodiques de troisième espèce, il faut avoir recours, pour l'élément simple, à des fonctions de deux variables dont l'une est la variable qui figure dans la fonction à décomposer et dont l'autre coïncide avec les pôles successifs.

L'idée géniale d'Hermite est d'une fécondité telle que chaque auteur, en étudiant des fonctions nouvelles, s'efforce de trouver un élément simple correspondant. C'est ce que Poincaré a fait pour les fonctions fuchsiennes en créant les fonctions zêtafuchsiennes comme il avait créé les fonctions thêtafuchsiennes.

En faisant l'exposé général de la théorie de la décomposition en éléments simples, je prendrai d'abord les fonctions rationnelles et celles qui s'y rattachent, puis, d'après Hermite, les fonctions doublement périodiques et la fonction zêta, les fonctions doublement périodiques de seconde et troisième espèce, puis les fonctions d'un point analytique, puis les fonctions automorphes d'Henri Poincaré, puis enfin les fonctions f dites triplement périodiques de trois variables réelles x, y, z vérifiant l'équation de Laplace

$$\frac{\partial^2 T}{\partial x^2} + \frac{\partial^2 T}{\partial y^2} + \frac{\partial^2 T}{\partial z^2} = 0.$$

APPELL. *Mém. Sci. Math.*, 1929, **36**, 1.

6

COURBE DE CORRÉLATION

Si on porte dans un tableau à double entrée, en coordonnées logarithmiques, les mesures individuelles des caractères de l'indice choisi, on obtient un nuage de points, de part et d'autre d'une ligne plus ou moins rectiligne, surtout dans la plus grande partie de sa région médiane. Cette ligne demeure sensiblement droite quel que soit le degré d'allométrie de l'indice.

En coordonnées arithmétiques, quand l'allométrie est très accusée, l'ensemble des rapports individuels se répartit en un nuage de points nettement incurvé. S'il n'y avait pas de variation dans les rapports individuels, ceux-ci s'aligneraient suivant la courbe de corrélation des deux caractères. Comme on l'a vu, pour exprimer l'allométrie la plupart des auteurs ont adopté la fonction $y = bx^{a}$. C'est l'approximation la plus commode. Cependant, il faut remarquer que la courbe correspondant à cette formule n'est certainement pas l'expression de la vraie corrélation entre deux caractères. Elle s'adapte évidemment bien à la partie moyenne du nuage, mais rien ne prouve qu'elle s'adapte également bien aux individus extrêmes. En effet, souvent les points correspondant aux tailles exceptionnelles, les très petites et les très grandes, ne se trouvent pas sur la ligne figurative. D'autre part, lorsqu'un individu se place sur cette ligne, à l'une ou à l'autre de ses extrémités, cela n'est pas une preuve en faveur de la bonne adaptation de la formule au phénomène biologique. L'alignement de cet individu peut être dû au hasard des variations individuelles, qui l'amènent accidentellement sur la ligne théorique.

Si on étudie une population abondante, représentant un grand nombre d'exemplaires extrêmes et si on divise cette population en tronçons successifs, dont on calcule les différents centres de gravité, on obtient, en reliant ces centres, une estimation approchée de la vraie courbe de corrélation qui, souvent, n'est pas une droite en coordonnées logarithmiques.

COUSIN. *Arch. Zool. Expt.*, 1939, **81,** 294.

7

THÉORIE DE POINCARÉ

La question peut, comme nous l'avons dit dès le début, être abordée par une voie toute différente qui a donné lieu à de nombreux travaux. On regarde alors le corps comme formé d'un nombre immense de points matériels agissant les uns sur les autres, et l'on cherche à calculer les effets produits par les actions mutuelles de ces points. Poincaré, dans ses leçons *Sur la théorie mathématique de l'élasticité*, a fait usage de cette méthode.

Le point de départ de Poincaré consiste à admettre l'existence d'un potentiel des forces intérieures, fonction des petits déplacements éprouvés par les points du système à partir d'un certain *état naturel*, et à développer ce potentiel par la formule de Taylor en s'arrêtant aux termes du second ordre. L'auteur s'abstient de supposer les forces extérieures nulles dans l'état naturel, parce qu'autrement, dit-il, il pourrait arriver que les déplacements correspondant à l'état final, appelé d'équilibre contraint ne fussent pas très petits. Pour éclaircir cette idée, il cite le cas d'un gaz qui ne peut, sans pression superficielle, occuper un volume fini. Après avoir ainsi posé le problème, Poincaré, pour continuer les calculs, suppose que les actions moléculaires deviennent insensibles quand les distances dépassent une très petite longueur, appelée rayon d'action moléculaire, et, après une discussion serrée, il néglige tout ce qui est effectivement négligeable. Finalement, distinguant deux cas, suivant que les forces intérieures sont ou non centrales, c'est-à-dire dirigées suivant les droites de jonction des couples de points, il aboutit au tableau suivant du nombre des coefficients distincts:

		Nombre de coefficients.
Forces intérieures non centrales	Forces extérieures non nulles dans l'état naturel	27
	Forces extérieures nulles dans l'état naturel	21
Forces intérieures centrales	Forces extérieures non nulles dans l'état naturel	21
	Forces extérieures nulles dans l'état naturel	15

Dans le cas particulier de l'isotropie, la conclusion de Poincaré est qu'il y a en général trois coefficients, se réduisant à deux ou à un seul quand les 27 coefficients de l'anisotropie se réduisent à 21 ou à 15. On a d'abord les deux coefficients λ et μ de Lamé. Le troisième coefficient υ s'annule si les forces extérieures sont nulles dans l'état naturel. La réduction à un seul coefficient se produit si, en outre, les forces intérieures sont centrales, car on trouve alors que μ ne diffère pas de λ.

LECORNU. *Mém. Sci. Math.*, 1929, **35**, 14.

8

ERREUR DIFFÉRENTIELLE. TOLÉRANCE

Le type élémentaire d'erreur est l'erreur différentielle. Un exemple excellent en est fourni par la fabrication en série des pièces mécaniques. Le prototype est caractérisé par une certaine dimension x_0, mais le « jeu » toléré permet d'accepter les pièces de dimension légèrement différente x; cette dimension restant cependant entre les limites x_1 et x_2, généralement fixées par des calibres:

$$x_1 \leqslant x \leqslant x_2. \tag{1}$$

Si x_1 et x_2 sont symétriques par rapport à x_0 on peut écrire:

$$\Delta x = x_2 + x_1, \tag{2}$$

$$x_1 = x_0 - \frac{\Delta x}{2} \qquad x_2 = x_0 + \frac{\Delta x}{2}, \tag{3}$$

$\frac{\Delta x}{2}$ est la *tolérance* ou l'*erreur* sur x. J'appellerai incertitude sur x l'intervalle Δx.

J'insiste sur les caractères de ce type d'erreur. Toutes les dimensions x sont supposées définies absolument. Dans la réalité physique cela veut dire qu'elles sont définies matériellement avec une incertitude δx très inférieure à la tolérance, de second ordre. Ainsi une lame de mica clivée suivant deux plans réticulaires sera définie à quelques Angströms près tandis que la tolérance pourra être supérieure à 0,1 mm. si cette lame sert à découper des rondelles isolantes.

Portées en abscisses sur l'axe ox, les dimensions x sont représentées par des points L, les limites L_1 et L_2 encadrant toutes les positions acceptables de L. La détermination exacte d'un très grand nombre de points L permet de tracer une courbe de dispersion des dimensions x, en portant en ordonnée au-dessus de chacune des valeurs *distinctes* des x le nombre des cas où elle a été constatée. Les courbes de dispersion analogues sont toujours figurées par une ligne de tracé *continu*, le tracé réel en est cependant essentiellement *discontinu*. D'abord parce que le nombre des mesures, N, aussi grand qu'il soit, reste toujours fini et que les ordonnées de la courbe ont une croissance discontinue. D'autre part l'incertitude de second ordre δx sur la définition ou la mesure de x nous oblige à n'admettre entre L_1 et L_2 qu'un nombre fini de valeurs distinctes de x. Cette conception discontinue de l'incertitude Δx permet seule la construction de la courbe continue. L'introduction des notations différentielles, plus ou moins altérées, ne doit faire aucune illusion. Quand on définit la probabilité γ par le quotient:

$$\gamma = \frac{1}{N}\frac{dN}{dx}$$

il est bien entendu que ni le numérateur ni le dénominateur ne peuvent devenir infiniment petits.

DUFFIEUX. *Ann. Phys.*, 1942, **17,** 210.

9

ONDES LIQUIDES DE GRAVITÉ

Pour commencer par un cas simple où nous trouverons immédiatement la solution, considérons un bassin de dimensions infinies tant en profondeur que dans le sens horizontal: le fluide occupe tout l'espace au-dessous du plan horizontal $z = 0$ de la surface libre. Dans cette hypothèse, il n'y a pas de parois (elles sont infiniment éloignées), et la condition d'annulation de la dérivée $\frac{d\phi}{dn}$ est remplacée par la suivante: la fonction ϕ doit s'annuler asymptotiquement lorsqu'on s'éloigne de l'origine des coordonnées, ce qui signifie que le mouvement est constamment nul à l'infini, qui n'est jamais atteint par une perturbation ayant son origine à distance finie.

Reprenons, dans le cas actuel d'un bassin infini, l'équation relative à la surface libre $z = 0$

$$\frac{\partial\phi}{\partial z} - \frac{\partial^2\phi}{\partial t^2} = 0,$$

et faisons, à son sujet, la remarque capitale suivante. *Cette équation est vérifiée non seulement à la surface* $z = 0$, *mais dans tout le fluide.* En effet, son premier membre est une fonction harmonique (puisque ϕ en est une), qui est nulle à la surface libre et qui s'annule aussi aux grandes distances, où ϕ tend partout vers zéro. Donc, ce premier membre est identiquement nul, et l'équation devient une équation *indéfinie* du problème.

Reprenons maintenant l'équation

$$\frac{p}{\rho} = z - \frac{\partial \phi}{\partial t},$$

qui a lieu partout à toute époque; dérivons-la par rapport à t en suivant une même particule fluide

$$\frac{1}{\rho}\frac{dp}{dt} = \frac{\partial \phi}{\partial z} - \frac{\partial^2 \phi}{\partial t^2} = 0;$$

ainsi, *une particule fluide supporte la même pression pendant tout le mouvement.*

Désignons par Z l'ordonnée de la particule à l'état final de repos, nous aurons

$$\frac{p}{\rho} = Z$$

et, par suite,

$$z - Z = \frac{\partial \phi}{\partial t}.$$

$(z - Z)$ est la *dénivellation h* de la particule à l'époque t.

VERGNE. *Mém. Sci. Math.*, 1928, **34**, 6.

10

THÉORIE DE L'ÉLASTICITÉ—APERÇU HISTORIQUE

La théorie mathématique de l'élasticité est une création du XIXe siècle; mais elle avait été amorcée au XVIIe et au XVIIIe par diverses recherches concernant les corps élastiques ou plastiques.

Galilée avait abordé le problème de la résistance d'une poutre encastrée dans un mur; il croyait pouvoir regarder comme inextensibles les fibres de cette poutre.

En 1678, Hooke fit connaître la loi d'après laquelle il y a proportionnalité entre la déformation et l'effort. Il y avait été conduit par l'étude des ressorts spiraux. Mariotte retrouva cette loi, en 1686, à propos du problème de la poutre;

il remarqua en outre que dans la flexion d'une poutre, une partie des fibres s'allonge tandis que l'autre se raccourcit, le passage des unes aux autres s'effectuant par la fibre neutre, qu'il plaçait à mi-hauteur de la poutre. Le problème de la résistance se trouvait ainsi relié à celui de la flexion. Il restait à calculer la courbure prise par la fibre centrale (lieu des centres de gravité des sections droites), question dont Jacques Bernoulli s'occupa à diverses reprises (1691, 1694, 1705). Puis J. Bernoulli et Euler consacrèrent divers travaux à l'Élastique plane. Euler, dans l'un de ses Mémoires, rechercha la forme de cette courbe en supposant que les fibres soient soumises à un effort de compression et il fonda ainsi la théorie de la résistance d'une colonne, théorie qui fut ensuite développée par Lagrange et par Euler lui-même. Euler entreprit également les premières recherches sur la stabilité des corps élastiques.

Bientôt les vibrations de ces corps appellent l'attention des savants. En 1700, Sauveur fait de l'acoustique une science spéciale. L'étude mathématique des vibrations des cordes, des tiges, des membranes, entreprise par Taylor, Jean et Daniel Bernoulli, d'Alembert et Lagrange, fournit l'occasion d'importants développements analytiques.

Les résultats obtenus ne s'étendaient pas sans difficulté au cas des plaques. A la suite des recherches expérimentales de Chladni, l'Académie des Sciences créa pour 1811 un prix qui provoqua les travaux de Sophie Germain et de Poisson.

Ces travaux inaugurent au XIXe siècle la théorie mathématique de l'élasticité. En même temps l'étude de la résistance des matériaux, abandonnée depuis Coulomb (1773), était reprise par Young et Navier en vue de répondre aux besoins des ingénieurs. Puis, le 14 mai 1821, Navier ingénieur des Ponts et Chaussées, lut à l'Académie des Sciences un Mémoire sur la théorie de l'élasticité, qui fut imprimé seulement en 1827. Dans ce Mémoire il établissait, le premier, les équations de l'équilibre et du mouvement des corps élastiques à trois dimensions; aussi doit-il être regardé comme le véritable fondateur de la théorie de l'élasticité.

LECORNU. *Mém. Sci. Math.*, 1929, **35**, 44.

11

La série hypergéométrique

La *série hypergéométrique*

$$F(\alpha, \beta, \gamma, x) = \sum_{n=0}^{n=+\infty} \frac{(\alpha, n)(\beta, n)}{(\gamma, n)(1, n)} x^n,$$

$$a_n = \frac{\alpha(\alpha+1)\ldots(\alpha+n-1)\,\beta(\beta+1)\ldots(\beta+n-1)}{\gamma(\gamma+1)\ldots(\gamma+n-1)\,1.2\ldots n}$$

dépend des quatre quantités α, β, γ, x que Gauss nomme respectivement: 1er, 2e, 3e, 4e élément; c'est une série entière en x, une série de polynomes en α et en β, une série de facultés en γ.

Les éléments α et β jouent évidemment un rôle symétrique. Le coefficient a_n de x^n a une valeur finie et non nulle, hormis les cas suivants:

I. *A partir d'un certain rang:* 1° *si γ est nul ou entire négatif, les a_n sont infinis;*

2° *si γ et α (ou β) sont simultanément nuls ou entiers négatifs, les a_n sont sous la forme indéterminée* 0 : 0;

3° *si α (ou β) est nul ou entier négatif, les a_n sont nuls; la série se réduit à un polynome en x de degré* $|\alpha|$ (ou $|\beta|$).

Ces valeurs exclues, pour examiner la convergence de la série, Gauss forme le rapport

$$\frac{a_{n+1}}{a_n} = \frac{n^2 + (\alpha+\beta)n + \alpha\beta}{n^2 + (\gamma+1)n + \gamma} = 1 + \frac{\alpha+\beta-\gamma-1}{n} + 0\left(\frac{1}{n^2}\right);$$

en supposant x complexe, mais α, β, γ réels, il en déduit que la série converge pour $|x| < 1$, diverge pour $|x| > 1$; pour $|x| = 1$, il y a doute: il n'examine que le cas ou $x = 1$.

K. Weierstrass a complètement résolu la question.

De Fériet. *Mém. Sci. Math.*, 1937, **85**, 4.

12

Équations de Dirac-Madelung

Nous utilisons le langage géométrique et les notations propres à la relativité restreinte. Un point quelconque de l'Univers ou espace-temps est défini par ses quatre coordonnées contrevariantes (réelles)

$$x_1, \quad x_2, \quad x_3 \text{ d'espace,} \qquad x_4 = ct \text{ de temps;}$$

le module carré du vecteur qui joint l'origine à ce point est

$$g^{ij}x_ix_j; \tag{1}$$

en pratique nous choisirons des axes rectangulaires ordinaires; le module carré précédent s'écrit alors

$$x_1^2 + x_2^2 + x_3^2 - x_4^2, \tag{2}$$

les vecteurs du genre temps ont donc un module carré négatif, c'est le cas des vitesses; les vecteurs du genre espace ont un module carré positif.

Un grand nombre de nos formules, établies en axes rectangulaires, sont aussi valables en axes obliques quelconques, pourvu toutefois que le changement de coordonnées résulte d'une transformation (linéaire) de déterminant égal à $+ 1$.

Nos notations demandent une justification. Il a paru commode, suivant l'exemple donné par beaucoup de physiciens, d'adopter les indices inférieurs pour les composantes contrevariantes et les indices supérieurs pour les composantes covariantes, tout simplement parce que les composantes contrevariantes sont d'un emploi plus courant que les autres. D'autre part le module carré (2) est souvent pris avec un signe différent: cette convention ne paraît préférable que pour les problèmes relatifs à la mécanique du point matériel.

Le volume d'espace-temps construit sur quatre vecteurs a_i, b_i, c_i, d_i, soit

$$\begin{vmatrix} a_1 & a_2 & a_3 & a_4 \\ b_1 & b_2 & b_3 & b_4 \\ c_1 & c_2 & c_3 & c_4 \\ d_1 & d_2 & d_3 & d_4 \end{vmatrix}$$

sera noté

$$\alpha^{ijkl}a_ib_jc_kd_l;$$

les α^{ijkl} sont les composantes covariantes d'un tenseur; leurs seules valeurs possibles sont $- 1, 0, + 1$. En particulier $\alpha^{1234} = + 1$.

YVON. *J. Physique*, 1940, **1**, 18.

BOTANIQUE

1

L'ÉTUDE DES ASSOCIATIONS SOUS-MARINS

Dans la recherche de la végétation sous-marine la méthode dont on se sert pour se procurer du matériel est de la plus grande importance. Pour des examens qualitatifs, tels que la composition floristique et faunistique d'une région, il suffit de se servir de divers filets, dragues, etc., qui nous amènent tout le matériel rencontré sur leur parcours. Mais ce matériel ne peut pas nous renseigner sur la distribution des diverses espèces dans l'espace et sur la composition quantitative de l'association. Un seul moyen nous le permet, c'est de voir de nos propres yeux le lieu d'habitation des algues et des animaux. Ce n'est pas difficile dans l'étage littoral, mais dès que nous voulons descendre plus bas, il faut nous servir du scaphandre. Peu de sociologues marins s'en sont servis, seul Torsten Gislen (1930) l'a fait d'une manière systématique. J'ai eu l'occasion d'étudier la végétation sous-marine avec le scaphandre et je veux apporter ici quelques résultats de mes observations. C'est mon collègue Ivan Kuscer qui a fabriqué le scaphandre et c'est lui également qui m'accompagnait dans mes descentes. Je me suis familiarisé avec les méthodes d'océanographie à l'Institut Océanographique de Split, où j'ai pu déterminer une partie du matériel recueilli. Je dois mes remerciements au directeur de l'institut M. Dr A. Ercegovic pour l'accueil aimable qu'il m'a réservé à cet institut.

Je dois justifier cet article, peut-être pas assez documenté, faute d'observations encore nécessaires, par le fait qu'il paraît impossible de continuer les recherches dans le proche avenir à cause des événements mondiaux. C'est ainsi que je veux donner aux fruits de mes observations de trois étés une forme pouvant servir à des recherches ultérieures et à une discussion constructive.

ZALOKAR. *Bull. Soc. Bot. Gen.*, 1942, **33**, 172.

2

RHIZOME DE L'HYDROSTACHYS GOUDOTIANA

Anatomiquement, il est bien visible tout d'abord que la souche qui fixe la plante au sol correspond à un rhizome. Elle est constituée, en effet, par un parenchyme dans lequel sont répartis sans ordre un certain nombre de faisceaux de forme et de grosseur très diverses, ce caractère étant en rapport avec la forme, elle-même inégale, du rhizome, due à la disposition également très contournée des racines.

Les plus grands faisceaux se trouvent vers le centre; sinueux, ils sont à section transversalement allongée. Les faisceaux de la périphérie sont à section irrégulière ou parfois elliptique. Toutes ces formes peuvent, du reste, être d'autant moins précises qu'elles dépendent beaucoup de l'orientation des racines sur le rhizome; un même cordon pourra ainsi, selon le niveau, prendre des aspects différents, car son parcours est influencé par l'inégalité de forme du rhizome; et la position des racines sur cette souche allongée et tourmentée n'est elle-même nullement régulière. A cela s'ajoute que, ça et là, sur leur trajet, plusieurs faisceaux peuvent se fusionner.

Au niveau où nous les considérons, le plus gros, qui est un des plus centraux, est à cellules périphériques lignifiées entourant une zone de 10 à 15 cellules à parois minces, qui bordent elles-mêmes une petite lacune centrale où devait se trouver un vaisseau spiralé disparu. Nous basons cette hypothèse sur le fait que, dans les autres faisceaux qui sont pleins, le centre est occupé par un petit groupe de ces vaisseaux spiralés. Ce groupe est alors enveloppé, dans les dits faisceaux, par un collenchyme dont les éléments sont surtout épaissis aux angles, l'ensemble de la paroi présentant un certain nombre de ponctuations.

Les faisceaux ainsi constitués sont plongés dans un parenchyme composé de grosses cellules de 150 μ en moyenne, remplies d'un amidon, dont les petits grains sont arrondis ou ovales, sans hile apparent, de 10 à 15 μ de diamètre. On trouve, en outre, dans de nombreuses cellules, tant au centre que sur les bords, de grosses mâcles échinées d'oxalate de calcium.

GAFFIER. *Ann. Fac. Sci. Marseille*, 1932, **5** (2), 125.

3

SYSTÈME LIGNEUX DES TIGES CHEZ LES UTRICULARIÉES

Les conditions qui marquent l'infériorité du système ligneux des tiges des Utriculariées, sont les suivants: une moindre différenciation des éléments ligneux secondaires, un moindre developpement de ce bois secondaire, une moindre condensation dans le groupement des éléments ligneux primères, l'absence des rayons limites ou leur très faible indication, l'absence de différenciation entre les régions réparatrices et les régions sortantes.

London B.Sc. (*General*).

4

LIPIDES

Nous avons mis en évidence par coloration par le Soudan III, le noir Soudan et le mélange nadi la présence de globules lipidiques, souvent très petits mais quelquefois relativement gros, leur diamètre pouvant atteindre la largeur du corps microbien. Ils augmentent généralement en taille et en nombre au cours de la vie de la cellule. Ces globules, situés dans le cytoplasme, existent probablement dans toutes les espèces bactériennes. Ils ont été signalés maintes fois par les Cytologistes. Nous avons pu préciser que chez les Beggiatoacées ils appartenaient au groupe des phospholipides.

DELAPORTE. *Rev. Gén. Bot.*, 1940, **52,** 113.

5

FLEUR DE L'ARISTOLOCHE

Les carpelles du pistil se développent tous d'ordinaire complètement et également. Pourtant dans l'Aristoloche les styles et les stigmates avortent et les six carpelles se réduisent à leurs ovaires. Ce sont alors les connectifs des anthères, épaissis, soudés lateralement en tube, développés et couverts de papilles vers le haut qui jouent le rôle de stigmates et du style. On pourrait croire que le style et le stigmate concrescents avec l'androcée, comme il vient d'être dit pour les Orchidées et les Stylidées.

London B.Sc. (*Special*).

6

Nature chimique des gamones et des termones

L'identification chimique des substances actives, présentes dans les sécrétions de gamètes, s'est développée à la suite de la remarque suivante: une préparation de caroténoïdes bruts, obtenue à partir de carottes et de feuilles, avait été abandonnée durant un certain temps à l'air et à la lumière. Or, voici que des extraits aqueux de cette préparation, agissent à la manière des secrétions de gamètes femelles de *Chlamydomonas eugametos.* Ces mêmes extaits aqueux, soumis à un éclairement plus long, manifestent ensuite les propriétés des substances inconnues libérées par les gamètes mâles.

La collaboration d'un chimiste spécialisé dans l'étude des caroténoïdes, Richard Kuhn, permit à Franz Moewus, de mettre à profit cette première remarque.

L'analyse directe des produits fournis par la cellule était impossible en raison de leur extraordinaire dilution. Un examen spectroscopique des humeurs secrétées par les gamètes, permit cependant d'y reconnaître la présence d'un pigment de constitution chimique analogue à celle du pigment tiré du safran (*Crocus sativus*): la crocine. Cette première vérification est complétée par une épreuve physiologique: une solution préparée à partir des cristaux de la crocine de safran, transforme en motogamètes les prégamètes de *Chlamydomonas.* Cette solution agit encore à la dilution fabuleuse de 1 : 250.000.000.000.000 ! Un simple calcul montre que cinq molécules de crocine suffisent pour activer quatre prégamètes.

Ces premières enquêtes délimitent nettement le territoire chimique à explorer; il s'agit des caroténoïdes et plus particulièrement de la crocine.

Chodat et De Siebenthal. *Bull. Soc. Bot. Gen.*, 1942, **33**, 101.

7

Le milieu marin

Le milieu marin est bien différent du milieu terrestret Les plantes sur la terre ont besoin du sol non seulemen.

comme base de fixation, mais aussi comme source de nourriture; dans la mer cette deuxième fonction de la base n'existe plus. Le milieu nutritif est l'eau de mer et les changements dans sa composition chimique peuvent apporter des changements de la végétation, tandis que les conditions chimiques du support n'ont pas une grande importance. Ainsi nous ne trouvons pas de différence appréciable entre la végétation des rochers granitiques ou des rochers calcaires dans la mer, pourvu que les conditions physiques soient les mêmes. Seuls les organismes endolithiques dépendent du chimisme du rocher, celui-là s'opposant à la formation des cavernes ou la facilitant.

L'eau de mer est l'élément qui joue le rôle du sol dans la formation des associations « édaphiques ». La végétation nitrophile rudérale trouve sa contrepartie dans la végétation des ports, qui abondent en matières organiques. Toutefois ici la végétation n'est pas riche, parce qu'à l'abondance de substances nitriques (provenant de matières organiques) s'associent une pénurie en oxygène et la présence de substances nuisibles (H2S). A la végétation halophile correspond la végétation marine des eaux saumâtres ou plutôt elle en est le complément.

Aux influences externes dans les associations terrestres nous pouvons comparer diverses conditions sous la mer. La question des précipités et de l'approvisionnement de l'eau ne se présente pas dans les régions profondes, mais elle devient actuelle dès que nous montons au-dessus du niveau de la marée basse (étage littoral et supralittoral). La température varie beaucoup suivant les endroits et la profondeur. De même la lumière, qui influence surtout les algues, mais dont l'influence sur certains animaux n'est pas nulle. Aux vents sur la terre nous pouvons comparer l'action des vagues, la houle et le ressac, qui jouent un très grand rôle dans la formation des associations littorales et infralittorales. On peut aussi considérer comme condition externe l'abondance en nourriture vivante planctonique que consomment beaucoup d'habitants sessiles du benthos.

De plus, une association marine est beaucoup plus liée à son ambiance qu'une association terrestre. Une phase seule de la vie des individus se déroule dans l'association, d'autres se déroulent ailleurs, surtout dans le plancton (p. ex. les larves, les gamètes, des stades jeunes de quelques algues). Comme la reproduction de beaucoup d'organismes est bornée à un temps très court les modalités de la propagation dépendent des conditions du moment. Il en peut suivre un banc de larves qui, emportées par le courant, vont coloniser un endroit et feront défaut ailleurs.

En examinant une association, nous devons en avoir l'ensemble devant nos yeux.

ZALOKAR. *Bull. Soc. Bot. Gen.*, 1942, **33**, 174.

8

MOUVEMENTS FOLIARES ET FLORAUX D'ASARUM EUROPAEUM

Au printemps, les colonies d'*Asarum europæum* ont l'air d'avoir été apprimées contre le sol; il y a là peut-être l'effet du poids de la neige ou bien l'effet d'un tropisme.

Les expériences que j'ai faites m'ont montré les faits suivants: Les jeunes pétioles des feuilles qui viennent de sortir du bourgeon, sont nettement négativement géotropiques. Si l'on donne à ces pétioles une position horizontale (l'expérience étant faite avec des plantes en pot), ils se redressent verticalement. Mais ce géotropisme négatif ne subsistera pas pendant toute la vie la feuille. Dans la forêt, les pétioles des feuilles restent dans la position dressée pendant tout l'été, cela leur permet d'émerger de ce qui les entoure, cela porte également le limbe vers la lumière. Les feuilles nées cette année passeront l'hiver et subsisteront jusqu'au mois d'août avant de tomber. Mais au printemps déjà, les pétioles des vieilles feuilles ne sont plus dressés, ils se tordent et touchent le sol, leur géotropisme négatif s'affaiblit donc à partir d'un certain moment et il est remplacé par une nastie. L'aspect aplati de la colonie, au printemps, s'explique donc par cette nastie.

Mes expériences m'ont montré qu'il s'agit ici, en effet, d'une nastie. J'ai placé mes plantes en pot, dans une position telle que les pétioles des anciennes feuilles étaient dans une direction plus ou moins verticale, l'un dressé, l'autre suspendu. Quelques jours après, les deux pétioles avaient repris une situation plus ou moins horizontale correspondant à leur manière d'exister dans la forêt. Dans cette expérience, l'une des feuilles s'abaisse, l'autre se relève pour occuper la position horizontale, mouvements qui indiquent une réaction géotropique. La lumière ne joue aucun rôle dans cette géonastie, car cette dernière se produit également à l'obscurite complète. C'est un phénomène lent, comparé à la réaction des jeunes feuilles. J'ai enregistré des cas cependant où les vieilles feuilles n'ont pas réagi du tout et sont restées dans la position anormale que je leur avais donnée.

Toutes les jeunes feuilles sont positivement héliotropiques. J'ai fait des expériences avec des plantes demi étiolées, portant de longs pétioles avec de petits limbes. Lorsqu'on coupe le limbe, la réponse phototropique ne se fait plus, même après un temps prolongé, le limbe est donc l'organe percépteur. La sensibilité phototropique s'affaiblit avec l'accroissement du limbe. Les vieilles feuilles, à gros limbe, sont presque complètement insensibles à la lumière.

Le pédoncule de la fleur en bouton est négativement géotropique; après l'anthèse et avant la pollinisation de la fleur, il est plagiotrope, presque positivement géotropique. Le pédicelle, dans ce dernier cas, est courbé près de son insertion sur la tige, l'axe floral et une partie du pédoncule sont dirigés suivant l'horizontale. Si l'on donne à l'axe floral une position verticale, l'ouverture de la fleur étant dirigée vers le haut ou vers le bas, le pédicelle fait une courbure telle que l'axe floral est toujours ramené dans une situation horizontale. Tout cela change avec la pollinisation. La fleur pollinisée est, en effet, insensible à la pesanteur, elle se maintient dans les positions anormales qu'on lui donne. Il n'y a donc pas de mouvements post-floraux.

Avant l'éclosion du bourgeon, la vernation a pour effet de plisser la feuille suivant la nervure médiane, la face inférieure étant extérieure. A ce moment, le pétiole est 2 à 3 fois plus court que le limbe et à son insertion sur ce dernier, il fait une courbure appréciable. A l'épanouissement, le pétiole s'allonge beaucoup et la courbure, dont nous avons parlé, s'accentue jusqu'à faire un angle droit entre le pétiole et la nervure médiane. La pointe de la feuille est tournée vers le bas et le pétiole fait une courbure qui ressemble à la crosse chez les jeunes plantules de beaucoup de Dicotylédonnées lorsqu'elles sortent de terre. Mais cet état ne dure pas longtemps et, par une croissance exagérée du côté inférieur de l'insertion du pétiole, jointe à un diagéotropisme du limbe, ce dernier se renverse et se déploie et expose maintenant sa face supérieure vers le ciel. La nervure médiane fait une rotation de presque 180° dans un plan vertical. Il semble que l'insertion du pétiole sur le limbe joue ici le rôle d'un organe moteur. Or, les plantes étiolées que j'ai eu l'occasion d'observer, ont conservé cette courbure pendant assez longtemps (1 semaine environ), tandis que les feuilles exposées à la lumière ouvrent leur limbe dans l'espace d'un jour après l'éclosion. Il semble donc que la présence de la lumière soit nécessaire pour déclencher les mouvements décrits.

LEEMANN. *Bull. Soc. Bot. Gen.*, 1927, **19**, 99.

9

ROUGISSEMENT D'UNE ALGUE VERTE

Dans notre étude systématique des conditions nutritives qui entraînent l'accumulation des caroténoïdes chez le *Dictyococcus cinnabarinus*, nous avons limité notre champ de travail aux aliments qui entrent dans la matière vivante comme éléments plastiques. Ces éléments sont le carbone, l'azote, le phosphore, le magnésium et le soufre.

Nous ne nous sommes pas occupés des éléments et d'éventuels facteurs qui agissent, à de très faibles concentrations, et sur la croissance et sur la pigmentation de l'Algue de nos expériences. Sous cette réserve, nous pouvons énoncer les principaux résultats obtenus.

Le glucose et, plus généralement, les sources carbonées assimilables, favorisent l'accumulation des caroténoïdes, particulièrement en lumière forte. La cause de la synthèse excessive de ces pigments semble résider dans le fait que l'Algue est incapable de se défendre contre une abondante offre de carbone. L'intervention du glucose n'est pas déterminante; ce sucre pousse seulement à l'accumulation des caroténoïdes. En effet, dès qu'il est épuisé, les cultures redeviennent vertes.

Une carence en azote, en phosphore ou en sulfate de magnésium provoque l'accumulation des pigments caroténoïdes. A chacune de ces trois alimentations carencées correspond un métabolisme particulier. Ces trois métabolismes sont accompagnés d'un phénomène commun: l'accumulation des caroténoïdes.

Il arrive qu'une culture soit atteinte d'une chlorose rapide qui aboutit à des cellules blanches et mortes. Ce phénomène, dont nous ne pouvons préciser le mécanisme, est accompagné d'une légère lyse générale et d'une phospholyse massive.

HAAG. *Bull. Soc. Bot. Gen.*, 1942, **33**, 35.

10

LA POLLINISATION D'ASARUM EUROPAEUM

Il s'agissait donc pour moi de démontrer quel est le véritable mode de pollinisation de l'*Asarum europæum*; il n'y avait que l'expérience qui pouvait me le montrer.

Sur les quelques centaines de fleurs que j'ai eues entre les mains, aucune ne renfermait un insecte. J'ai examiné un certain nombre de fleurs où les étamines étaient toutes couchées, le stigmate était mûr, mais sur aucun de ces stigmates examinés au microscope, je n'ai trouvé du pollen étranger. Mais on trouve toujours du pollen sur les papilles au moment où les grandes étamines font leur déhiscence. Notons encore cette particularité que la maturité du stigmate se prolonge au-delà de la maturité des étamines et que le pollen est lourd et peu adhérant.

Pour arriver à la certitude il fallait donc: 1. assister à la déhiscence des étamines; 2. exclure rigoureusement tous les insectes. J'ai pu démontrer que les insectes ne jouent aucun rôle dans la pollinisation de la fleur de l'*Asarum europæum*. J'avais d'abord mis du papier gluant autour des fleurs pour capturer les insectes qui auraient voulu s'y introduire. Aucun insecte n'est tombé dans le piège. Jugeant ce moyen insuffisant, j'ai changé de méthode. J'ai soigneusement nettoyé une dizaine de pots dans lesquels j'avais planté des pieds d'*Asarum europæum*. Les fleurs étaient toutes en bouton au début de l'expérience. Pour empêcher les insectes de sortir de la terre j'ai couvert cette dernière d'une couche de plâtre de 2 cm. d'épaisseur. Ce plâtre se moulait autour des tiges et en se durcissant fermait hermétiquement le dessus du pot. Les plantes n'ont pas souffert de ce procédé. Tous ces pots ont été placés dans une baignoire avec 10 cm. d'eau à la chambre de bain. Pendant tout le temps que durait l'expérience j'ai surveillé toutes les fleurs étroitement. J'ai ainsi la certitude d'avoir exclu rigoureusement tous les insectes.

J'ai examiné toutes les fleurs, il y en avait vingt, qui ont mûri dans ces conditions et toutes étaient pollinisées. Donc je puis conclure de cette expérience que les fleurs d'*Asarum europæum* s'autopollinisent et que les insectes ne sont pas du tout nécessaires pour opérer la pollinisation.

J'ai encore une dernière preuve que l'*Asarum europæum* s'autopollinise: c'est d'avoir assisté moi-même au microscope à cette pollinisation. Il faut se représenter la fleur dans sa position horizontale. Les grandes étamines sont dressées contre le style, les anthères entre les lobes du stigmate. Tout cela forme une espèce de cylindre dont l'axe est horizontal. Les loges des anthères, au lieu d'être vraiment extorses, sont à 45° et les papilles des stigmates les touchent très souvent. Lors de la déhiscence de l'anthère il se fait alors une fente unique pour deux loges. Les bords de cette fente se retroussent au fur et à mesure qu'elle s'allonge et il y a finalement un mouvement qui ressemble à celui quand on retourne un sac. Souvent le pollen est emporté par les bords qui se retournent et il est déposé directement sur les papilles les plus proches. Le reste du pollen roule par grains ou par paquets en dehors de l'anthère et il ne peut tomber ailleurs que sur le stigmate.

Ainsi se confirme l'hypothèse de Kirchner qui est vérifiée maintenant par mes expériences.

Dans une position verticale la fleur ne pourra pas s'autoféconder aussi facilement que dans la position horizontale, cette dernière semble donc avoir un sens biologique.

LEEMANN. *Bull. Soc. Bot. Gen.*, 1927, **19**, 154.

11

CONCRÉTIONS DE SILICE CHEZ LES PODOSTÉMACÉES

L'anatomie des Podostémacées a fait déjà le sujet de diverses recherches, dont les premières, dues à Cario, en 1881, ont porté moins sur l'ensemble de l'anatomie du *Tristicha hypnoides* Spreng. que sur les curieuses concrétions de silice que l'on rencontre très fréquemment chez les Podostémacées et qui avaient été déjà vues par Tulasne. Ces éléments siliceux se trouvent dans les régions périphériques de la racine de la tige et de la feuille. Surtout abondants dans l'épiderme des feuilles et des tiges, ils se recontrent aussi autour des faisceaux vasculaires.

La taille et les dispositions sculpturales de ces concrétions sont très variées dans une même espèce. Elles remplissent généralement la cellule qui les contient. Leur surface est souvent creusée de cannelures, de sillons en spirale, ou diversement corrodée. Leur partie périphérique est d'aspect homogène et transparent, mais l'intérieur contient de petites inclusions liquides. L'abondance de ces concrétions est telle, parfois, dans le revêtement épidermique, que la plante conserve après dessiccation le même aspect qu'à l'état frais. Ces concrétions ont été parfois regardées comme un dispositif protecteur contre les dégâts mécaniques. Cette opinion ne correspond pas toujours aux faits observés: MM. Chodat et Vischer ont constaté au Paso de Yagui, au Brésil, que l'abondance du revêtement de silice est fonction de la situation des plantes. Ainsi le *Podostemon atrichus*, qui se développe sous la corniche des cascades, est cuirassé par 3 à 4 couches de cellules à silice et l'*Apinagia Yguazuensis*, qui n'en possède pas, et, de ce fait, est beaucoup plus flexible, se développe dans l'eau courante.

GAFFIER. *Ann. Fac. Sci. Marseille*, 1933, **5**, 110.

12

La Pléomorphisme parmi les Protococcacées

Dans son travail sur les « Protococcacées et Desmidiées récoltées à Java », Ch. Bernard dit en traitant le genre *Ankistrodesmus*: « C'est à propos de ces types si pléomorphes qu'il convient d'insister sur la nécessité de faire des cultures avant de créer de nouvelles dénominations ».

A Genève, on suit déjà depuis longtemps la méthode des cultures pures. Il s'est montré que dans plusieurs genres, presque toutes les formes qui n'ont presque aucune différence morphologique apparente, qui ont été isolées, présentaient des caractères bien définis, par rapport à chaque milieu de culture. M. Chodat a montré cela pour plusieurs genres de Protococcacées, en particulier pour *Chlorella* et *Scenedesmus*. De plusieurs espèces conjecturales de ce dernier genre, il a isolé une dizaine d'espèces élémentaires, clairement définies. Ce travail-ci est une confirmation de ces observations. J'ai étudié quatre formes d'*Ankistrodesmus* dont les cellules ont le même aspect et dont on ne peut distinguer, dans des conditions normales, qu'une forme à cause de l'arrangement particulier de ses cellules, tandis que je ne puis pas distinguer les trois autres formes les unes des autres. Nous avons vu qu'en les plaçant dans certains milieux, toujours rigoureusement identiques pour les quatre formes, leurs colonies montraient des différences remarquables, quant à leur grandeur et leur couleur; et les courbes qui expriment la variation de la taille des cellules de chaque forme étaient nettement différentes les unes des autres et cela dans des conditions qui ne différaient pas trop de celles qui peuvent se trouver dans le milieu naturel de ces algues. Il s'agit donc bien de quatre formes différentes et nous leur avons donné des noms différents. Nous les avons appelées des espèces. Ce ne sont évidemment pas des espèces linnéennes, mais de « petites espèces », des espèces élémentaires. De même que chez les plantes phanérogames, la conclusion qu'une espèce linnéenne peut embrasser un grand nombre d'espèces élémentaires, ce qui paraissait ridicule pendant longtemps, commence à être admis d'ordinaire, de même il faudra s'habituer à tirer également cette même conclusion pour des plantes aussi primitives et petites que les Protococcacées et d'entendre, par exemple par *Ankistrodesmus Braunii*—dans le sens que ce terme avait jusqu'à maintenant—un nom collectif pour un grand nombre d'espèces élémentaires. J'ai évité les termes « variétés physiologiques » ou « races biologiques ». Par ces termes, on peut entendre des formes qu'on ne peut distinguer les unes des autres que par des caractères physiologiques, par exemple par une capacité différente d'assimiler une nourriture donnée, mais qui ne présentent point de différences morphologiques. Or, j'ai montré que sur un même milieu, les cellules des quatre formes étudiées, présentent presque toujours des différences morphologiques. Il est vrai que ces différences ne peuvent souvent être saisies que par la méthode statistique, mais elles existent néanmoins. Un milieu donné—et j'insiste qu'il peut s'agir d'un milieu très semblable à ceux qui peuvent se trouver aux stations naturelles de ces algues—ne peut non seulement avoir une valeur différente pour le métabolisme des quatre *Ankistrodesmus*; il peut aussi influencer différemment la forme de leurs cellules et la morphologie de leurs colonies. Le fait que, par rapport à un milieu donné, les quatre *Ankistrodesmus* peuvent différer aussi bien par des caractères physiologiques que par des caractères morphologiques, me paraît justifier l'appellation d'espèces en entendant, par ce terme, des espèces élémentaires.

ŒTTLI. *Bull. Soc. Bot. Gen.*, 1927, **19**, 80.

13

ERICA VAGANS À ENTREVERNES

Du point de vue floristique par contre, la présence à Entrevernes de cette bruyère pose un problème de solution délicate. Espèce essentiellement atlantique, dont l'aire s'étend de l'Irlande au Portugal, *Erica vagans* est abondante dans les Pyrénées occidentales, disséminée dans l'Ouest de la France, avec quelques rares stations éparses dans le bassin parisien et le Massif central. Dans le bassin du Rhône, on lui connaît des stations isolées, sur le plateau de Chambaran à Roybon et, non loin de Vienne, à Eyzin-Pinet. Reuter a signalé près de Genève, à la lisière du bois de Jussy, une tache d'*Erica vagans*, et Christ considère qu'il s'agit du « dernier avant-poste d'un territoire des plus irréguliers ». Il n'est pas déraisonnable de supposer que des stations de très faible étendue existent ou ont existé ailleurs dans le bassin du Rhône et ont pu échapper aux investigations.

Comment, dans ces stations éparses et, en particulier, dans celle d'Entrevernes, la plante a-t-elle pu arriver et s'installer ? Les graines des *Erica* sont de faibles dimensions. Le vallon d'Entrevernes, orienté sensiblement du nord au sud, et, grâce à des synclinaux de même orientation, en connection, à travers les Bauges, avec la Combe de Savoie et le Graisivaudan, peut être une voie de migration. Ainsi que l'a remarqué déjà M. Beauverd, ce couloir peut être parcouru par de violents courants atmosphériques. Compte tenu des courants ascendants et du cheminement des semences à haute altitude que de récentes observations ont révélé, le transport par le vent de graines de bruyère à partir d'une station éloignée est admissible. L'hypothèse d'un transport par adhérence aux plumes d'oiseaux peut être aussi invoquée.

L'installation d'*Erica vagans* dans la station d'Entrevernes est récente. En 1908, M. Beauverd a consigné les déclarations précises d'un « montagnard d'Entrevernes connaissant à fond la localité pour l'avoir journellement parcourue en tous sens durant une trentaine d'années ». Il lui affirmait « qu'aucune bruyère autre que la Callune n'avait été autrefois observée en cette station; il n'y aurait que vingt ans à peine qu'une petite touffe d'*Erica vagans*, maintenant très développée, apparut au sommet de la tache naissante, qui depuis lors ne fit que s'agrandir ». Or, au moment de la découverte, la bruyère occupait une surface assez grande en bordure du sentier suivant la crête, sans quoi elle aurait sans doute échappé à celui qui, le premier, l'a remarquée. Marc Le Roux, en 1907, indiquait qu'elle formait une tache d'environ 100 m. de diamètre: cette indication doit être comprise comme s'appliquant à la surface totale de la station car la présence d'une barre rocheuse exclut la possibilité d'une extension uniforme de la plante. Personnellement, en

1909, sans avoir fait de mensurations exactes, j'ai noté la densité du peuplement de Bruyère. En 1924, j'ai constaté que la station comportait deux taches, distantes d'une vingtaine de mètres, que j'ai repérées: l'une à 30 m. au nord du sentier mesurant environ 3 m. 50 sur 1 m. 50; l'autre plus au nord couvrant une surface de 10 m. sur 4 m. En 1933 j'étais frappé de la réduction de ces deux taches. Au lieu de croître en masse compacte, à l'exclusion de toute autre espèce, la Bruyère se présentait sous forme de brins épars, intercalés entre des pousses d'autres plantes telles que Molinie et Callune et nettement localisés à la périphérie de la tache primitive, dont le centre était dégarni. En 1938, malgré la précision du repérage, il m'a été impossible de retrouver une tige d'*Erica vagans* sur l'emplacement de la première tache; la seconde tache, très appauvrie, réduite aux dimensions de 2 m. 50 sur 0 m. 50, comprend des brins épars de Bruyère entremêlés de Molinie et Callune, entourés d'un épais tapis d'*Hypnum Schreberi*. Ainsi, au cours d'une cinquantaine d'années, dans cette station d'Entrevernes, *Erica vagans* s'est installée, s'est étendue, puis a régressé et est à la veille de disparaître. L'extension de la plante, il a été facile de le constater, résulte essentiellement d'un processus végétatif par ramification progressive et enracinement à la périphérie des brins couchés sur le sol; le fait est conforme avec ce que l'on peut observer sur des touffes d'*Erica vagans* en culture.

Comment expliquer, après une phase d'extension, la régression et, mieux, la disparition de la Bruyère? On pourrait invoquer l'action de froids hivernaux rigoureux, auxquels cette espèce atlantique résisterait mal. De fait, dans le jardin botanique de l'Ecole des Eaux et Forêts une touffe d'*Erica vagans* provenant de brins arrachés à Entrevernes en 1924, et en pleine prospérité, a été complètement détruite par les froids de l'hiver 1928-29. Mais on ne peut préciser quels ont été, à Entrevernes, les abaissements de température et, en montagne, il faut tenir compte éventuellement de l'effet protecteur d'une couche de neige. La cause principale de la régression paraît être l'impossibilité pour la Bruyère de résister actuellement à la concurrence vitale d'autres espèces. Le fait semble évident quand, à la place des touffes compactes et étendues de jadis, on considère ces brins distants, enlisés parmi les mousses, entremêlés de jeunes pousses vigoureuses de Molinie, dominés par des tiges de Callune ou d'*Arctostaphylos Uva-ursi.* D'autre part il est significatif que l'*Erica vagans* fleurissant abondamment et, suivant toutes probabilités, fructifiant, ne se propage pas et n'occupe pas de nouveaux emplacements aux environs.

GUINIER. *Bull. Soc. Bot. Gen.*, 1942, **33**, 139.

14

LA GERMINATION ET LE DÉVELOPPEMENT DES SPORES D'ASPERGILLUS NIGER

Rappelons d'abord qu'il existe une différence entre le mode de sporulation des microbes et celui des champignons.

Tandis que, chez les premiers, les spores représentent un état de résistance et d'intermittence, chez les champignons, les spores représentent une formation permanente de la reproduction. D'après les recherches de Buchner (cité par Kruse), et ses élèves, la forme de sporulation des bactéries traduit un manque de substances nutritives du milieu de culture. Selon d'autres auteurs, elle représenterait une forme de résistance contre l'action toxique de certains produits du métabolisme intermédiaire.

En faveur de la théorie de Buchner, plaide le fait que les bacilles normaux et bien développés, ensemencés dans l'eau ordinaire ou dans une solution saline, donnent immédiatement naissance à la forme sporulée, alors que les mêmes bactéries laissées à se développer dans un milieu nutritif riche et spécifique se développent sans passer par la forme sporulée. Si, de ces cultures, ces bacilles sont passés en d'autres cultures similaires, le fait se répète sans qu'apparaisse la forme sporulée. Mais dans le cas où les cultures deviennent trop vieilles et où la substance nutritive a été épuisée par le processus du métabolisme général des bactéries, on observe parfois l'apparition de formes sporulées. W. Kruse et ses élèves reprenant cette question, arrivent à la conclusion que le manque de substances nutritives ou d'oxygène conduit à la forme de sporulation des bactéries.

D'après l'opinion de Pfeffer, la lumière n'exerce aucune action sur la germination des spores de Champignons. R. Kolkwitz a observé que les spores de différents Myxomycètes introduites dans une mince couche d'eau ordinaire, germent après quelques heures, tandis que, dans une solution de saccharose dans l'eau distillée, elles ne germent pas; ceci prouve qu'en dehors des aliments organiques, la présence des substances inorganiques est abolument nécessaire.

Knight et Fildes ont pu isoler, de la levure, une vitamine qui n'est pas identique à l'auxine et qui est nécessaire pour la croissance des bactéries de la classe *Sporogenes* et de la classe *Betulinum*, à la dose de 0,000.02%.

L. Nattan-Larrier affirme que certaines spores peuvent germer dans l'eau distillée, mais que la plupart demandent un milieu favorable; il ne précise pas combien de temps ces spores peuvent continuer à croître en eau distillée.

Brefeld a obtenu pour la première fois (1895) une germination des spores de *Ustilago-Cynodonties* dans l'eau et dans d'autres milieux nutritifs. Mais il n'a pas pu obtenir de germination dans l'eau distillée. McAlpine, en 1910, a étudié de même la germination des spores qui, selon lui, se développent tout aussi bien dans l'eau que dans les milieux nutritifs.

En 1937, M. Yen Wen-Yu reprend, en France, l'étude de la germination de la même spore dans différents milieux liquides. Dans l'eau ordinaire, les spores ne manifestent aucune activité germinative.

Ceci est en contradiction avec les recherches de McAlpine et, par contre, confirme entièrement celles de Brefeld. Dans le même travail, il montre qu'il a pu obtenir la germination des spores de certains champignons dans l'eau ordinaire stérilisée, mais que cette germination a été retardée, se produisant au bout de 24 heures au lieu de 3-6 heures que dure la germination de la même spore dans un milieu nutritif. Habituellement, la germination de la spore en eau distillée n'est pas suivie d'un développement de l'embryon. Nous trouvons extrêmement intéressantes les expériences classiques de Guilliermond sur l'action toxique et inhibitrice de divers colorants sur la croissance des Champignons.

KAUFFMANN-COSLA *et al.* *Rev. Gén. Bot.*, 1940, **52**, 97.

ZOOLOGIE

1

DÉVELOPPEMENT DE L'ECTODERME CHEZ LES ANNÉLIDES

Le type primitif de développement des formations ectodermiques se laisse observer chez les Archiannélides d'aujourd'hui et chez les *Polychaeta.* Ils ont l'ébauche commune des formations ectodermiques se développant de la même manière; on y peut apercevoir une sorte de nouveau centre d'organisation sous la forme de dérivés du blastomère *d* 2. Ainsi, c'est le mode épiblastique de développement des formations ectodermiques qui doit être considéré ici comme général et fondamental; ce mode-là n'est qu'une esquisse du nouveau centre de développement des dérivées ectodermiques et du nouveau mode de leur développement (plaque somatique).

Comme résultat d'une segmentation spirale régulière sur le pôle animal de l'œuf, un groupe de micromères ectodermiques apparaît, formant trois quartettes. Ce petit groupe fournit de nombreux dérivés: mésenchyme larvaire et protonéphridies, revêtement dermique du trochophore et du ver lui-même, le centre nerveux syncipital. Le rôle des micromères du premier quartette doit attirer notre attention particulièrement: leurs dérivés participent à la formation du revêtement dermique de l'hémisphère supérieur du trochophore, c'est-à-dire du revêtement dermique prospectif du lobe céphalique du ver.

SCHMIDT. *Arch. Zool. Expt.*, 1939, **81,** 319.

2

LE SYSTÈME NERVEUX DE PLEUROBRANCHAEA

Si ce genre n'est pas représentatif de tout le groupe des Notaspides, il n'y a pas lieu de s'en préoccuper, car les divergences sont d'une importance secondaire. Plus ou moins de rapprochement des ganglions, de concentration, quelques anastomoses, quelques filets nerveux plus ou moins prolongés ne constituent pas des différences importantes. Nous voyons, par exemple, dans un même petit groupe de Mollusques un nerf apparemment impair partir de la commissure buccale, ou bien naître de deux racines droite et gauche, qui se réunissent pour se disjoindre presque aussitôt, ou enfin deux nerfs bien distincts partant des deux ganglions, remplir le même office: différences sans aucune importance, si l'on ne perd pas de vue que tout organe impair doit être considéré chez les Mollusques comme remplaçant deux organes latéraux pairs fusionnés.

Si nous prenons comme point de départ les études de Lacaze-Duthiers sur le Pleurobranche orangé (1859), nous voyons que son étude du système nerveux, excellente pour l'époque n'a plus aujourd'hui qu'un intérêt historique; qu'elle est même entachée de quelques erreurs. En particulier, le nerf pénial avec son ganglion est représenté comme partant du ganglion cérébroïde et donnant une branche qui s'anastomose avec le nerf génital d'origine pédieuse. En réalité, le nerf pénial part du ganglion pédieux, et ceci paraît bien être le cas général.

PRUVOT-FOL. *Arch. Zool. Expt.*, 1939, **81,** 10.

3

LA THÉORIE DE LA RÉCAPITULATION

Tout le monde connaît la loi biogénétique fondamentale dite de Haeckel, l'ontogénèse est une récapitulation de la phylogénèse, c'est-à-dire que le développement de l'individu est une rapide répétition de l'évolution de l'espèce, un court résumé de la chaîne infinie de ses ancêtres.

Les faits les plus caractéristiques qui démontrent désormais cette loi d'une manière, peut-on-dire, irréfutable, sont tellement connus qu'il ne nous paraît point nécessaire de les rappeler ici; quelques exemples, la solipédation se produit graduellement chez le cheval et seulement aux derniers stades de son développement; nombre de cétacés qui plus tard au lieu de dents possèdent ce qu'on nomme des fanons, ont des dents aux mâchoires, tant qu'ils sont à l'état de fœtus et qu'ils ne peuvent prendre aucune nourriture.

London B.Sc. (*Special*).

4

La vessie natatoire des Polyptères

La vessie natatoire des Polyptères diffère de celle de tous les poissons, les Dipreustes exceptés, et ressemble aux poumons des Vertébrés terrestres, en ce qu'elle consiste en deux sacs cylindriques (celui du côté droit plus long que l'autre), se réunissant en avant pour former un tube très court qui communique avec la face ventrale de l'œsophage par une fente longitudinale, comparable à la glotte. Mais, contrairement à ce qui se voit chez les Dipreustes, la vessie natatoire ne reçoit que du sang artériel.

London B.Sc. (*Special*).

5

Œufs du Condor

Généralement la ponte est de deux œufs, d'un seul quelquefois. Plus ou moins ovoïdes, ils sont blancs: les uns sans tache, les autres avec une légère jaspure de gris et de lilas, surtout au bout pointu. La coquille est assez rugueuse, et les pores sont prononcés. Les dimensions sont inégales de l'un à l'autre, comme en témoignent les quatre mesures suivantes: 108 mm. × 72, 130 mm. × 60, 132 mm. × 63, 135 mm. × 71.

L'incubation, faite presque exclusivement par la femelle, dure entre quarante-deux et quarante-cinq jours. Desmurs rapporte qu'en 1847 une Poule couva un œuf de Condor dans le Regent's Park de Londres, y employant cinquante-quatre jours: 7 mai au 30 juin. Cette différence tendrait à prouver que la chaleur de celle-ci était insuffisante pour le développement normal du poussin.

Il arrive que sur deux œufs un seul est fécondé. C'est ainsi que l'oiselet, élevé par moi il y a quinze ans, à San

Bernardo, fut trouvé solitaire sur son rocher, à côté d'un œuf stérile.

HOUSSE. *Ann. Sci. Nat.*, 1939, **2**, 125.

6

LE GANGLION GASTRIQUE DU POULPE

Organe impair et asymétrique, ce ganglion est situé à droite du carrefour où se joignent l'estomac, l'intestin et le cæcum spiral. La taille, moderée, atteint 5 à 6 mm. chez les gros Octopus. Il a la forme d'un ovoïde à grand axe perpendiculaire à la direction de l'intestin. Les nerfs issus de ce ganglion peuvent être, ainsi que le montrent nos dissections, répartis en 3 faisceaux ou pédicules:

1° le pédicule supérieur comprend en général 3 nerfs émanés du pôle supérieur, qui se dirigent vers l'estomac et l'œsophage. Ce sont le nerf gastrique antérieur; le nerf gastro-œsophagien qui se scinde en une branche gastrique et une branche œsophagienne; celui-ci après avoir donné des rameaux à l'œsophage, s'anastomose avec les nerfs œsophagiens du ganglion buccal, et le nerf gastrique postérieur.

2° le pédicule dorsal comprend un faisceau de nerfs disposés en éventail, qui se rendent dans le cæcum spiral; ils sont en général au nombre de 6.

3° le pédicule ventral est formé de 3 nerfs émanés du pôle postérieur du ganglion, qui se dirigent vers la droit le long de l'intestin. Ce sont: le nerf hépato-pancréatique droit qui longe le canal hépato-pancréatique droit, lui abandonne un rameau (nerf du canal hépato-pancréatique droit); le tronc commun des nerfs intestinaux, qui longe l'intestin et se divise rapidement en plusieurs branches, et le nerf hépato-pancréatique gauche, d'où provient le nerf du canal hépato-pancréatique gauche.

Cette description correspond dans ses grandes lignes à celle des auteurs (Chéron 1866, Pfefferkorn 1915), dont nous précisons les données en fixant le nombre et la destinée de ces nerfs.

BOGORAZE et CAZAL. *Arch. Zool. Expt.*, 1946, **84**, 117.

7

RECHERCHES SUR LES « HYMENOLEPIS DIMINUTA »

Avant d'aborder les anomalies que nous avons relevées, rappelons en quelques mots l'anatomie normale des organes génitaux d'*Hymenolepis diminuta.* Il existe trois testicules dont un poral, c'est-à-dire situé du côté du pore génital, les deux autres antiporaux, situés de l'autre côté et séparés du poral par les glandes femelles. Cette position est la même dans tous les anneaux, car un des traits caractéristiques d'*H. diminuta* est d'avoir les pores génitaux unilatéraux. Les testicules ont chacun un canal par lequel ils se déversent. Les trois canaux se réunissent en un seul dit canal déférent, qui se dilate bientôt en une vésicule séminale externe. Puis le canal déférent entre dans la poche du cirre, se dilate encore pour former une vésicule séminale interne qui se continue par le cirre proprement dit. L'appareil femelle consiste en un ovaire multilobé dont la partie postérieure communique avec une glande vitellogène massive. Il existe également une petite glande coquillière. Ces glandes communiquent par un fin canal: l'oviducte, avec l'extrémité proximale du vagin dilatée qui est dite réceptacle séminal. Le vagin a un trajet à peu près parallèle au canal déférent et débouche près de la poche du cirre, formant avec celle-ci le pore génital. Les spermatozoïdes, formés dans les testicules, se rendent dans les conduits mâles, sont injectés par le cirre dans le vagin voisin, s'accumulent dans le réceptacle séminal; ils vont finalement féconder les ovules formés dans l'ovaire, et ayant reçu les liquides des glandes vitellogène et coquillière. Puis, ces ovules fécondés s'engagent dans un nouvel organe d'existence virtuelle jusque là, l'utérus, qui se dilate, prend une importance énorme, devient sacciforme et finalement remplit tout l'anneau mûr. Pendant ce temps, l'ovule s'est segmenté et l'embryon hexacanthe s'est formé.

PALAIS. *Ann. Fac. Sci. Marseille*, 1933, **6**, 121.

8

Analyse d'une hybridation

On a vu que certains indices morphométriques donnent la possibilité d'identifier des individus d'espèces ou de races différentes. Le meilleur indice morphométrique pour caractériser deux espèces est généralement celui dont les valeurs numériques dans ces deux espèces seront les plus éloignées l'une de l'autre. Cet indice est évidemment celui qui comprend les deux caractères dont les moyennes sont en plus grande opposition métrique en sens contraire dans les deux espèces.

Par exemple: *Gr. campestris* a une tête relativement grosse et un tibia court et *Gr. bimaculatus*, une tête relativement petite et un tibia long. L'indice *Th/Ti* pourra donc être un rapport susceptible de présenter une valeur taxonomique.

Mais, souvent il est difficile, par la simple observation, de juger de l'opposition des dimensions des caractères chez les deux espèces, en raison même de l'allométrie de certains indices. Ainsi, s'il est vrai que *Gr. campestris* a une très grosse tête, cela est surtout exact pour les exemplaires de très grande taille. Les petits individus ont une tête petite, si bien qu'à tailles égales, les grands exemplaires des deux espèces se distinguent aisément par le volume respectif de leurs têtes, les petits, au contraire, présentent un contraste beaucoup plus atténué. Pour certains caractères, ce fait est encore beaucoup plus accusé. Ainsi, on lit dans les diagnoses même des deux espèces de Gryllides: *Gr. campestris* mâle, pronotum plus large en avant qu'en arrière; *Gr. bimaculatus* mâle, pronotum également large en avant et en arrière. Or, ce caractère de détermination ne permet de différencier que les individus de grande taille; car, plus le *Gr. campestris* est grand, plus le pronotum s'évase à l'avant. Par contre, lorsqu'on observe un petit individu, on remarque que son pronotum répond à la diagnose donnée pour *Gr. bimaculatus*: pronotum également large en avant et en arrière. Dans une note antérieure, j'ai donné une analyse de la variabilité de la forme du pronotum chez les deux espèces de Gryllides, ainsi que chez leurs hybrides. Je ne reviendrai donc pas ici sur la nécessité de la prise en considération de l'allométrie de taille.

COUSIN. *Arch. Zool. Expt.*, 1939, **81**, 293.

9

CAS DE CANCER MALIN DES POISSONS

Le problème du cancer chez les Vertébrés à sang-froid possède sa vaste littérature et il devrait inspirer un vif intérêt au point de vue pratique, aussi bien que théorique. Avant tout, il rend possible le développement des recherches concernant l'éthiologie du cancer sur le matériel autre que celui que l'on emploie d'habitude, dans des conditions de milieu qui différent au point de vue chimique et physique de celles que nous trouvons dans les cas de cancer chez les animaux à sang chaud et chez l'Homme.

Ensuite, s'il s'agit spécialement des poissons, nous rencontrons ici des cas de cancers malins, si l'on peut ainsi dire « épidémiques » ou mieux en masses, lorsque toute une culture des poissons est atteinte de cette maladie. Les recherches sur des cas pareils pourraient fournir quelques indications, concernant soit l'influence libératrice de certaines conditions du milieu dans le cas du cancer, soit d'autres causes de la propagation des néoplasmes malins.

Cependant, si les descriptions de néoplasmes malins chez les poissons sont nombreuses, les recherches méthodiques concernant ce problème ne sont pas abondantes. Quoique parfois très détaillées, elles se rapportent presque exclusivement à la morphologie de cas rassemblés par hasard. Ainsi Plehn (1909), J. Wolff (1913), Roussy et Wolf (1929), Takahashi (1929), Thomas (1931), Williams (1931), Haddov et Blake (1933). Les autres résument les cas de cancers chez les poissons, connus dans la littérature, ou bien ceux rassemblés par eux.

Ce ne sont que des cas d'épizootie, paraissant souvent dans les élevages de truites et de saumons et causés par l'épithélioma de la glande thyroïde, qui ont donné lieu à des recherches plus détaillées sur la virulence de ce néoplasme, son anatomie et sur son caractère clinique (Harway, Gaylord et Cloves 1910 et les autres).

Il s'agit ici d'un cas très particulier de cancer malin des poissons. En outre, il se rapporte aux poissons, élevés dans les aquariums du Musée des Colonies à Paris, où il paraît sporadiquement. Il pourra donc, probablement, devenir le sujet de nouvelles recherches.

RAABE. *Arch. Zool. Expt.*, 1939, **81**, 1.

10

LES MONOCYSTIDÉES

Les Monocystidées des Oligochètes comprennent de nombreux genres dont certains, en particulier le genre Monocystis, sont très riches en espèces. Cependant, étudiant les Lumbriciens des environs de Monthellier, nous les avons vu parasités par des Monocystidés dont la plupart n'ont pas été décrites. Cela tient à ce qu'aucun travail n'a été fait dans notre région sur ce sujet et la faune Grégarinienne a l'air d'être très variable suivant l'habitat du ver.

Le polymorphisme des Monocystidées des Oligochètes est à rapprocher de celui des Flagellés des Termites. La cause dans l'un et l'autre cas serait probablement la même: l'abondance de nourriture déterminerait la variation.

TUZET et LOUBATIÈRES. *Arch. Zool. Expt.*, 1946, **84**, 132.

11

CONSTITUANTS DES CORPS GRAS CHEZ LES INSECTES

L'acide *butyrique* CH^3-$(CH^2)^2$-COOH et son homologue en C^6 l'acide *caproïque* existent très probablement dans les beurres d'Aphidiens; je ne les ai pas isolés, mais leur présence est rendue très vraisemblable par l'indice de Reichert-Meissl élevé que fournit le beurre de *Pemphigus*.

Même remarque pour les acides *caprylique* et *caprique* dont la présence doit être déduite avec beaucoup de vraisemblance de l'indice de Polenske donné par le même *Pemphigus*.

L'acide *laurique* CH^3-$(CH^2)^{10}$-COOH qui constitue une portion importante du beurre de Coco, du suif de Tangkallak, de la graisse de *Litsœa sebifera*, etc. existe aussi chez les *Pemphigus* où j'ai pu établir sa présence en l'isolant et en le caractérisant par son point de fusion et son poids moléculaire (200). Il fond à 43°, 5.

L'acide *myristique* CH^3-$(CH^2)^{12}$-COOH existe en abondance dans la graisse de cochenilles (*Coccus cacti* L.) où il a été trouvé par Liebermann (1885). On sait que parmi les végétaux il se trouve en forte proportion dans le beurre de muscade. dans la graisse de *Virola venezuelensis* et le beurre de Dika,

L'acide *palmitique* CH^3-$(CH^2)^{14}$-COOH a été signalé par Tsujimoto (1908) dans l'huile de chrysalides de vers à soie. C'est un acide des plus répandus dans le règne animal et qui existe certainement chez un grand nombre d'insectes.

L'acide *stéarique* CH^3-$(CH^3)^{16}$-COOH se trouve dans l'huile de *Pyrausta nubilalis* Hübn., dans la graisse de Cantharides, etc.

Les termes supérieurs de cette série n'ont pas encore été signalés dans les graisses d'insectes (je laisse entièrement de côté tout ce qui a trait aux cires). Il serait intéressant de rechercher l'acide arachidique chez les larves de Coléoptères qui se nourrissent de graines d'arachide.

TIMON-DAVID. *Ann. Fac. Sci. Marseille*, 1930, **4**, 41.

12

LA LOI D'HOMOLOGIE PARALLÈLE

On a remarqué depuis longtemps que certains caractères des pattes, par exemple ceux d'une griffe dissymétrique, se retrouvent d'une patte à l'autre, à condition d'intervertir les faces antiaxiale et paraxiale quand on passe du groupe I-II au groupe III-IV. Si les 4 pattes, de chaque côté, étaient parallèles et orientées perpendiculairement au corps, toutes les faces antérieures (ou toutes les faces postérieures) seraient homologues les unes des autres. On peut appeler cela l'*homologie parallèle*.

Je me suis demandé s'il s'agissait d'une véritable loi. Les exemples cités par les auteurs étaient en petit nombre. N'en existait-il pas d'autres, beaucoup plus nombreux et importants, incompatibles avec l'explication donnée ? Fallait-il, d'autre part, étendre la loi à toute la patte ou la réserver aux seules extrémités distales ?

Après beaucoup d'observations, je crois être en mesure d'affirmer que la loi d'homologie parallèle s'applique à tous les caractères des pattes. C'est une loi générale et primitive, très importante. Il faut en tenir compte dans les notations.

J'ai rencontré dans la présente étude beaucoup d'exemples d'homologie parallèle. Je ne les ai pas tous cités. J'appelle ici seulement l'attention sur ceux qui ne dépendent pas de la griffe, ni de la chætotaxie. La loi s'applique à la division des fémurs et à des détails d'ornementation.

Pour d'autres Acariens, les exemples d'homologie parallèle sont aussi nombreux et évidents, de sorte qu'il faudrait s'attacher surtout, aujourd'hui, aux exemples contraires, s'il y en a, c'est-à-dire à ceux où l'on retrouverait, à la face antiaxiale d'une patte postérieure, par exemple, et non à la face paraxiale, un caractère exclusivement antiaxial d'une patte antérieure.

L'homologie parallèle est beaucoup plus qu'une façon d'expliquer des caractères. Elle nous démontre qu'à l'origine les pattes étaient parallèles. Un Acarien comme *Anystis*, où elles sont restées presque parallèles, surtout aux premiers états, est primitif à cet égard. La division des pattes en deux groupes de sens opposés, en rapprochant du corps les faces para, a introduit plus tard des conditions nouvelles pour ces faces. C'est peut-être parce qu'ils frottaient constamment et inutilement l'idiosoma que certains poils para des pattes postérieures, au fémur et au génual en particulier, ont disparu.

GRANDJEAN. *Ann. Sci. Nat.*, 1939, **2**, 40.

13

REMARQUES SUR LE GENRE « LORDALYCUS »

Les caractères de ce genre peuvent être résumés de la manière suivante, en distinguant les primitifs et ceux qui témoignent d'une évolution régressive ou progressive.

Les caractères primitifs sont avant tout ceux de la mandibule. Il faut y ajouter ceux de la chætotaxie des pattes avec la disposition fréquente par paires des poils et des solénidions et leur alignement d'un article à l'autre, ou sur le tarse; le grand nombre de poils, particulièrement sous les tarses, même aux tarses IV; le grand nombre de solénidions, en ce qui concerne les baculiformes et les fusiformes pour lesquels il ne paraît pas y avoir néotrichie. Au palpe on peut noter comme caractère primitif la présence de 4 poils au tibia, au lieu des 3 habituels et peut-être celle de 2 solénidions au dernier article. La régression relativement faible de la prélarve est un caractère primitif très important.

Parmi les caractères d'évolution régressive, l'un des plus notables est la déficience de nombreux poils dorsaux et postérieurs. Le prodorsum a perdu 2 paires de poils. Quant au-dessus de l'hysterosoma, jusqu'à la région anale, c'est le plus pauvrement pileux que je connaisse chez les *Endeostigmata.* L'animal a perdu aussi sa 3e paire de verrues génitales, tous ses poils eugénitaux chez la femelle, sauf un seul, ses fissures lyriformes et ses yeux.

Les caractères progressifs sont ceux de la néotrichie solénidionale et des trichobothries. Celles-ci, avec leurs grandes bothridies et leurs poils minces, sont assez perfectionnées. La réunion des deux bothridies de la paire antérieure dans une fossette commune est un caractère de spécialisation très exceptionnel. Il faut ajouter la forme insolite du corps, les tubercules de la surface, en particulier celui qui borde en arrière la fossette commune, les cils fourchus des ongles.

GRANDJEAN. *Ann. Sci. Nat.*, 1939, **2**, 119.

14

BIOLOGIE DES JUMEAUX

Dans cet ouvrage, l'auteur nous donne un exposé extrêmement clair et, comme toujours, parfaitement documenté des faits relatifs à la gemellité. Après avoir distingué des vrias jumeaux provenant d'un même œuf, des faux jumeaux provenant d'œufs distincts, il attire l'attention sur les nombreux problèmes de biologie générale que pose et permet de résoudre, au moins en partie, le cas des premiers. C'est ainsi qu'on a pu établir que la formation des jumeaux uniovulaires se ramène, à l'origine, à une perturbation dans les débuts du développement de l'œuf, entraînant un dédoublement des ébauches. Quand chez les mammifères cette subdivision de l'ébauche de l'individu se fait sur le blastocyte, comme chez les Tatons, les embryons sont monochorioniques, mais il est probable que dans le cas des jumeaux uniovulaires dichorioniques ce sont les tout premiers blastomères qui ont été isolés, comme on a pu le réaliser expérimentalement chez les Tritons, l'Amphioxus et divers Invertébrés. Analysant un des caractères les plus frappants de ces vrais jumeaux, leur ressemblance, M. Caullery remarque qu'il ne s'agit pas d'une identité absolue; des particularités, des différences se font jour qui en fixent des limites. C'est pourquoi, d'ailleurs, ils constituent un matériel précieux pour l'étude des variations du génotype en présence des circonstances extérieures diverses. La gémellité normale conduit à l'étude des monstres doubles qui n'en sont que des cas particuliers. Un fait singulier sur lequel M. Caullery attire l'attention est que, contrairement à ce qui a lieu pour la majorité des problèmes biologiques qui intéressent l'Homme, on est beaucoup mieux documenté sur les humains que sur les autres mammifères.

L.F. *Arch. Zool. Expt.*, 1946, **84**, 169.

15

Cellules cirières

Ce sont des leucocytes spécialisés dans un rôle de mise en réserve et sans doute homologues aux cellules adipeuses des autres insectes. La cire est l'élément principal de réserve chez les Coccides et on retrouve même cette substance dans le vitellus des œufs.

Les cellules cirières semblent dériver de jeunes eucocytes à la suite d'une diminution progressive du rapport caryo-cytoplasmique en même temps que le cytoplasme, devenant moins basophile, se charge de vacuoles. Celles-ci, d'abord fines et périphériques, renferment chacune un granule très réfringent, dont l'ensemble donne à la cellule un aspect granuleux. Le noyau est encore central au début de cette évolution, mais bientôt une ou deux vacuoles plus grandes se forment et le refoulent à la périphérie. Ces dernières vacuoles apparues sont soudanophiles; elles contiennent de la cire qui, par fixation à l'alcool faible, apparaît sous forme de cristaux biréfringents. Ces cristaux perdent alors, dans ces conditions, leur affinité pour le colorant des graisses et fondent à une faible température (approximativement 30-33°); ils se reforment par refroidissement. Ces vacuoles s'accroissent, se fusionnent et ces processus déterminent la formation d'une cellule cirière typique.

Les jeunes cellules cirières en formation conservent pendant un certain temps les propriétés phagocytaires des leucocytes qui leur ont donné naissance. Elles hébergent, en outre, fréquemment des symbiotes. La présence de ces derniers, dans de telles cellules, qui dans des conditions expérimentales peuvent également renfermer des granulations d'encre de Chine, témoigne à l'encontre de l'opinion de divers auteurs, que ces symbiotes sont phagocytés. Ceux-ci conservent néanmoins leur pouvoir de bourgeonnement, tandis que la cellulehôte se vacuolise de plus en plus, tendant ainsi vers une forme prématurée de cellule cirière adulte; mais cette cellule paraît déficiente ainsi que semble bien l'indiquer l'aspect de son cytoplasme qui devient de plus en plus hyalin. Elle prend finalement l'aspect d'une sphère translucide, à paroi extrêmement mince et à noyau difficilement reconnaissable, car il perd ses propriétés de colorabilité. De nombreux symbiotes s'y observent, notamment des formes jeunes ayant pris naissance par bourgeonnement intracellulaire. Il est vraisemblable qu'à un moment donné ces sortes de cellules éclatent et recèdent au plasma les symbiotes qu'elles y avaient prélevés en même temps que les jeunes formes nées par bourgeonnement.

POISSON et PESSON. *Arch. Zool. Expt.*, 1939, **81**, 25.

16

Drepanoptera Gabonensis

Tête, thorax, abdomen et couleur du fond des ailes, uniformément brun terne (antennes absentes).

Ailes antérieures falquées avec ocelle réniforme allongé; la partie hyaline très mince entourée d'un filet blanc, d'une bande brun-jaune, et d'un serti noirâtre. La rayure interne forme une large surface triangulaire blanchâtre au contact du bord interne de l'aile et elle se fond dans la rayure externe. Cette dernière occupe exactement le milieu de l'aile; elle part de la côte pour venir diagonalement se poser sur le bord supérieur de l'ocelle, et elle ressort en dessous dans sa partie concave, qu'elle longe pour venir se fondre avec la rayure interne. Dans la cellule une bande diffusément plus claire semble contourner aussi l'ocelle. L'aire externe est parsemée d'atomes blancs dans sa moitié proximale. La bande submarginale plus jaunâtre est séparée de la marge gris-clair par une ligne onduleuse brunâtre formant des lobes internervulaires très peu profonds et réguliers. La tache apicale pseudo-ocellaire noire forme un ovale très allongé, la moitié proximale à peine est cernée de blanc. A l'apex une ligne fulgurante blanche délimite une zone apicale brun-rouge. La nervure radiale antérieure se détache du pédoncule des autres radiales.

Ailes postérieures avec un ocelle réniforme très court et plus large que l'antérieur, ayant les mêmes couleurs. La rayure interne blanchâtre est large et contourne l'ocelle presque à son contact; la partie basale de l'aile est d'une teinte brunâtre à peine plus foncée. La rayure externe commence sous l'ocelle même et va presque en ligne droite au bord anal de l'aile. L'aire externe est modérément parsemée d'atomes blanchâtres. La bande submarginale jaunâtre est séparée de la marge grisâtre par une ligne irrégulière brune et elle renferme, dans les espaces internervulaires, des taches assez épaisses et régulièrement réunies.

Le dessous des ailes est identique comme dessin et comme couleur; mais à l'aile postérieure la rayure externe se prolonge jusqu'à la subcostale, dans une direction parallèle au bord anal de l'aile.

Envergure: 138 millimètres.

TESTOUT. *Ann. Soc. Linn. Lyon*, 1935, **79**, 45.

17

ANATOMIE GÉNÉRALE DES DIPTÈRES

Les glandes endocrines post-cérébrales des insectes contractent des rélations étroites avec le système nerveux stomatogastrique. Celui-ci présente, chez Chironomus plumosus, une configuration typique. De la région tritocérébrale de chacun des ganglions cérébroïdes part, vers l'avant, un nerf labro-frontal; les deux nerfs labro-frontaux aboutissent à un ganglion frontal triangulaire de situation très antérieure. Issu du ganglion frontal, un nerf récurrent parcourt vers l'arrière la face dorsale de l'oesophage; sur une grande partie de son trajet, ce nerf est inclus dans l'aorte d'où il sort au niveau du ganglion hypocérébrale situé à l'antérieur du collier oesophagien et accolé à la face ventrale de l'aorte. Le ganglion hypocérébrale, enfin, envoie un nerf stomacal à chacun des deux ganglions stomacaux symétriques situés latéralement, sur la paroi du tube digestif, à la limite de l'oesophage et du proventricule.

POSSOMPÈS. *Bull. Soc. Zool. France*, 1946, **71**, 100.

18

LA TÉRATOLOGIE DES INSECTES

Les monstruosités entomologiques offrent au biologiste un vaste champ d'investigations encore bien insuffisamment exploré, et leur étude lui apporte mainte occasion d'exercer ses facultés de synthèse. Il s'en aperçoit dès le début car, devant préciser son sujet, il se voit constraint de remettre en question la définition même de la monstruosité. Les tératologistes classiques s'occupaient à peu près exclusivement des Vertébrés superieurs et ils avaient ainsi la partie belle. Lorsqu'on s'adresse à des Invertébrés, dont l'organisation en diffère si profondément, les définitions courantes s'avèrent insuffisantes ou imprécises: il en faut tirer les éléments et même faire appel à des éléments nouveaux. La tâche est si malaisée que c'est de justesse que l'on peut éviter de dénier toute existence à la tératologie: tiraillée entre la génétique et la pathologie elle risquerait fort d'être démembrée si deux raisons ne venait à son secours. La première est qu'il reste malgré tout un noyau de faits qu'on serait assez embarrassé de loger dans la pathologie, qui ne sont pas de nature héréditaire et qui sont plus que de simples phénomènes de variation. Puisqu'on ne peut détruire complètement la tératologie, peut-être est-il de meilleure politique de la laisser vivre sa vie et jouer le rôle de science-tampon. La seconde raison est qu'il est impossible d'isoler le rôle tératogénétique de l'hérédité de celui du milieu, et qu'à vouloir le faire on tomberait de quelque chose d'artificiel dans quelque chose de plus artificiel encore.

Définir les monstruosités comme des modifications d'origine embryonaire, c'est préjuger dans notre définition d'une époque d'origine qu'il s'agit précisément de déterminer après étude morphologique. Une telle erreur de méthode, sans conséquence grave chez les vertébrés supérieurs dont l'organisation est, à la fin de la vie embryonnaire, très voisine de celle de l'âge adulte, devient intolérable lorsqu'il s'agit d'insectes pour qui la fin de la vie embryonnaire n'est que le prélude de la vie larvaire et qui devront subir de profondes métamorphoses avant d'atteindre l'état parfait. Chez eux tout concourt à montrer que des anomalies morphologiquement identiques peuvent avoir une origine totalement différente.

BALAZUC. *Bull. Soc. Zool. France*, 1946, **71**, 96.

GÉOLOGIE

1

Le Calcaire portlandien dans la région marseillaise

Le Calcaire portlandien, représenté en différents points de la région marseillaise, se présente généralement sous la forme d'une importante masse de calcaire dur et très blanc en surface, mais à cassure variable, grise, jaune ou même rougeâtre. Au point de vue statigraphique, il représente le Jurassique supérieur tout à fait terminal; ainsi défini, il va de la dolomie jurassique aux marnes valanginiennes.

Nous avons pensé, connaissant le travail de Dejean, sur le Crétacé marin de Provence, qu'il pourrait y avoir quelque intérêt à préciser la pétrographie de ce calcaire qui, le plus souvent, est entièrement dépourvu de fossiles déterminables. Par ailleurs, ce calcaire portlandien est très semblable, à l'œil nu, aux calcaires immédiatement supérieurs et, en particulier, à l'Urgonien; et la complexité géologique de la Provence ne permet pas toujours au géologue de distinguer ces deux formations par les méthodes normales de la stratigraphie; c'est là, nous semble-t-il, que la pétrographie peut lui venir en aide, c'est de là qu'est venue notre idée de faire l'étude pétrographique du calcaire portlandien pour faciliter la solution du problème dans les cas stratigraphiquement douteux.

Les gisements qui ont fait l'objet d'une étude approfondie sont les suivants:

1° La Nerthe: jurassique supérieur du Vallon de la Cloche.

2° Montredon: jurassique supérieur du Mont Rose.

3° Mazargues: jurassique supérieur de l'Arénas.

En dernier lieu, nous avons coupé quelques échantillons recueillis à l'Ouest et à l'Est d'Allauch; nous verrons plus loin que ce dernier gisement, par les coupes qu'il nous a fournies et par sa stratigraphie, pose un problème important dont la solution demandera de nouvelles recherches.

Enfin, en cours d'année, nous avons effectué quelques coupes dans des échantillons recueillis en divers points du Var (*Nans*, *Régusse*, *Draguignan*, *Brue-Auriac*) ; ces coupes, si elles ne donnent pas de résultats importants, parce qu'elles sont restées isolées, viennent cependant confirmer les conclusions que nous tirerons de nos études plus systématiques et plus complètes.

Il nous paraît bon d'indiquer ici la façon dont nous avons procédé pour récolter les échantillons : un gisement étant choisi, nous recherchions sa base (contact avec la dolomie) et nous ramassions les échantillons en nous élevant dans la série et en les numérotant avec soin ; et cela, jusqu'à ce que nous atteignions le sommet du gisement (contact avec les marnes valanginiennes). Les échantillons étaient ensuite coupés dans leur ordre de récolte, et les plaques minces numérotées dans le même ordre. Cette méthode nous permettra d'affirmer que telle coupe par exemple provient d'un échantillon inférieur ou supérieur à tel ou tel autre. De plus, nous aurons la certitude d'avoir récolté à des niveaux différents et d'avoir obtenu, non pas certes tous les faciès et tous les organismes existant dans le calcaire blanc, mais bien le maximum d'entre eux.

GAUTIER. *Ann. Fac. Sci. Marseille*, 1930, **4**, 253.

2

LA DISPERSION CRISTALLINE

La dispersion cristalline produit en lumière blanche certaines modifications dans les figures d'interférence, notamment en ce qui concerne la disposition des couleurs sur les courbes isochromatiques, qu'il convient d'examiner. En effet, comme à un cristal donné correspond, pour chaque radiation, un ellipsoïde optique distinct, qui peut en outre avoir dans certains cas une orientation différente d'une couleur à l'autre, il s'ensuit nécessairement que pour chaque radiation du spectre, nous aurons également des isogyres et des courbes isochromatiques différentes. En lumière blanche, la figure d'interférence qu'on observe, resulte de la superposition de ces différentes figures colorées, de sorte qu'en un point où une radiation quelconque est détruite soit par interférence soit par l'analyseur il ne saurait y avoir l'obscurité, cette condition ne s'appliquant pas simultanément à toutes les radiations. En conséquence la teinte qui apparaîtra au point considéré est le complémentaire de celle qui correspond à la radiation éliminée.

London B.Sc. (*Special*).

3

Magnitude et Intensité des séismes

Les progrès des études séismologiques permettent aujourd'hui de présenter de la séismicité du globe un tableau plus fidèle en toutes ses parties. En se basant sur les enregistrements obtenus par les séismographes, il est possible d'attribuer à chaque séisme un nombre qui est fonction de l'énergie développée au foyer. Ce chiffre, la magnitude, est un élément constant pour un séisme par opposition à l'intensité qui varie avec la distance à l'épicentre et qui est mesurée en utilisant les degrés d'une échelle empirique.

Gutenberg et Richter ont défini la magnitude comme étant le logarithme de l'amplitude maxima, mesurée en microns, du mouvement inscrit par un séismographe standard à une distance donnée (100 km.) de l'épicentre. L'échelle de magnitude est donc une échelle physique indépendante de l'observateur et du lieu d'observation. Elle peut néanmoins être comparée aux échelles d'intensité employées depuis longtemps dans les études macroséismiques: une secousse de magnitude 3 est généralement ressentie à l'épicentre; celle de magnitude 4,5 peut causer des dégâts légers; les plus grands séismes ont une magnitude qui va de 7 à 8,5. La distance à laquelle les ondes d'un tremblement de terre peuvent s'inscrire sur les séismographes actuellement en service dépend de sa magnitude: les grands séismes (magnitude comprise entre 7 et 8,5) sont inscrits par les observatoires séismologiques du monde entier, les séismes dont la magnitude est comprise entre 6 et 7 sont bien inscrits entre 5.000 et 10.000 km. de distance; ceux de magnitude inférieure à 5,3 ne sont en général pas inscrits au delà de 1.000 kilomètres.

L'énergie E mise en jeu dans un séisme est liée à la magnitude M par une formule empirique simple, valable plus particulièrement pour la Californie:

$$\log E = 11{,}3 + 1{,}8\ M$$

Les séismes les plus faibles qu'on puisse enregistrer ($M = 0$) correspondent à la mise en jeu d'une énergie de 2×10^{11} ergs; les séismes les plus violents ($M = 8{,}5$), à une énergie de 4×10^{26} ergs.

L'introduction de cette nouvelle grandeur, la magnitude, permet donc désormais des comparaisons entre les séismes qui sont ressentis dans les différents pays du monde entier, et permet ainsi de dresser de la séismicité des diverses régions un tableau beaucoup plus exact que celui qu'on pouvait faire jusqu'ici en se basant uniquement sur les renseignments macroséismiques.

ROTHÉ. *Rev. Étude. Calam.*, 1945, **8**, 119.

4

ÉTUDE PÉTROGRAPHIQUE DE LA HAGUE

L'objet principal de ce travail est la description des roches éruptives de la Hague, soit la région comprise entre le Houguet (anse de Vauville) à l'Ouest et la ville de Cherbourg à l'Est. Pour établir l'âge de ces roches il a été nécessaire d'étudier leurs contacts avec les terrains sédimentaires. A l'époque où j'ai commencé cette étude, en 1921, on considérait ces contacts comme métamorphiques. L'étude sur le terrain m'a amenée à la conviction que ce n'était pas exact: les contacts se présentaient tantôt normaux, tantôt mécaniques, sans aucun métamorphisme de contact. Leur interprétation m'a donc obligée, dans quelques cas, à discuter les conditions stratigraphiques et tectoniques du pays, ainsi qu'à étudier au microscope quelques-unes des roches sédimentaires.

On appelle la Hague l'étroite péninsule qui s'étend au Nord Ouest de Cherbourg. Les terrains cambriens et siluriens, dont ce territoire est composé, ont une direction générale W 20° N et forment un petit synclinal, qu'on désigne habituellement sous le nom de synclinal de Jobourg. On trouvera la description détaillée de la structure de ce pli dans de nombreuses publications de M. Bigot. Pour le but que je cherche à atteindre, il importe seulement d'indiquer que ce synclinal n'est qu'un petit détail du vaste synclinal de Siouville, situé plus au Sud (voir la Feuille de la carte géologique au 1 : 80.000 : Les Pieux). Les couches siluriennes et dévoniennes de l'extrémité du flanc sud de ce dernier sont coupées comme à l'emporte-pièce par le granite de Flamanville qui les a en partie digérées et en partie transformées en cornes et autres roches métamorphiques. Ce granite est donc nettement d'âge *post-dévonien.*

Il était beaucoup moins facile d'avoir une opinion sur l'âge des massifs granitiques qui forment à la Hague le socle du flanc nord du synclinal de Jobourg. M. Bigot a établi depuis longtemps l'âge précambrien du granite à amphibole d'Aurigny (îles Anglo-Normandes). Considérant que la continuation du massif d'Aurigny se retrouvait dans le granite à amphibole près d'Herqueville, à la Hague, M. Bigot attribuait à ce dernier également l'âge précambrien. Pour les autres granites de la Hague, en se basant sur quelques vagues (et nullement vérifiées) apparences de métamorphisme de contact avec les couches siluriennes et cambriennes, on admettait jusqu'en 1923 qu'ils sont postérieurs au Silurien et, par analogie avec Flamanville, on les croyait post-dévoniens. J'ai pu démontrer en 1923 que ni les grès, ni les schistes cambriens ou siluriens de la Hague ne sont touchés par le métamorphisme de contact. D'autre part, les granites et les roches filoniennes acides qui les traversent, se rencontrent sous forme de galets dans le conglomérat de la base du Cambrien. Les granites de la Hague sont donc *précambriens.*

En poursuivant mes études j'ai dû reconnaître que les contacts des roches précambriennes avec le Cambrien, que j'avais cru d'abord anormaux, sont souvent au contraire normaux, mais pouvant parfois présenter un léger déplacement des masses en présence. Si on observe sur le terrain les relations des deux roches qui se touchent, on voit parfois que sous le plan de contact, qui peut être plus ou moins redressé, existe une zone qui se différencie de la roche plus profonde; ce qui pouvait laisser croire que cette zone spéciale, sans limites bien définies, était le résultat d'actions mécaniques. C'est particulièrement sous l'influence de M. Maurice Lugeon, qui m'a fait l'honneur de m'accompagner sur le terrain, en examinant attentivement et de nouveau quelques-uns de ces contacts, que j'ai admis qu'on avait affaire bien souvent à la superposition normale du grès cambrien sur une vieille surface pénéplainée et en partie désagrégée et reconsolidée.

JÉRÉMINE. *Bull. Soc. Géol. France*, 1930, **30,** 3.

5

GISEMENT DE TUNGSTÈNE DE MANDELESSE

Les premières fouilles executées dans le gisement de Mandelesse, près de Limoges, remontent à 1917. On reconnut, dès les premiers travaux, la présence de quatre filons quartzeux mineralisés en wolfram. Sur l'un de ces filons, d'allure à peu près verticale, deux puits d'une vingtaine de mètres de profondeur furent creusés, ainsi que, partant d'eux, 440 mètres de galeries. L'épaisseur du quartz variait de 25 à 45 centimètres avec une minéralisation irregulière; l'exploitation de ces galeries a permis d'extraire 320 tonnes de quartz renfermant 30 kgs de wolfram par tonne. Par suite, peu après la fin de la guerre, de la chute brusque du prix du tungstène, qui en 1939 avait atteint 300 francs le kg, les travaux ont été abandonnés.

KOPENHAGUE. *Ann. Mines Carb.*, 1946, **4,** 1262.

6

OBSERVATIONS SUR LE GENRE OSTREA

L'*O. bogharensis*, qu'elle soit considérée au sens de Nicaise ou des géologues actuels, ne diffère du type plissé mésogéen, *O. strictiplicata* que par l'enroulement du crochet, c'est-à-dire que l'*O. bogharensis* est nettement exogyroïde. Mais dans les terrains éocènes des régions du Caire (Basse-Egypte), les individus d'une même localité sont tantôt légèrement opisthogyroïdes, tantôt exogyroïdes francs; et entre ces deux formes il existe des formes de passage dont la progressivité d'un type à l'autre est tout à fait insensible. A mon avis les formes exogyroïdes sont simplement des variations individuelles du type mésogéen (*O. strictiplicata*). On ne doit pas les considérer comme appartenant à une espèce distincte, ni à une variété *bogharensis* Coq. comme l'a fait Thomas.

Les différences établies par Astre entre l'*O. multicostata* Des. et l'*O. moussoulensis* Astre (ou *strictiplicata* R. et D.) sont basées sur le fait que le plateau cardinal surplombe une partie de la cavité de la valve gauche chez la première et qu'il ne la surplomble pas chez la seconde. Or cette constatation n'a rien d'absolu, et ce caractère ne fournit donc pas une bonne diagnose. Car l'*O. multicostata* du Bassin de Paris montre un grand nombre d'individus chez lesquels le plateau cardinal ne surplombe pas la cavité de la valve gauche. Par contre, l'*O. moussoulensis* (ou *strictiplicata*) a souvent des individus chez lesquels le plateau cardinal surplombe une partie de la cavité de la valve. Par exemple, dans la Basse-Egypte, les échantillons présentent une forme générale et une ornementation identiques au type de la Montagne-Noire; seulement parfois la coquille est moins épaisse, de sorte que le crochet est moins puissant et la cavité de la valve gauche prolonge une petite partie sous le plateau cardinal. Ils montrent d'autre part que la disparition de la petite partie de la cavité prolongée sous le plateau cardinal est en relation avec l'augmentation d'épaisseur du test, de manière que cette dernière croît et la partie de cavité prolongée sous le plateau cardinal diminue de plus en plus; finalement il ne reste plus rien quand l'épaisseur du test atteint le degré maximum.

D'autre part, MM. Douvillé et Astre ont constaté que les plis des formes mésogéennes sont plus gros, plus réguliers, moins nombreux et moins dichotomes. A première vue, ces caractères paraissent bien évidents; mais par l'examen précis d'un grand nombre d'échantillons de ces deux types, on voit que la dissemblance d'ornementation n'est pas si vigoureuse; au contraire, il y a passage graduel. Les formes spéciales à plis très gros du type mésogéen différent autant de ses formes normales que celles du Bassin de Paris; par contre, certains individus de cette dernière région présentent de gros plis non seulement aussi vigoureux que ceux d'*O. moussoulensis* mais encore plus forts que ceux des formes jeunes de ce dernier type. En outre, Astre a constaté que le nombre de plis d'*O. moussoulensis* est très variable, de 30 à 60; bien que ceux d'*O. multicostata* soient très nombreux, le nombre ne semble pas être de beaucoup supérieur à cette limite.

Les différences que l'on peut observer se traduisent tout au plus par une coquille plus bombée ou plus globuleuse et un test plus épais chez l'*O. strictiplicata* que chez l'*O. multicostata*. Il serait plus logique de considérer la première comme une variété de la seconde; c'est-à-dire que l'*O. multicostata* du Bassin de Paris s'est adaptée au régime mésogéen, plus chaud, plus calcaire et a occupé des zones moins littorales où l'eau était relativement plus tranquille. De sorte que, à part les individus très littoraux et ceux vivant exceptionnellement dans des conditions défavorables où se trouvent les types du Bassin de Paris, les formes mésogéennes sont beaucoup plus développées et présentent surtout des types gryphoïdes et exogyroïdes.

YANG KIEH. *Bull. Soc. Géol. France*, 1930, **30,** 80.

7

L'ÂGE DE LA TERRE

Après une critique concernant l'incertitude des méthodes employées, sans oublier que l'unité de mesure elle-même, qui est l'année, est une grandeur inconstante et que les processus d'érosions et de dépôts ne peuvent avoir une allure constante durant des milliers d'années, l'auteur constate la concordance frappante dans le résultat des calculs qui assignent un peu plus que 1.700 millions d'années à l'âge de la terre. Les calculs les plus probants concernent le rapport soit entre l'uranium et le plomb, soit entre le rubidium et le strontium dans les minéraux différents. Ce qui est plus remarquable encore, c'est qu'on retrouve un chiffre assez voisin quant à l'âge de l'Univers par des calculs établis selon les théories de Milne et d'Arnolt et selon la formule empirique de Huble pour la vitesse de l'expansion cosmique.

MALFITANO. *J. Physique*, 1939, **10**, 291D.

8

LA GÉOLOGIE DE LA RÉGION MARSEILLAISE

Une des principales difficultés de la géologie provençale et en particulier des terrains qui nous occupent, est la pauvreté ou l'absence complète de fossiles dans certains étages, ou la mauvaise conservation de ceux-ci quand ils se rencontrent au sein d'une roche très dure d'où il est impossible de les dégager. L'étude stratigraphique ne peut se faire, dans ce cas, qu'en suivant les quelques intercalations marneuses fossilifères qui servent de repère ou par des déterminations parfois peu précises, basées sur l'aspect extérieur de la roche.

On conçoit donc l'intérêt que pourrait présenter un procédé de détermination microscopique, le calcaire le plus pauvre en fossiles renfermant presque toujours en effet une faune microscopique pouvant être aisément étudiée en lames minces. De plus outre la faune, la texture, les matériaux constituants et l'état sous lequel ils se présentent sont autant d'éléments de détermination.

Malheureusement cette détermination est beaucoup plus difficile qu'il ne paraît. En effet, nous avons vu que les faciès microscopiques étaient considérablement plus variés que ne le laissait supposer un simple examen à l'œil nu et que dans une couche qui paraît homogène, le microscope révélait de nombreuses modalités. La faune et les matériaux constitutifs ne sont aussi la plupart du temps nullement caractéristiques et parmi les foraminifères, le plus grand nombre: Rotalidées, Textularidées, Miliolidées, sont tout à fait banaux et se rencontrent sans modifications apparentes dans tous les terrains. Seuls certains foraminifères très complexes (Miliolidées trématophorées, Orbitolines) pouvant être déterminés avec précision et caractérisant certaint terrains, seuls certains matériaux très particuliers, peuvens être utilisés.

Deux étages peuvent être déterminés avec précision : ce sont l'Urgonien et le Sénonien.

Urgonien.—L'association algues calcaires-Orbitolines est, dans notre région, caractéristique de l'Urgonien et on est à peu près certain de rencontrer dans une coupe mince au moins un de ces éléments.

Le Sénonien est également caractérisé par la présence des Foraminifères : Orbitolites, Miliolidées trématophorées, Rotalidées à test hyalin surtout. Ces Foraminifères à test blanc sont le plus souvent déterminables à la loupe.

Les autres étages du Crétacé sont moins nettement caractérisés. Les caractères spéciaux à chacun d'eux peuvent être résumés ainsi :

Valanginien.—Calcaires riches en foraminifères (Textularidées surtout) parfois de grande taille et en gastropodes (Nérinées).

Hauterivien.—Calcaires détritiques ou graveleux à ciment cristallin très pauvres en foraminifères.

Aptien.—Calcaires marneux grisâtres très finement granuleux azoogènes ou grès riches en esquilles de test avec Orbitolina conoïdea-discoïdea.

Cénomanien.—A la base : grés rougeâtre à Orbitolina plana pétri de grains de quartz.—Calcaire à Caprines : Calcaire jaunâtre très cristallin donnant des coupes très transparentes : prismes d'Inocérames, fragments d'Orbitolines ; cristaux de quartz.

Turonien.—Grès zoogène à bryozoaires et, surtout, échinodermes (piquants d'oursins ou articles de Crinoïdes) grains de quartz. Ou bien : Calcaire à Durania avec de nombreux petits foraminifères et débris d'Orbitolites.

DEJEAN. *Ann. Fac. Sci. Marseille*, 1930, **4**, 248.

9

Principaux produits minéraux de la Turquie

Les réserves des gisements de charbon de l'Anatolie et des côtes de la Mer Noire ont été évaluées à 1 milliard de tonnes: la mine la plus importante s'étend sur plus de 170 km et fournit un charbon bitumineux à pouvoir calorifique élevé: sa teneur en matières volatiles est d'environ 30-35%. Il y a d'autres gisements qui, jusqu'à présent, ont été relativement peu exploités près des Dardanelles et en Asie Mineure. Aux environs de Kutahya il existe un important gisement de lignite (réserves plusieurs millions de tonnes): le combustible extrait est, entre autres emplois, utilisé pour la production de l'énergie électrique fournie à la Turquie méridionale.

L'existence de minerais de chrome a été décelée dans plus d'une certaine de localités. Les gisements récemment decouverts de la région de Guleman fournissent un minerai de qualité superieure (49-52% de Cr^2O^3). Les réserves seraient très importantes. Les riches gisements de minerai de cuivre de la province d'Ergani qui s'étendent sur 1200 km renferment environ 1,2 million de tonnes de minerai. Parmi les autres richesses minérales il y a lieu de citer l'antimoine, le plomb et le zinc dont il existe pour ces deux derniers métaux d'assez nombreux gisements. Le solfatare de Ketschi-Burlu fournit un soufre dont la qualité est appréciée. La Turquie dispose également de minerai de mercure.

Kopenhague. *Ann. Mines Carb.*, 1946, **4**, 1243.

10

La vie dans les lagunes de salure variable

La vie animale dans ces lagunes de salure variable n'était évidemment possible qu'à des êtres spécialement adaptés: tels devaient être les Gigantostracés et les Placodermes, auxquels nous ne connaissons point d'analogues actuels. Mais pour d'autres Poissons, la comparaison avec leurs congénères vivants est très significative. Ainsi le genre dévonien Dipterus se rattache intimement aux Dipreustes actuels, et par exemple au genre Neuceratodus qui vit dans les lacs temporaires des déserts australiens (ceux mêmes que nous venons d'evoquer) et est capable, grâce à sa respiration pulmonaire, de continuer à vivre envasé pendant les périodes de sécheresse. De même les Holoptychins appartiennent à un groupe de Ganoïdes (les Crossoptérygiens) représenté actuellement par les Polyptères du Tchad. Enfin le Phyllopode Estheria, connu dans les Vieux-grès-rouges, et que nous retrouverons dans les lagunes permotriasiques, est biologiquement voisin des Artemia qui peuplent encore les lagunes des steppes russo-asiatiques.

London B.Sc. (Special).

GÉOGRAPHIE

1

CHAMP DE LAPIÉ

Partout où le sol calcaire n'est pas couvert de vegetation il presente des formes déchiquetées pour lesquelles on a adopté le nom de lapié employé dans le Jura. Un champ de lapié typique a l'aspect d'une surface sillonnée de canaux irreguliers, larges de quelques centimetres, profonds de 1 à 2 mètres, et séparés par des arêtes étroites ou arrondies. Le terrain est comme labouré par ces crevasses, rendant la marche difficile ou même dangéreuse. Leur formation est due surtout a l'œuvre de la décomposition chimique des diaclases, secondée par l'action des plantes, par la décomposition mecanique elle-même, et par le ruissellement qui entraine les produites de décomposition.

London B.Sc. (*Special*).

2

LE PLATEAU DE LANNEMEZAN

Il convient de dire un mot du cône édifié au debouché de la Neste, le plateau de Lannemezan. . . . « Il est formé par une plaine unie couverte de bruyères, d'ajoncs, de fougères, sans aucune habitation. Plus au Nord, quelques vallonnements très doux marquent la source de la plupart des cours d'eau qui descendent en divergeant de cette région. Ils ne tardent pas à se creuser plus profondément et alors apparaissent des villages, des cultures, de bosquets de chênes ou de hêtres. Ces petites vallées offrent des paysages riants, tandis que la lande se retrouve à la surface des plaines élevées qui les séparent ». On voit au flanc des tranchées du chemin de fer des argiles rouges à gros blocs à demi decomposés avec des lits de graviers ou de sable: une couche d'humus acide brune ou noirâtre couronne le tout. Un déboisement imprudent a laissé ces sols meubles sans défense contre l'érosion; là se trouve l'origine des troubles transportés par toutes les rivières qui divergent en éventail du plateau. Les eaux de pluie s'égouttent en un temps tres court, et, tôt après l'ondée les torrents meurent de soif. Pour soutenir leur débit et fertiliser les valles, on a commencé dès 1859 à faire appel aux eaux de la Neste dérivées par un canal. Depuis, pour augmenter la dotation de l'irrigation, le Ministere d'Agriculture avait entrepris l'aménagement des réserves du Neouvieille. L'industrie hydro-électrique y a trouvé son compte.

London B.Sc. (*Special*).

3

L'INDUSTRIE EN SAVOY

C'est de l'agriculture, de la fertilité du sol, de ses paturages et du travail de ses bras que le Savoyard dut tirer ses principaux moyens d'existence,—Mais l'industrie ne chômait pas. Bien au contraire. La transformation des matières premières, soit indigènes, soit venues du dehors, était une nécessité dans un pays qui ne pouvait importer que très peu de produits fabriqués, n'étant riche ni en monnaie pour payer ni en marchandises d'échange. Une multitude d'industries familiales se créa donc sur tous les points du duché; on en conserve aujourd'hui le souvenir et les traces et même, pour beaucoup d'entre elles, la survivance. Le souverain et son gouvernement se montraient attentifs à susciter les entreprises et, une fois crées, à les proteger contre l'âpreté de la concurrence. La rançon de cette protection était le plus souvent le contrôle officiel des prix de revient et des prix de vente. Une marge de bénéfice était laissée au fabricant et le consommateur n'avait pas à redouter qu'aucune spéculation se fit à son detriment.

London B.Sc. (*Special*).

4

LE PLATEAU D'EN-PARIS

Mais, si l'accès du plateau d'En-Paris est aisé à pied, par le sentier qui conduit de Huez au col de Sarenne, il est plus facile encore d'y accéder, en suivant d'abord, en autocar, la Romanche, depuis le Bourg-d'Oisans jusqu'au Fréney et l'embouchure du Ferrand près du barrage du Chambon; de ce barrage un transbordement s'impose pour gagner Mizoën en un petit autocar et traverser cet étroit mais pittoresque village. Ce fut ce chemin que notre caravane de la Société Linnéenne adopta pour atteindre Mizoën, Clavans et Besse. Route longeant les ravins du Ferrand, impressionnante et sauvage, sans parapets et parfois en voie d'éboulement. Terrains schisto- marneux friables du Lias, très peu favorables au passage de l'autocar.

La route de Besse se greffe sur celle de Clavans, traverse le pont du torrent le Ferrand, et se développe en plusieurs tournants. Elle franchit ensuite un petit affluent du Ferrand qui tombe en une jolie cascade sur des schistes lustrés et des micaschistes, laissés à découvert, par l'érosion du Lias. Ce point de bifurcation du pont du Ferrand, qui commande les quartre chemins (Chambon-Mizoën, Clavans, Huez, Sarenne et Besse), aura plus tard une importance touristique de premier ordre, car la route de Clavans communique déjà avec la Maurienne, par un bon sentier muletier, qui suit le vallon synclinal de Clavans, longeant ainsi la base des glaciers des Grandes-Rousses (glaciers du Sablat, des Quirelies, du Sauvage, etc.). Clavans sera un jour le carrefour des quatre plus belles Alpes de l'Oisans pour les sports d'hiver (alpes de Sarenne, d'Huez, du plateau d'En-Paris et de Mont-de-Lans-le-Chambon).

Bonnamour *et al.* *Ann. Soc. Linn. Lyon*, 1935, **79**, 8.

5

Après les avalanches de Goncelin

« J'ai suivi le cours du torrent partout où il a été accessible . . . Dans la partie inférieure, le cours des eaux est borné par des escarpements en roche vive. Elles ne peuvent ni se dévier, ni en corroder les parois.

« Arrivé à mi-hauteur de la montagne, le torrent se divise en une multitude de ramifications; de nouvelles gorges se sont formées en si grand nombre que je n'ai pu les compter.

Toute cette montagne, parée naguère d'une riche verdure, est sillonnée à des profondeurs généralement assez considérables pour faire concevoir de vives inquiétudes sur notre sort à venir.

« Les gorges sont taillées à pic, elles ne peuvent subsister longtemps dans l'état où elles se trouvent, le terrain qui les avoisine est crevassé à une certaine distance, et une pluie assez considérable, le gel et le dégel en hiver occasionneront infailliblement de nouveaux déchirements dont le résultat sera d'amener à la plaine d'immenses matériaux.

« Pour vous donner une idée du nombre de ces gorges, j'en ai compté 41 sur le versant appelé *Maillardat* dans un espace que j'ai jugé ne pas excéder 60 à 70 hectares.

« Du reste, les pertes en récoltes, en forêts et en terres qui ne sont plus susceptibles de culture est immense. La restauration des chemins de communication est presque impossible et l'avenir des malheureux habitants du hameau de *Soulières* tout aussi sinistre que celui des habitants de la plaine. . . .»

GAUTIER. *Rev. Étude. Calam.*, 1945, **8,** 147.

PHYSIOLOGIE

1

La membrane protoplasmique

Si nous nous demandons quelle est la nature de cette membrane protoplasmique la réponse sera aujourd'hui très malaisée; car la théorie d'Overton, règnant jusqu'à ces temps derniers, n'est plus suffisante. Overton considerait la membrane protoplasmique comme une membrane lipoïde; de sorte que seules les substances solubles dans les lipoïdes peuvent la traverser. En réalité, il est demontré que ce n'est nullement le cas. On a donc cherché à modifier cette théorie. J. Traube et Czapek considerent la membrane comme une emulsion des substances grasses cette théorie a trouvé une base experimentale dans les recherches de Clowes qui par l'action des sels de $CaCl_2$ et autres ions polyvalents a pu renverser une emulsion d'huile dans l'eau en emulsion d'eau dans l'huile, c'est-à-dire de rendre cette membrane permeable, soit pour les solutions aqueuses, soit pour les solutions huileuses. D'autres experimentateurs considerent la membrane comme une substance proteique ou même anorganique, non-lipoïde.

London B.Sc. (*Special*).

2

Pourpre rétinien

Un cas tout a fait comparable est celui du déplacement de la sensibilité maxima de la vision quand varient la valeur absolue de l'éclairement et la position de l'image sur la retine. On sait qu'il faut distinguer dans la rétine l'appareil de vision diurne et l'appareil de vision crépusculaire. Ce dernier paraît constitué par le système de bâtonnets impregnés de pourpre rétinien. Ces bâtonnets manquent au centre de la rétine. Ils interviennent pour les faibles éclairements et dans la vision périphérique. Or le pourpre rétinien, d'après

Koenig, a son maximum d'absorption au voisinage de 500 et l'éclairement avec de faibles lumières a pour effet de porter le maximum de visibilité de $\lambda = 550\mu$ à $\lambda = 505\mu$, c'est-à-dire dans la région du spectre où le pourpre absorbe le plus fortement.

London B.Sc. (*Special*).

3

Rôle physiologique des Porphyrines de l'œuf

Quelle peut-être la signification physiologique de ces porphyrines ?

En ce qui concerne l'hématoporphyrine, commune à tous les œufs, où elle est peu abondante, on pourrait admettre qu'elle joue là un rôle analogue à celui de la porphyrine que l'on rencontre dans les parties du squelette en voie d'ossification, dans les piquants du jeune hérisson ou les plumes du jeune pigeon, d'une porphyrine peut-être indispensable à la formation des os, comme à la calcification de la coquille. En réalité, il paraît certain que les porphyrines ne se trouvent dans les concrétions calcaires que si les organismes possèdent de l'hémoglobine dans leurs tissus. On est donc amené aux deux hypothèses suivantes :

a) Ou bien, la porphyrine ne représente qu'un produit d'excrétion et ne joue aucun rôle dans la calcification.

b) Ou bien les organismes, qui peuvent produire du calcaire sans hémoglobine et sans porphyrine, le font par un mécanisme tout différent de celui qu'utilisent dans ce but les animaux supérieurs.

J'ai recherché systématiquement la présence des porphyrines dans un grand nombre d'organismes animaux ou végétaux en voie de calcification. C'est ainsi que j'ai examiné notamment en lumière de Wood :

1° Des rameaux d'algues calcaires du genre Lithothamnium.

2° Le squelette calcaire en voie de formation de la seiche (Sepia).

3° Les piquants de l'oursin (Paracentrotus lividus).

4° La carapace du crabe et de la langouste.

5° La coquille de nombreux mollusques marins gastéropodes et lamellibranches appartenant aux genres: Pectunculus, Tellina, Halyotis, Cardium, Arca, Murex, Natica, Cypræa, Mitra, Trochus, etc.

Dans aucune de ces parties calcaires, choisies intentionnellement chez des sujets dépourvus d'hémoglobine, il n'a été possible de découvrir la moindre trace de porphyrine, ce qui montrerait bien que, dans certains groupes zoologiques tout au moins, elle n'est pas indispensable à la formation des tissus calcifiés.

Si on envisage maintenant l'ooporphyrine vraie, renfermée en quantité relativement grande dans la couche externe des œufs bruns, son rôle apparaît plus réduit encore et limité vraisemblablement à un simple produit d'excrétion. Elle ne peut être en effet nécessaire à la calcification de la coquille, puisque certains oiseaux pondent toute leur vie des œufs parfaitement blancs. Bien mieux, on trouve chez la poule tous les intermédiaires entre l'œuf blanc et l'œuf à coquille brune, et cela parfois dans la même basse-cour. Il ne semble donc pas que la nourriture soit cause de cette excrétion porphyrique.

Peut-être la race et le climat interviennent-ils ? C'est un fait connu que les poules de race asiatique pondent de préférence des œufs teintés. D'autre part, c'est surtout dans l'Ouest de la France, en Vendée ou dans les Deux-Sèvres, que l'on trouve des œufs à coquille très brune, recherchés, paraît-il, des consommateurs en Angleterre.

Mais il est aussi un fait que de nombreuses statistiques faites chez des aviculteurs m'ont permis de constater: c'est que les œufs à coquille foncée paraissent provenir surtout de poules à plumage clair ou même entièrement blanc. Il est rare qu'une poule noire ponde des œufs très pigmentés alors que le fait est fréquent chez les blanches. On pourrait dès lors admettre que le plumage agit par sa plus ou moins grande perméabilité à la lumière et que la sécrétion d'ooporphyrine ne serait qu'une sorte d'état pathologique provoqué par une photosensibilisation plus ou moins intense de la poule.

GOUZON. *Ann. Fac. Sci. Marseille*, 1934, **7**, 131.

4

ACTION DE L'OOPORPHYRINE SUR LES CELLULES SEXUELLES DE L'OURSIN

Les cellules sexuelles de l'oursin (Paracentrotus lividus) constituent un réactif physiologique sensible et d'emploi très commode. Elles ont été utilisées par Loch pour l'étude de l'action photosensibilisatrice de l'éosine. Je me suis adressé à elles pour expérimenter l'influence analogue de l'ooporphyrine sur la fécondation et la segmentation de l'œuf.

On prépare une suspension de cellules mâles et femelles en délayant dans un peu d'eau de mer un fragment des glandes génitales correspondantes (jaunes chez le mâle, orangées chez la femelle). L'ooporphyrine est solubilisée de la manière suivante. On mêle au mortier:

Ooporphyrine cristallisée	0 gr. 01
Co^3NaH	0 gr. 05
Eau de mer	10 cc.

(L'addition de bicarbonate de sodium n'apporte aucune modification appréciable dans la tonicité de l'eau de mer qui demeure inoffensive pour les cellules non sensibilisées.)

La solution obtenue est employée à la dose de 1 partie pour 9 d'eau de mer, ce qui porte la concentration en porphyrine à 1 p. 10.000. On ajoute à ce mélange placé en pleine lumière 1 goutte de la suspension cellulaire, que l'on observe aussitôt au microscope:

a) Les spermatozoïdes, après 5 minutes d'exposition au soleil, perdent leur mobilité et deviennent inaptes à la fécondation;

b) Les ovules placées dans les mêmes conditions se montrent beaucoup plus résistants à l'action de l'ooporphyrine: il faut une heure d'irradiation pour qu'ils ne puissent plus être fécondés par des spermatozoïdes actifs. L'influence photosensibilisatrice a cependant provoqué chez quelques-uns d'entre eux un début de segmentation parthénogénétique, limité à trois ou quatre cellules.

La fécondation artificielle des œufs d'échinodermes conduit au bout de dix heures à des embryons composés d'une cinquantaine de cellules. Avec des soins particuliers, il est possible de les conserver vivants une semaine et même plus.

La même opération a été effectuée à la lumière en présence d'ooporphyrine, dans de l'eau de mer renfermant ce pigment à une concentration de 1 p. 5.000. Si on ajoute au liquide une goutte de suspension des deux cellules sexuelles, la fécondation se produit aussitôt. L'œuf s'entoure bien d'une membrane protectrice, mais la segmentation est rapidement entravée. Au bout de dix heures, l'embryon se compose seulement de 4 à 8 cellules qui sont rapidement détruites par une exposition un peu plus longue à la lumière.

Si les expériences précédemment décrites sont effectuées dans une pièce faiblement éclairée par une lampe à incandescence, la fécondation se produit normalement et donne un embryon viable.

Conclusion.—L'ooporphyrine exerce, en présence de la lumière, une action nocive sur les cellules sexuelles de l'oursin, mais l'ovule résiste beaucoup plus longtemps que le spermatozoïde, tué presque aussitôt.

Elle n'empêche pas la fécondation, mais arrête très rapidement la segmentation de l'œuf et le développement de l'embryon.

GOUZON. *Ann. Fac. Sci. Marseille*, 1934, 7, 199.

PSYCHOLOGIE

1

La parole

La parole exprime, avons nous dit, l'état de conscience du sujet qui parle. Mais c'est trop général. L'état de conscience, à un moment donné, est un ensemble extrêmement complexe de sensations, d'images, de sensations corporelles, d'émotions ou de sentiments, et ce n'est pas cet ensemble complet qui est exprimé. Dans cet ensemble il y a un point central, et c'est à celui-ci que correspond le langage. L'enfant qui crie pour obtenir son biberon n'exprime pas tout ce qu'il ressent; il exprime ce qui l'intéresse, ce qui l'émeut, ce qui attire avant tout son attention. Nous rappellerons que la pensée active tend toujours vers un choix, et c'est ce choix que la parole rend manifeste, en même temps que les éléments musicaux du langage et de la mimique générale rendent manifeste l'émotion concomitante et l'attitude générale de la personne qui parle, devant ce qu'elle exprime.

London B.Sc. (*General*).

2

Colère passive

Existe-t-il une colère passive, comme il y a une joie passive, une tristesse passive et une peur passive ? Nous craindrions, en l'affirmant, de céder à ce besoin de symétrie qui fait, en psychologie comme ailleurs, imaginer les oppositions et les ressemblances aux dépens de la vérité; et cependant il y a des colères froides qui se traduisent psychiquement, non par une agitation centrifuge, mais par une concentration douloureuse de la pensée et de l'affectivité. La psychologie n'en est pas faite.

London B.Sc. (*General*).

3

Sentiment et sensation

Si le sentiment n'est autre chose qu'une sensation, rien n'est plus légitime que de rapprocher le plaisir de la joie et la peine de la tristesse. Si le sentiment n'est pas une sensation, il faut, au contraire, le séparer de l'émotion. Telles sont les deux solutions entre lesquelles il est permis d'opter, mais dans l'état de nos connaissances, bien difficile de choisir. N'oublions pas, au reste, que rapprocher n'est pas identifier. Que le plaisir soit l'ébauche d'une joie, et la peine, l'indication d'une tristesse, c'est là peut-être ce que l'on demontrera un jour. La tristesse et la joie n'en demeurerons pas moins des émotions, c'est-à-dire les réactions très complexes, le plaisir et la peine des réactions très simples et tout au plus des rudiments d'émotion.

London B.Sc. (General).

ANTHROPOLOGIE

1

Traits de caractère des races

On peut dire d'une façon générale que les races américaines, mongoloïdes, etc., sont graves, méditatives, un peu obtuses, mélancoliques; et qu'au contraire, les races nègres ou mélanesiennes sont enjouées, rieuses, vives et superficielles comme les enfants; mais que d'exceptions à cette règle générale ! Et puis, chaque voyageur, chaque observateur juge à sa manière un peuple donné, suivant la nature des rélations (pacifiques ou hostiles, etc.) qu'il a eues avec lui. On ne peut donc rien affirmer, une fois que l'on veut sortir des généralités banales consistant à dire que les « sauvages » manquent de prévoyance et d'idées générales; qu'ils sont cruels; que leurs facultés imitatives sont très développées, etc.; certains traits de caractère sont particulièrement marqués dans certaines races: ainsi la versatilité et l'inconstance des Bochimans sont affirmés par tous ceux qui les ont etudiés et retentit même dans leur langage.

London B.Sc. (General).

2

La danse et les différences sexuelles

Quiconque enseigne les sports, la gymnastique, la danse, sait que les hommes et les femmes s'y comportent d'une manière essentiellement opposée. L'homme préfère les mouvements amples, les grands pas, la projection des membres, le saut et tout ce qui est centrifuge. La femme aime les mouvements resserrés, les petits pas, le repliement des membres, le contour fermé, et tout ce qui est centripète. En remontant dans le développement de l'humanité cette différence croît sensiblement. Plus un peuple est primitive, plus les différences sexuelles sont accentuées. Le type moteur des deux sexes est reflété le plus visiblement par la danse, mais le chant aussi est le miroir de cette différence; l'homme, en chantant préfère les mouvements amples; le chant feminin, au contraire, a généralement l'ambitus petit, les distances moindres, et un retour centripète au point de départ, qui sera plus tard la tonique.

London B.Sc. (General).

ÉTUDES MILITAIRES

1

Constitution et emploi des réserves

L'emploi des réserves est le principal moyen permettant au chef de modifier ou de rétablir l'économie de son système de forces, et d'intervenir personnellement dans la lutte.

Les réserves existent à la compagnie, au bataillon et au régiment.

Les réserves des compagnies engagées ont un rôle très limité et fixé d'avance: dans le combat offensif, elles sont destinées à se fondre dans l'échelon de feu; dans le combat défensif, elles concourent à la défense d'une partie du terrain confié à la compagnie. Lorsque ces réserves ont été dépensées, il est souvent difficile d'en reconstituer d'autres.

Par contre, dans le bataillon et le régiment, les unités en réserve peuvent recevoir des missions variées, et leur importance est telle que le chef a le devoir de les reconstituer le plus tôt possible après leur emploi. Il ne doit cependant pas hésiter, s'il le juge nécessaire, pour rétablir une situation compromise, à dépenser toutes ses réserves, sans souci de leur reconstitution immédiate.

London B.Sc. (*General*).

2

L'infanterie

Protégée et accompagnée par ses propres feux et par les feux de l'artillerie, éventuellement précédée et appuyée par les chars de combat, l'aviation, etc., l'infanterie conquiert le terrain, l'occupe, l'organise et le conserve.

Sa tâche est particulièrement rude, mais glorieuse entre toutes.

Au combat, c'est elle qui dépense le plus; tous les efforts du commandement doivent donc tendre à la ménager, à maintenir ou à exalter son moral et son energie.

L'infanterie agit par le feu et par le mouvement.

Les feux de l'infanterie sont caracterisés par leur précision, par leur diversité et par leur puissance aux courtes et moyennes distances.

Ils peuvent également exercer des actions retardatrices et d'interdiction jusqu'aux distances maxima que permet l'exécution méthodique des tirs.

L'infanterie est pourvue:

—d'armes à tir tendu, en général automatiques et à grand débit, dont les actions rasants sont particulièrement efficaces;

—d'armes à tir courbe, de petit et de moyen calibres, qui lui permettent d'attaquer l'ennemi abrité ou defilé;

—d'armes anti-chars et d'armes contre avions:

—d'armes blanches.

On distingue:

—les armes légères, portées par un seul homme;

—les armes lourdes, décomposées pour leur transport en plusieurs fardeaux ou remorquées.

Dans la défensive, son armement confère à l'infanterie une puissance d'arrêt considérable même sur les fronts étendus. Cette puissance prend toute sa valeur par une organisation du terrain aussi poussée que possible.

London B.Sc. (*General*).

3

LE LOGEMENT ET LE CANTONNEMENT DES TROUPES

Pour compléter l'étude que nous avons publiée sur cette question dans notre numéro de janvier (pages 8 et s.), nous croyons utile de reproduire les renseignements suivants:

M. Callet, député, a demandé au ministre de la Défense nationale et de la Guerre:

1° quelles sont les obligations des propriétaires vis-à-vis des officiers et hommes de troupe logés chez eux en ce qui concerne:

a) la question couchage, draps, couvertures, chauffage;

b) la lumière, bureaux, garages et chambre à coucher;

c) quelles sont les formalités à accomplir pour percevoir l'indemnité, le cas échéant, si celle-ci est due.

Voici la réponse du ministre (« J. O. » *Débats Parlementaires*, Ch. des députés N° 15, 17 février 1940, page 210):

1° *Obligation des propriétaires en ce qui concerne le couchage.*—Pour avoir droit aux indemnités de logement prévues à l'article 33 du décret du 2 août 1877 dont le taux a été fixé en dernier lieu par le décret du 26 mai 1939, l'habitant doit fournir au moins, par lit d'officier:

a) un lit pourvu d'un sommier, d'un matelas, d'un traversin ou oreiller, de draps, d'une ou plusieurs couvertures. En outre, l'officier doit disposer d'un cube d'air minimum de 12 mètres cubes. Si ces conditions ne sont pas remplies, l'indemnité à payer par lit sera celle prévue pour un homme de troupe. (Circulaire du 8 décembre 1916);

b) les sous-officiers doivent être pourvus d'un lit et les soldats d'un lit ou au moins d'un matelas et d'une couverture pour deux. (Art. 23 du décret du 2 août 1877.)

2° *Obligation des propriétaires en ce qui concerne la lumière, les bureaux, garages, etc.*—Les militaires en billet de logement ont droit au feu et à la chandelle. Sur ce dernier point, il ne peut s'agir que de l'usage de l'éclairage dans les conditions normales dans l'armée. Pour les locaux qui ne sont pas à usage exclusif d'habitation et pour lesquels un décret ne prévoit pas d'allocation forfaitaire (bureaux, hangars, etc.), une instruction du 3 mars 1915 prévoit un taux d'indemnité de location dont le rajustement est à l'étude. Dans ces locaux, les frais de chauffage et d'éclairage restent à la charge de l'Etat.

3° *Formalités à accomplir pour percevoir l'indemnité de logement.*—Chaque maire doit adresser en double expédition à l'intendant militaire, des états récapitulatifs appuyés par des états d'effectifs qui lui ont été remis par l'officier commandant la troupe logée sur billet de logement. (Art. 30 à 32 du décret du 2 août 1877.)

Vie Comm., 1940, **17**, 88.

4

L'ORGANISATION GÉNÉRALE DE LA CIRCULATION EN TEMPS DE GUERRE

Garantir le secret des opérations militaires, réprimer l'espionnage, tels sont les principes dominants de la réglementation de la circulation en temps de guerre.

Les mesures pratiques qui découlent de ces principes sont doubles :

1° contrôler d'une manière permanente les personnes qui se déplacent sur les diverses parties du sol français ;

2° restreindre la circulation et le séjour, notamment dans la zone des Armées, tout en tenant compte des nécessités imposées, soit par la continuation normale de la vie économique, soit par des considérations familiales.

L'étude de ces mesures nous conduira à examiner successivement :

1° les zones de circulation ;

2° les modes de déplacement ;

3° les titres de circulation.

A.—*Les zones de circulation*

Au point de vue de la circulation, le territoire français métropolitain (2) est divisé en *deux zones principales*, elles-mêmes subdivisées en *zones secondaires*, savoir :

1° *Zone des armées* comprenant :
1. une *zone avancée ;*
2. une ou plusieurs *zones réservées ;*
3. une *zone non réservée.*

2° *Zone de l'intérieur* comprenant :
1. une ou plusieurs *zones spéciales* pouvant être :
   a) non soumises à l'état de siège ;
   b) soumises à l'état de siège ;
2. la *zone de l'intérieur proprement dite.*

La délimitation de ces diverses zones est fixée par arrêtés ministériels, sauf la zone avancée et la (ou les) zone réservée des armées qui sont fixées par le général commandant en chef sur chaque théâtre d'opérations. Les modifications sont portées à la connaissance des maires par les préfets.

B.—*Les modes de déplacement*

En matière de circulation, la réglementation distingue nécessairement les modes de déplacement. Le contrôle d'un voyageur en motocyclette est, en effet, beaucoup plus difficile que celui de la personne utilisant un moyen de transport public ou circulant à pied ou en voiture hippomobile.

D'autre part, certaines personnes ont besoin de circuler d'une manière continuelle dans une région déterminée pour les nécessités de leur commerce ou de leur métier, alors que d'autres n'ont besoin de faire, pour une raison de famille ou d'intérêt, qu'un simple déplacement plus ou moins éloigné et pour une durée limitée.

Pour ces diverses raisons, les déplacements peuvent être classés en deux groupes principaux, savoir:

1° dans un *premier groupe*, on peut comprendre ceux effectués:

a) soit à pied, à cheval, à bicyclette, en voiture hippomobile ou en chemin de fer;

b) soit en utilisant un véhicule automobile ou un bateau affectés à un transport public;

c) soit, enfin, en aéronef, à titre de passager;

2° rentrent dans le *second groupe*, les déplacements effectués en motocyclette, en automobile ou en bateau.

C.—*Les titres de circulation*

Pour tenir compte des principes qui viennent d'être exposés, l'Administration a institué une certain nombre de titres pour la circulation des civils dans les zones réglementées, parmi lesquels nous distinguerons les plus usuels, savoir:

1° le *sauf-conduit* destiné aux personnes devant faire un voyage limité dans le temps, dans un lieu déterminé plus ou moins éloigné de leur résidence;

2° la *carte de circulation temporaire* réservée aux personnes appelées à circuler d'une manière constante pour les besoins de leur commerce ou de leur profession dans une région limitée;

3° le *permis M.A.B.* permettant d'utiliser pour un déplacement, quel qu'il soit, dans une zone déterminée, un voiture automobile, une motocyclette ou un bateau.

Vie Comm., 1940, **17**, 61.

VOCABULARY

ABBREVIATIONS

a., active (*i.e.* transitive).	esp., especially.	part. { participle. participial.
adj., adjective.	f., feminine.	phys., physics.
adv., adverb.	geol., geology.	prep., preposition.
art., article.	indef., indefinite.	pron., pronoun.
bot., botany.	inf., infinitive.	r., reflexive.
chem., chemistry.	lit., literally.	s., substantive.
conj., conjunction.	m., masculine.	sing., singular.
cpd., compound.	math., mathematics.	v., verb.
def., definite.	n., neuter (*i.e.* intrans.).	zoöl., zoology.
Eng., English.	p., plural.	

Prepositions.—The commonest uses are briefly given. It is impossible to deal with them exhaustively in a Vocabulary. The sense can generally be deduced from the context.

Verbs.—Attention is drawn to the article on **se**. Reflexive uses of verbs are not given individually unless there seems special reason to do so.

Proper Names.—These are given in a separate list at the end.

A

A (à), prep., *at, in, on, to, with,* etc.; — ce que, *by the fact that.* Forms the equivalent of Eng. cpd. s., *e.g.* pierre à bâtir, *building-stone*, canne à sucre, *sugar-cane;* and of Eng. cpd. adj., *e.g.* à raies vertes, *green-striped.*

abaissement, s. m., *lowering.*

abaisser, v. a., *lower;* (math.) let . . . *fall.* S'—, *sink; shrink.*

abandonner, v. a., *abandon; give up; lose.* — à l'air, *leave in the air.* Je l'abandonne à elle-même, *I let go.*

abdomen, s. m., *abdomen.*

abdominal, e, adj., *abdominal.*

abeille, s. f., *bee.*

abondance, s. f., *abundance.*

abondant, e, adj., *abundant, rich, copious.*

abord, s. m. Au premier —, *at first.* D'—, tout d'—, *first; first of all; at first.*

aborder, v. n., *arrive, reach; commence.*

aboutir, v. n., *end.*

abréviation, s. f., *abbreviation.*

abri, s. m., *shelter.* A l'—, *sheltered.* A l'— de, *protected from.*

abscisse, s. f., *abscissa, x co-ordinate.*

absence, s. f., *absence.*

absolu, adj., *absolute.*

absolument, adv., *absolutely.*

absorber, v. a., *absorb.*

absurde, adj., *absurd.* Preuve —, *proof by reductio ad absurdum.*

accélération, s. f., *acceleration.*

accélérer, v. a., *accelerate.*

acception, s. f., *meaning.* Dans toute l'— du mot, *in the full sense of the word.*

accessoire, adj., *accessory, subsidiary.*

accidentel, le, adj., *accidental.*
accoler, v. a., *to fasten together.*
accommodation, s. f., *adjustment.*
accompagner, v. a., *accompany.*
accomplir, v. a., *accomplish, carry out.*
accord, s. m., *agreement.* Être d'—, *agree, be agreed.*
accorder, v. a. n. — sur, *attune to.*
accroissement, s. m., *increase.*
accroître, v. a. n. S'—, v. r., *increase.*
accueil, s. m., *welcome.*
accumuler, v. a., *accumulate.*
accuser, v. a., *reveal, acknowledge.*
achever, v. a., *finish; complete.*
acide, s. m., *acid.*
aciduler, v. a., *acidulate.*
acier, s. m., *steel.*
acoustique, s. t., *acoustics.*
acquérir, v. a., *acquire.*
acti-f, -ve, adj., *active.*
action, s. f., *action; influence.*
activité, s. f., *activity.*
actuel, le, adj., *present; present-day; extant.*
actuellement, adv., *at present.*

adaptation, s. f., *adaptation.*
addition, s. f., *addition.*
additionner, v. a., *add up.* Additionné de, *to which . . . has (have) been added.*
adhérer, v. n., *adhere.*
adipeux, adj., *adipose.*
adjacent, e, adj., *adjacent; contiguous.*
admettre, v. a., *admit; admit of; acknowledge; let in.*
adopter, v. a., *adopt.*
adorer, v. a., *worship.*
adulte, adj., *full-grown.*
adventi-f, -ve, adj., *adventitious.*

aéré, e, part. adj., *airy.*
aérien, ne, adj., *aerial.*

affaire, s. f. Avoir — à, *have to deal with.*
affaissement, s. m., *subsidence.*
affecter, v. a. Affecté de, *subjected to.*
affirmer, v. a. n., *assert.*
afin, conj. — de (+ inf.), *in order to.* — que, *in order that.*

âge, s. m., *age.* Moyen âge, *Middle Ages.*
agent, s. m., *agent.*
agglomération, s. f., *agglomeration.*
agilité, s. f., *nimbleness.*
agir, v. n., *act.* Il s'agit de . . . , *it is a matter (question) of . . .*
agitation, s. f., *agitation; activity; shaking.*
agiter, v. a., *shake.*
agrandir, v. a., *enlarge.*
agréable, adj., *pleasant.*
agriculteur, s. m., *tiller of the soil; farmer.*

aide, s. f., *help.* A l'— de, *with the help of.*
aigle, s. m., *eagle.*
aigu, ë, adj., *acute.*
aiguille, s. f., *needle, spike.*
aiguillon, s. m., *sting.*
aile, s. f., *wing.*
ailé, e, adj., *winged.*
ailleurs, adv. D'—, *besides; in other respects.*
aimant, s. m., *magnet.*
aimantation, s. f., *magnetisation.*
aimanté, e, part. adj., *magnetic.*
aimanter, v. a., *magnetise.* S'—, v. r., *be magnetised.*
aimer, v. a., *love; like.* — beaucoup, *be very fond of.*
ainsi, adv. conj., *so, thus.* Et — de suite, *and so on.* — que, *as; like; as well as.*
air, s. m., *air.* Avoir l'— de, *look as if, seem to.*
aire, s. f., *area.*
aisé, e, adj., *easy.*
aisément, adv., *easily, readily.*

ajonc, s. m., *whin, furze.*
ajouter, v. a., *add.*

albumen, s. m. (bot.), *albumen.*
albumine, s. f. (chem.), *albumen.*
albuminoïde, adj., *albuminoid.*
alcali, s. m., *alkali.*
alcalin, e, adj., *alkaline.*
alchimie, s. f., *alchemy.*
alchimiste, s. m., *alchemist.*
alcool, s. m., *alcohol.* — éthylique, *ethyl alcohol.*
alcoolate, s. m., *alcoholate.*
alcoolique, adj., *alcoholic.*
alcoomètre, s. m., *alcoholometer.*
algébrique, adj., *algebraical.*
algues, s. f. pl., *algae.*
aligner, v. a. Être aligné(s), *stand in a row, in rows.*

aliment, s. m., *food; sustenance.*
alimentaire, adj., *alimentary.* Matières —s, *foodstuffs.*
alimentation, s. f., *feeding; food.*
alléger, v. a., *lighten.*
allemand, e, adj. s., *German.*
aller, v. n., *go; be going.* Il va de soi que . . ., *it is a matter of course that . . .*
alliage, s. m., *alloy.*
alligator, s. m., *alligator.*
allométrie, s. f., *allometry.*
allongé, e, part. adj., *elongated; long.*
allonger, v. a. S'—, *lengthen; expand; taper.* — en pointe, *taper into a point.*
allumer, v. a., *light.*
allumette, s. f., *match.*
allure, s. f., *pace; ways.*
alors, adv., *then.*
alpin, e, adj., *Alpine.*
altération, s. f., *change* (esp. for the worse).
alterner, v. a. n., *alternate.*
altitude, s. f., *altitude.*
aluminium, s. m., *aluminium.*
alun, s. m., *alum.*
alvéole, s. m. (zoöl.), *socket.*

amadou, s. m., *tinder.*
amande, s. f., *almond; kernel.*
amandier, s. m., *almond-tree.*
amas, s. m., *heap; pile.*
amateur, s. m., *lover* (may need expansion, e.g. "*flower-lover*").
ambiant, e, adj., *surrounding.*
ambiguité, s. f., *ambiguity; uncertainty.*
ambitus, s. m., *orbit, sphere of influence.*
ambulacre, s. m., *ambulacrum.*
améliorer, v. a., *improve.*
amendement, s. m., *improvement; manuring.*
amener, v. a., *bring; bring about; lead.*
am-er, -ère, s. f., *bitter.*
amertume, s. f., *bitterness.*
ami, s. m., *friend.*
amibe, s. f., *amoeba.*
amidon, s. m., *starch.*
ammonia-c, -que, adj. Gaz —, *ammonia.*
ammoniacal, e, adj., *ammoniated.*
ammoniaque, s. f., *ammonia.*
amoindrir, v. a. m., *lessen.*
amorcer, v. a., *to bait; decoy; feed.*
amorphe, adj., *amorphous.*
amphibie, s. m., *amphibian.*
amphibole, s. f., *horneblende.*
amphotère, adj., *amphoteric.*
amplitude, s. f., *amplitude; swing.*
amylacé, e, adj., *amylaceous, starchy.*

analogue, adj., *analogous.*
analyse, s. f., *analysis.*
analyser, v. a., *analyse.*
ancien, ne, adj., *ancient; former.* L'— monde, *the Old World.* Les —, s. m. pl., *the ancients.*
anciennement, adv., *formerly.*
ancienneté, s. f., *antiquity.*
androcée, s. f., *androecium.*
âne, s. m., *ass.*
anémone, s. f., *anemone.*
anglais, e, adj., *English.*
angle, s. m., *angle.* Prendre l'—, *read off the angle.*
anhydre, adj., *anhydrous.*
anhydride, s. m., *anhydride.*
animal, e, adj., *animal.*
animal, s. m., *animal.*
animalcule, s. m., *animalcule.*
animé, e, part. adj., *animate.*
anneau, s. m., *ring; annular joint.*
année, s. f., *year.*
annelé, e, adj., *annulated.*
annélides, s. m. pl., *annelids.*
annuel, le, adj., *annual.*
annuler, v. r. S'—, *vanish, become zero.*
anse, s. f., *handle; lobe.*
antédiluvien, ne, adj., *antediluvian.*
antenne, s. f., *antenna; feeler; aerial.*
antérieur, e, adj., *anterior; front.* Très — à, *long previous to.*
anthère, s. f., *anther, tip.*
anthéridie, s. m., *antheridium.*
anthérozoïde, s. m., *antherozoid.*
anthèse, s. f., *anthesis; full bloom.*
anthracite, s. m., *anthracite.*
antimoine, s. m., *antimony.*
antipode, adj., *antipodal.*

antirefléchissant, adj., *low-reflecting.*
anus, s. m., *anus.*

aorte, s. f., *aorta.*
août, s. m., *August.*

apercevoir, v. a., *perceive; notice.* S'— de, *perceive.*
aplatir, v. a., *flatten.*
aplatissement, s. m., *flattening.*
aplomb, s. m., *perpendicular.* D'—, *perpendicular(ly).*
apparaître, v. n., *appear; make one's appearance.*
appareil, s. m., *apparatus.* — photographique, *camera.*
apparence, s. f., *appearance.*
apparent, e, adj., *apparent.*
apparition, s. f., *appearance.*
appartenir, v. n., *belong.*
appât, s. m., *bait.*
appauvrir, v. a., *impoverish.*
appeler, v. a., *call.* S'—, *be named.*
appendice, s. m., *appendage.*
applicable, adj., *applicable.*
application, s. f., *application; purpose.* Point d'— de la pesanteur, *point where the weight acts.*
appliquer, v. a., *apply; lay; set.* S'—, *lie; apply.*
apporter, v. a., *bring.*
apprécier, v. a., *estimate.*
apprivoiser, v. a., *tame; make sociable.*
approché, e, part. adj., *approximate.*
approcher, v. a. n., *bring near; draw near; get nearer.*
approprié, e, part. adj., *suitable.*
approprier, v. a. S'—, *appropriate (to oneself).*
approximati-f, -ve, adj., *approximate; rough.*
approximation, s. f., *approximation.*
approximativement, adv., *approximately; roughly.*
appui, s. m. Point d'—, *fulcrum.*
appuyer, v. a. n. — . . . à, *lean . . . against.* S'— sur ce que, *rely on the fact that.*
après, prep., *after.* With perfect inf., *after . . . -ing.* D'—, *according to.*
âpreté, s. f., *roughness.*
apte (à), adj., *fit.*

aquatique, adj., *aquatic; water-.*

arabe, adj. s. m. f., *Arab; Arabian.*
arachide, s. f., *groundnut.*
arachnide, s. m., *arachnid.*
arbitraire, adj., *arbitrary.*
arbre, s. m., *tree.* — de Diane, *arbor Dianae; arborescent silver.*
arc, s. m., *arc.*
archégone, s. m., *archegonium.*
archet, s. m., *fiddlestick.*
archipel, s. m., *archipelago.*
ardoise, s. f., *slate.*
arête, s. f., *edge.*
argent, s. m., *silver.*
argile, s. f., *clay.*
argileu-x, -se, adj., *clayey.*
argon, s. m., *argon.*
arithmétique, s. f., *arithmetic.*
armature, s. f., *armature; framework; coating.*
armer, v. a. n., *equip.*
arracher, v. a., *tear away; pluck.*
arranger, v. a., *arrange.* S'—, *contrive.*
arrêt, s. m., *stop (-ping, -page).* Point d'—, *stationary point.* — de développement, *arrested development.*
arrêter, v. a., *stop; bring to a stop.*
arrière, s. m., *back (part).* En —, *backward.* En — de, *behind.*
arriéré, e, adj., *backward.*
arrivée, s. f., *arrival.*
arriver, v. n., *arrive; come; happen; succeed.* — dans, *reach.*
arrondi, e, part. adj., *rounded.*
art, s. m., *art.* Objets d'— imitant le bronze, *imitation bronze art wares.*
artère, s. f., *artery.*
artériel, le, adj., *arterial.*
arthropode, s. m., *arthropod.*
article, s. m., *article;* (zoöl.) *joint.*
articulé, e, adj., *articulated, jointed;* s. m., *articulate.*
artificiellement, adv., *artificially.*
artiozoaire, s. m., *artiozoön* (pl. *-zoa*).

ascension, s. f., *rise.*
aspect, s. m., *appearance.*
aspiration, s. f., *inhalation; suction.*
assécher, v. a. S'—, *dry up.*
assembler, v. a., *put together; join.*
assertion, s. f., *assertion.*
assez, adv., *enough; fairly; rather.*
assiette, s. f., *plate.*
assimilable, adj., *assimilable.*
assimiler, v. a., *assimilate; compare.*
association, s. f., *association.*
assujettir, v. a., *compel.*
assurer, v. a., *ensure.*
astérie, s. f., *starfish.*
astronome, s. m., *astronomer.*
astronomie, s. f., *astronomy.*

atelier, s. m., *workshop, laboratory.*
Atlantique, adj., *Atlantic.*
atmosphère, s. f., *atmosphere.*
atmosphérique, adj., *atmospheric.*
attacher, v. a., *attach; fasten.*
attaquer, v. a., *attack; strike.*
atteindre, v. a., *attain; reach; affect.*
atteler, v. a., *harness.*
attendre, v. a. n., *wait; wait for.* — que, *wait for . . . to . . . , wait until.*
attente, s. f., *waiting.*
attention, s. f., *attention.*
attirer, v. a., *attract.* S'—, *attract each other.*
attouchement, s. m., *touch; contact.*
attracti-f, -ve, adj., *attractive.*
attraction, s. f., *attraction.*
attraper, v. a., *catch.*
attribuer, v. a., *attribute.*

aubier, s. m., *sapwood.*
aucun, e, adj., *any* (+ negative, etc.); *not . . . any, no* (+ ne).
augmentation, s. f., *increase.*
augmenter, v. a. n. S'—, v. r., *augment; increase* (de, *by, in*).
aujourd'hui, adv., *today; nowadays.*
auparavant, adv., *before.*
auprès de, prep., *in comparison with.*
auprès, adv., *close by.*
aussi, adv., conj., *also* (initially, + inversion) *so; accordingly.*
aussitôt, adv., *immediately.*
austral, e, adj., *southern; south.*
autant, adv., *as much; as many.* D'— plus, *all the more.*
automne, s. m., *autumn.*
autoriser, v. a., *authorise.* Rien n'autorise à penser, *there is nothing* to *warrant the belief.*
autour de, prep., *round.*
autre, adj., pron., *other; different.*
autrefois, adv., *formerly.*
autrement, adv., *otherwise.*
autruche, s. f., *ostrich.*
aux, contraction of à les (art.).
auxiliaire, adj., *auxiliary.*
avaler, v. a. n., *swallow.*
avance, s. f. D'—, *beforehand.*
avancé, e, part. adj., *advanced.*
avancer, v. a. n., *advance; go forward; put forward.* — . . . de deux rangs, *move . . . forward two places.*
avant, prep., *before.* — de (+ inf.), *before . . . -ing.* En —, *forward; ahead.*
avec, prep., *with,* etc.
avenir, s. m., *future.*
avertisseur, s. m., *informer.* Sonnerie —, *alarm-bell.*
avide, adj., *greedy; eager.* — de (chem.), *having an affinity for.*
avis, s. m., *opinion; notice.*
avoine, s. f., *oats.*
avoir, v. a., *have.* Il y a, *there is, there are.* Il a le corps couvert de . . . , *his body is covered with . . .*
avouer, v. a. n., *confess.*
axe, s. m., *axis.*
aye-aye, s. m. (zoöl.), *aye-aye.*
azotate, s. m., *nitrate.*
azote, s. m., *nitrogen; azote.*
azotique, adj., *nitric.*

B

bacille, s. m., *bacillus.*
baie, s. f., *bay.*
baigner, v. a., *bathe.*
baignoire, s. f., *bath, bath-tub.*
bain-marie, s. m. (chem.), *water-bath.*
baisse, s. f., *lowering (in the level of); fall.*

balayage, s. m., *sweeping.*
baleine, s. f., *whale.*
ballon de Claisen, s. m., *Claisen flask.*
banal, adj., *commonplace.*
banc, s. m., *bench; bank.*
bande, s. f., *strip.*
baquette, s. f., *rod.*
barbe, s. f., *beard.*
barrage, s. m., *dam.*
barre, s. f., *bar.*
barreau, s. m., *bar.*
baryte, s. f., *baryta (protoxide of barium).*
baryum, s. m., *barium.*
bas, se, adj., *low.* Adv., plus —, *lower down.*
bas, s. m. En —, *down; downward.*
base, s. f., *base; basis.*
basophile, adj., *basophilic.*
bassin, s. m., *basin.*
bateau, s. m., *boat.* — pêcheur, *fishing-smack.*
bâtir, v. a., *build.*
bâton, s. m., *stick; rod.*
battement, s. m., *flap (-ping).*
battre, v. a. n., *beat; thrash.*

beau, bel, belle, adj., *beautiful; fine.*
beaucoup, adv., s. m., *much, many, a great deal* (+ de if s. follows). De —, *far; by far.*
besoin, s. m., *need.* Au —, *if need be.*
betterave, s. f., *beet, beetroot.* Sucre de —, *beet-sugar.*
beurre, s. m., *butter.*

biberon, s. m., *excessive drinker; child's feeding-bottle.*
biconvexe, adj., *biconvex.*
bien, adv., *well; quite; indeed; very.* — des, *many.* — que, conj., *although.*
bientôt, adv., *soon; presently.*
bifurquer, v. n. Se —, v. r., *bifurcate.*
bilan, s. m., *balance sheet; list; schedule.*
bilatéral, e, adj., *bilateral.*
bile, s. f., *bile.*
binôme, s. m., adj., *binomial.*
bioxyde, s. m., *dioxide.*
bismuth, s. m., *bismuth.*
bitume, s. m., *bitumen; asphalt.*
bitumineu-x, -se, adj., *bituminous.*
bivalve, adj., s. m., *bivalve.*

blan-c, -che, adj., *white.*
blanc, s. m., *white.*
blé, s. m., *corn; wheat.*
blesser, v. a., *wound; injure.*
bleu, e, adj., *blue.*
bleuâtre, adj., *bluish.*
bloc, s. m., *block; lump.*

bœuf, s. m., *ox.*
bois, s. m., *wood.*
boisson, s. f., *drink; beverage.*
boîte, s. f., *box.*
bon, ne, adj., *good.*
bord, s. m., *border; edge; bank; shore.*
border, v. a., *border.*
boréal, e, adj., *northern.*
borique, adj., *boric.*
borner, v. a. Se —, *confine oneself.*
bosquet, s. m., *grove; thicket.*
botanique, s. f. adj., *botany; botanical.*
botaniste, s. m., *botantis.*
bouche, s. f., *mouth; orifice.* —volcanique, *crater.*
bouchon, s. m., *stopper.*
bouclier, s. m., *shield.*
boue, s. f., *mud.*
boueu-x, -se, adj., *muddy.*
bougie, s. f., *candle.*
bouillant, e, part. adj., *boiling.*
bouillir, v. a. n., *boil.* Faire —, *boil.*
bouillonnement, s. m., *bubbling.*
boule, s. f., *ball.* — de neige, *snowball.*
bourgeonnement, s. m., *budding.*
boussole, s. f., *compass.* — de déclinaison, *box compass.* — d'inclinaison, *dip compass.*
bout, s. m., *end.* Au — de, *after.* Au — de quelque temps, *after a time.*
bouteille, s. f., *bottle.*
bouton, s. m., *button; stud; pimple; bud on a tree.*
boutonnière, s. f., *buttonhole.*

brachine, s. m., *brachynus* (species of coleoptera).
branche, s. f., *branch.*
branchial, e, adj. (zoöl.), *branchial.*
branchies, s. f. pl., *branchiae; gills.*
bras, s. m., *arm.* — de mer, *arm of the sea.*
braver, v. a., *face.*
brièvement, adv., *briefly.*
brillant, e, adj., *brilliant; bright; sparkling.*
briller, v. n., *shine.*
brin, s. m., *stalk, spray, sprig.*

briser, v. a. n., *break.*
britannique, adj., *British.*
brochet, s. m. (zoöl.), *pike.*
bronze, s. m., *bronze.*
broyer, v. a., *crush; pound.*
broyeu-r, -se, s. m. as adj., *mandibulate.*
bruit, s. m., *noise; rumour.*
brûler, v. a. n., *burn.*
brume, s. f., *fog.*
brun, e, adj., *brown.*
brunâtre, adj., *brownish.*
brusque, adj., *abrupt; sudden.*
brusquement, adv., *abruptly.*
brut, e, adj., *raw; crude.* Diamant —, *rough diamond.*
bruyère, s. f., *heather.*

buccal, adj., *buccal.*
bûche, s. f., *log.*
buée, s. f., *mist.*
buffle, s. m., *buffalo.*
buisson, s. m., *bush.*
bulbe, s. m., *bulb.* Plante à —s, *bulbous plant.*
but, s. m., *aim; goal.*
butyrique, adj., *butyric.*

C

cabane, s. f., *hut.*
cadre, s. m., *frame.*
cadu-c, -que, adj. *See* Feuille.
caillebotter, *clot; curdle.*
cailler, v. a. Se —, v. n., *clot, curdle.*
caisse, s. f. — de résonance, *resonance-box.*
calcaire, adj., *calcareous.*
calcaire, s. m., *limestone.*
calcium, s. m., *calcium.*
calcul, s. m., *calculation.*
calculer, v. a. n., *calculate.*
calibre, s. m., *calibre, diameter.*
calice, s. m., *calyx.*
calorie, s. f., *calorie.*
calorifique, adj., *calorific.* Effets —s, *heating effects.*
calorimétrie, s. f., *measurement of heat.*
calotte, s. f., *cap.*
camarade, s. m., *comrade.*
cambium, s. m., *cambium.*
campagne, s. f., *country (-side).*
canal, s. m., *canal; channel.*
canalicule, s. m., *canaliculus.*
canine, s. f., *canine.*
canne, s. f., *walking-stick.* — à sucre, *sugar-cane.*
cannelure, s. f., *channelling; groove.*
cap, s. m., *cape.*
capacité, s. f., *capacity, capability.*
capillaire, adj., s. m., *capillary.*
capital, s. m., *capital.*
capitale, s. f., *capital (city).*
capitaliser, v. a., *capitalise.*
capuchon, s. m., *hood.*
capucine, s. f., *nasturtium.*
car, conj., *for.*
caractère, s. m., *character; characteristic.*
caractériser, v. a., *characterise.*
caractéristique, adj., *characteristic.*
carapace, s. f., *shell of turtle or tortoise.*
carbonate, s. m., *carbonate.*
carbone, s. m., *carbon.*
carbonifère, adj., as s. m., *carboniferous.*
carbonique, adj., *carbonic.*
carboniser, v. a., *carbonise; char.*
carbure, s. m., *hydrocarbon.*
carence, s. f., *deficiency, lack of assets.*
carnivore, adj., *carnivorous.*
carotte, s. f., *carrot.*
carpelle, s. f., *carpel.*
carré, e, adj., *square.*
carré, s. m., *square.*
carrefour, s. m., *crossroads; junction; cross.*
carte, s. f., *map.*
cartilagineu-x, -se, adj., *cartilaginous; gristly.*
carton, s. m., *cardboard.*
cas, s. m., *case.* C'est le — de, *this is the case with.* Ce qui est le — de, *as is the case with.*
caséine, s. f., *casein.*
cassant, e, adj., *brittle.*
casser, v. a. n., *break.*
cassure, s. f., *fracture; crack.*
catégorie, s. f., *category.*
cause, s. f., *cause; reason.* A — de, *because of.* A — même de, *by very reason of.*
causer, v. a., *cause; produce.*
caustique, adj., *caustic.*
cave, adj., *hollow.* S. f., *cellar.*
caveau, s. m., *cellar.*
caverne, s. f., *cave.*

ce, cet, m., **cette,** f. adj., *this; that.* (Often with -ci, -là, attached to noun.)

ce, c', pron. (See Grammar for uses with être.) — qui, — que, *that which, what.* En — que, *in that.*
ceci, pron., *this.*
céder, v. a. n., *yield.* — la place à, *give place to.*
ceinture, s. f., *girdle; belt.*
cela, pron., *that.* Par — même, *for that very reason.* Pour — même, *for this reason, purpose.*
célèbre, adj., *celebrated.*
celle, f. of celui.
cellulaire, adj., *cellular.*
celui, pron. m. (+ de or relative clause complement), *that,* etc. —-ci, *the latter.* — là, *the former; that one.*
cément, s. m., *cement* (of teeth).
cendre, s. f., *ash.*
cent, adj., *hundred.*
centaine, s. f., (*about a*) *hundred.*
centésimal, adj., *centesimal.*
centième, adj., s., *hundredth.*
centigramme, s. m., *centigramme.*
centilitre, s. m., *centilitre.*
centimètre, s. m., *centimetre.*
central, e, adj., *central.*
centre, s. m., *centre.*
cependant, adv., *however; nevertheless.*
céphalopode, s. m., *cephalopod.*
cercle, s. m., *circle; hoop.*
céréale, s. f., *cereal.*
cérébroïde, adj., *cerebroid; brainlike.*
cérébro-spinal, e, adj., *cerebrospinal.*
certain, e, adj., *certain.*
certainement, adv., *certainly.*
cerveau, s. m., *brain.*
ces, pl. of ce, adj.
cesse, s. f. Sans —, *incessantly.*
cesser, v. a. n., *cease;* (+ de + inf.) *stop . . . -ing.*
cétacé, s. m., *cetacean.*
ceux, m. pl. of celui.

chacun, e, pron., *each;* (m. alone), *everybody.*
chaîne, s. f., *chain; range.*
chair, s. f., *flesh; meat; pulp.*
chaleur, s. f., *heat.* — de fusion, *latent heat of fusion.*
chambre, s. f., *room.*
chameau, s. m., *camel.*
champ, s. m., *field.*
champignon, s. m., *fungus, mushroom.*
changement, s. m., *change.*
changer, v. a. n., *change* (+ de = *in,* or untranslated).
chanter, v. a. n., *sing.*
chanvre, s. m., *hemp.*
chapitre, s. m., *chapter.*
chaque, adj., *each; every; either.*
char, s. m., *car; cart.*
char de combat, s. m., *armoured car.*
charbon, s. m., *charcoal; coal; carbon.* — de terre, *coal.* — en poudre, *pulverised carbon.* — des cornues, *retort carbon.*
charge, s. f., *charge.*
charger, v. a. n., *charge.* Se —, *take upon oneself.* Ils sont chargés de, *their duties are to.*
charnière, s. f., *hinge; pivot.*
charnu, e, adj., *fleshy; pulpy.*
charpente, s. f., *framework.*
charrier, v. a., *carry along.*
charron, s. m., *wheelwright.*
charrue, s. f., *plough.*
chasser, v. a. n., *hunt; drive away; remove.*
chat, s. m., *cat.*
chaud, e, adj., *hot; warm.* Tenir — à, *keep warm.*
chauffe, s. f., *heating.*
chauffer, v. a., *heat.* — au rouge, *make red-hot.*
chauve-souris, s. f. (zoöl.), *bat.*
chaux, s. f., *lime.* — vive, *quicklime.* Eau de —, *lime water.*
chemin, s. m., *road; path.*
cheminée, s. f., *chimney.*
cheminer, v. n., *travel.*
chêne, s. m., *oak.* — liège, *cork-tree.*
chenille processionaire, s. f., *the precessing caterpillar.*
chercher, v. a. n., *seek; look for.*
cheval, s. m., *horse.*
chevalet, s. m., *bridge* (of violin, etc.).
chevaline, adj. f. (+ race), *equine.*
chevelu, e, adj., *hairy.*
chevelure, s. f., (*head of*) *hair; scalp.*
chèvre, s. f., (*she-*) *goat.*
chevron, s. m., *strip.*
chez, prep., *among; with; in.*
chiendent, s. m., *couch-grass.*

chiffre, s. m., *cipher; figure.*
chimie, s. f., *chemistry.*
chimique, adj., *chemical.*
chimiquement, adv., *chemically.*
chimiste, s. m., *chemist (scientific).*
chlorate, s. m., *chlorate.*
chlore, s. m., *chlorine.*
chlorhydrate, s. m., *hydrochloride.* — d'ammoniaque, *ammonium chloride.*
chlorhydrique, adj., *hydrochloric.*
chlorophylle, s. f., *chlorophyll.*
chlorure, s. m., *chloride.*
choc, s. m., *shock; impact; collision.*
choisir, v. a., *choose.*
chomer, v. n., *to be lacking.*
chose, s. f., *thing.* Pas grand'—, peu de —, *not much; little.*
chou, s. m., *cabbage.*
chute, s. f., *fall.*
ci, adv. pron. — -dessus, *above; abovementioned.*
cidre, s. m., *cider.*
ciel, s. m., *sky.*
cil, s. m., *eyelash; filament.*
cilié, e, adj., *ciliated.*
cime, s. f., *top; summit.*
ciment, s. m., *cement.*
cimetière, s. m., *cemetery.*
cinétique, adj., *kinetic.*
cinq, adj., *five.*
cinquante, adj., *fifty.*
circonférence, s. f., *circumference.*
circonscrire, v. a., *enclose; circumscribe.*
circonstance, s. f., *circumstance.*
circonvolution, s. f., *circumvolution.*
circuit, s. m., *circuit.* Coupe-—, fuse; *circuit-breaker.*
circulatoire, adj., *circulatory.*
cire, s. f., *wax.* — à cacheter, *sealing-wax.*
cirière, s. f., *working bee.*
cirque, s. m., (of mountains) *cirque.*
cirre, s. m., *cirrus; tendril.*
citer, v. a., *quote; mention.*

claire, adj., *clear; bright.*
classe, s. f., *class.*
classer, v. a., *class.*
cliché, s. m., *photographic negative; stereotype plate.*
climat, s. m., *climate.*
climatérique, adj., *climatic.*
cloaque, s. m. (zoöl.), *cloaca.*
cloche, s. f., *bell; bell-jar.*
cloison, s. f., *partition.* — de séparation, *septum, dissepiment.*
clore (past part. **clos**), v. a., *close.*

coagulable, adj., *coagulable.*
cœfficient, s. m., *coefficient.*
cœlentérés, s. m. pl., *coelenterata.*
coerciti-f, ve, adj., *coercive.*
cœur, s. m., *heart;* (bot.) *heartwood.*
coffret, s. m., *chest; box.*
cohésion, s. f., *cohesion.*
coïncider, v. n., *coincide.*
col, s. m., *collar; pass, defile; neck.* A long —, *long-necked.*
collection, s. f., *collection.*
coller, v. a., *stick . . . together.*
collerette, s. f., *small collar.*
collet, s. m., *collar.*
collier, s. m., *collar.*
colonie, s. f., *colony.*
colonne, s. f., *column.* — vertébrale, *spinal column.*
coloration, s. f., *colouring.*
coloré, e, part. adj., *coloured; deep-coloured.*
colossal, e, adj., *colossal.*
combien, adv. (+ de), *how much; how many.*
combinaison, s. f., *combination.*
combiner, v. a., *combine.*
combler, v. a., *fill.*
combustible, adj., *combustible;* s. m., *fuel.*
combustion, s. f., *combustion.*
commande, s. f., *order, command; control of motor,* etc.
comme, adv., *as; like;* (exclamatory) *how.*
commencement, s. m., *beginning.*
commencer, v. a. n., *begin.* Pour —, *at the outset; to begin with.*
comment, adv., *how.*
commerce, s. m., *commerce.*
commode, adj., *convenient.*
commodité, s. f., *convenience.*
commun, e, adj., *common.*
communication, s. f., *communication.* Sans — avec, *cut off from.*
communiquer, v. a. n., *communicate; impart; give.*

compact, e, adj., *compact.*
compagnon, s. m., *companion.*
comparable, adj., *comparable.*
comparativement, adv., *comparatively.*
comparer, v. a., *compare.*
compenser, v. a., *compensate; counterbalance.*
compl-et, -ète, adj., *complete; entire.*
complètement, adv., *completely.*
compléter, v. a., *complete.*
complexe, adj., *complex.*
complication, s. f., *complexity.*
compliqué, e, part. adj., *complicated; intricate.*
comporter, v. a., *admit of.* Se —, *behave.*
composant, s. m., *component.*
composé, e, part. adj., *compound.*
composé, s. m., *compound.*
composer, v. a. n., *compose; make up.*
composition, s. f., *composition.*
comprendre, v. a., *comprise; contain; understand.* Faire —, *make clear.*
comprimer, v. a., *compress.*
compte, s. m., *account.* Tenir — de, *take into account.*
compter, v. a. n., *reckon.*
compteur, s. m., *indicator; meter.*
comté, s. m., *county.*
concamération. *See* ventre de vibration.
concave, adj., *concave.*
concentré, e, part. adj., *concentrated; undiluted.*
concentrique, adj., *concentric.*
concernant, prep., *relating to.*
concerner, v. a., *concern.* En ce qui concerne, *so far as . . . is (are) concerned.*
concevoir, v. a., *conceive; understand.*
conclure, v. a. n., *conclude.*
conclusion, s. f., *conclusion.*
concourir, v. n., *compete; tend* (+ à).
concours, s. m., *meeting.*
concrescent, adj., *coalescing.*
concurrence, s. f., *competition.* — vitale, *struggle for life.*
condamner, v. a., *condemn.*
condenser, v. a., *condense.*
condition, s. f., *condition.* A — de (+ inf.), *provided that we,* etc.
conducteur, s. m., *conductor;* adj., *conducting, -ive.*
conductibilité, s. f., *conductivity.*
conduire, v. a., *conduct.*
conduit, s. m., *duct.*
cône, s. m., *cone.*
confection, s. f., *construction; manufacture.*
confluent, s. m., *confluence.*
confluer, v. n., *meet* (of streams).
confondre, v. a., *blend; confuse.* Se —, *converge.*
conforme, adj., *conformable.*
conformément, adv. — à, *in accordance with.*
conformer, v. a. Se — à, *comply with.*
congeler, v. a., *congeal.*
conique, adj., *conic(al).*
connaissance, s. f., *knowledge.*
connaître, v. a. n., *know.*
connu, e, part. adj. Les plus —s, *the best-known.*
consécuti-f, -ve, adj., *consecutive.*
conseil, s. m., *advice.*
conséquence, s. f., *consequence.* En —, *consequently.*
conséquent, s. m. Par —, *consequently; therefore.*
conservation, s. f., *preserving.*
conservatoire, s. m. — des Arts et Métiers, *Institute of Arts and Crafts.*
conserver, v. a., *preserve; keep; retain.*
considérable, adj., *considerable; large.*
considérer, v. a., *consider; examine.*
consister, v. n. — en, *consist of;* — à, *consist in.*
consommation, s. f., *consumption.* Faire une — considérable de, *consume in large quantities.*
consommer, v. a., *consume.*
constamment, adv., *constantly.*
constant, e, adj., *constant.*
constante, s. f., *constant.*
constatation, s. f., *statement.*

constater, v. n., *state; verify.* — que, *verify the fact that.*
constituant, s. m., *constituent.*
constituer, v. a., *constitute; form.*
constituti-f, -ve, adj. Elément —, *constituent.*
constitution, s. f., *composition.*
constructeur, s. m., *constructor; builder.*
construction, s. f., *construction; building; manufacture.*
construire, v. a., *construct; build; make.*
contact, s. m., *contact.* Au — de, *in contact with.*
contenir, v. a., *contain.*
contenter, v. a. Se —, *be satisfied* (de, *with*).
contestation, s. f., *dispute.*
continent, s. m., *continent.*
continental, e, adj., *continental.*
continu, e, adj., *continuous.*
continuer, v. n. a., *continue; prolong.*
contour, s. m., *contour.*
contourner, v. a., *pass round; wind round.*
contracter, v. a. n., *contract.*
contractile, adj., *contractile.*
contraction, s. f., *contraction.*
contractivité, s. f., *contractility.*
contraire, s. m., *reverse; opposite.* Au —, *on the contrary; on the other hand.*
contre, prep., *against.*
contribuer, v. n., *contribute.*
contrôler, v. a., *to check.*
convection, s. f., *convection.*
convenable, adj., *suitable.*
convenir, v. n., *agree;* (+ à) *suit.* Convenant mieux, *being more suitable.* On est convenu de, *we agree to.*
convention, s. f., *convention.*
conventionel, le, adj., *conventional.*
convergent, e, adj., *converg-ent, -ing.*
convexe, adj., *convex.*
coordonner, v. a., *coordinate.*
coquillage, s. m., *shell.*
coquille, s. f., *shell.*
coquillière, adj., *conchiferous.*
corail, s. m., *coral.*
corde, s. f., *cord; rope;* (math.) *chord.* — sonore, *harmonics of a string.*
cordon, s. m., *cord, bundle.*
coriace, adj., *leathery; tough.*
corne, s. f., *horn.*
corné, e, adj., *horney.*
cornée, s. f., *cornea.*
cornue, s. f., *retort.*
corolle, s. f., *corolla.*
corps, s. m., *body; substance.*
corpuscule, s. m., *corpuscle.*
correspondant, -e, part. adj., *corresponding.*
correspondre, v. n. Se —, v. r., *correspond.*
corroder, v. a., *corrode.*
cosinus, s. m., *cosine.*
côte, s. f., *rib; coast.*
côté, s. m., *side.* De tous (les) —s, *on all sides.* Du — de, *towards.* D'un autre —, *on the other hand.* Laisser de —, *neglect.*
cotylédon, s. m., *cotyledon.*
cou, s. m., *neck.*
couchage, s. m., *lodging; bedding.*
couche, s. f., *layer; stratum.*
coulée, s. f., *flow.*
couler, v. n., *flow.*
couleur, s. f., *colour; dye.*
coup, s. m. — d'œil, *glance.*
coupe, s. f., *cutting; section.* — transversale, *cross-section.*
couper, v. a. n., *cut; intersect.*
couplage, s. m., *coupling, interaction.*
courant, e, part. adj., *running.*
courant, s. m., *current.* Au — de, *informed of.*
courbe, adj., *curved.* S. f., *curve.*
courbure, s. f., *curvature.*
courir, v. n., *run.*
couronne, s. f., *crown.*
couronner, v. a., *crown; surround.*
cours, s. m., *course.* — d'eau, *stream.*
course, s. f., *race.*
court, e, adj., *short.*
coussin, s. m., *cushion.*
coût, s. m., *price, cost.*
couvent, s. m., *monastery.*
couver, v. n., *incubate, hatch.*
couvercle, s. m., *cover, lid.*
couverture, s. f., *covering; roofing.*
couvrir, v. a., *cover.*

craie, s. f., *chalk.*
crampon, s. m. (bot.), *crampon.*

cramponner, v. a. Se —, *cling.*
crâne, s. m., *skull; cranium.*
craquement, s. m., *crackling.*
cratère, s. m., *crater.*
crayon, s. m., *pencil.*
crépusculaire, adj., *twilight.*
crésol, s. m., *cresol.*
crétacé, adj. as s. m., *cretaceous.*
crête, s. f., *crest.*
creuser, v. a. n., *hollow out.*
creu-x, -se, adj., *hollow.*
crevasse, s. f., *crevasse.*
crinière, s. f., *mane.*
cristal, s. m., *crystal.*
cristallin, s. m., *crystalline.*
cristalliser, v. a., *crystallise.*
croc, s. m., *fang.*
crochet, s. m., *hook.*
crochu, e, adj., *hooked; crooked.*
crocodile, s. m., *crocodile.*
crocus, s. m., *crocus.*
croire, v. a. m., *believe; think.*
croissant, e, part. adj., *increasing.*
croître, v. n., *grow.*
crosse, s. f., *butt-end.*
croûte, s. f., *crust.*
croyance, s. f., *belief.*
crustacé, s. m., *crustacean.*
cryptogame, s. m., *cryptogam.*

cube, s. m., *cube.*
cubique, adj., *cubic.*
cuir, s. m., *leather.*
cuisse, s. f., *thigh.*
cuivre, s. m., *copper.* — jaune, *brass.*
cul-de-sac, s. m., *cul-de-sac, blind alley.*
culte, s. m., *worship.*
cultiver, v. a., *cultivate; grow.*
curieu-x, -se, adj., *curious.*
cuscute, s. f., *dodder.*
cutané, e, adj., *cutaneous.*

cycadées, s. f. pl., *cycadaceae.*
cycas, s. m., *cycad.*
cylindre, s. m., *cylinder.*
cylindrique, adj., *cylindrical.*

D

dahlia, s. m., *dahlia.*
danger, s. m., *danger.*
dangereu-x, -se, adj., *dangerous.*
dans, prep., *in, into, through, on,* etc.
dater, v. a. n., *date.*
dattier, s. m., *date-palm.*
dauphin, s. m., *dolphin.*
davantage, adv., *more.* Et . . . pas —, *nor . . . either.*

de, d', prep., *of, from, with, by, on, to,* etc.; (after plus, + number) *than; some* (often untranslated).
débarrasser, v. a., *rid; disembarrass.*
débit, s. m., *market; sale; debit side of ledger.*
débris, s. m., *debris; rubbish; flotsam.*
début, s. m., *beginning* (often pl.).
débuter, v. n., *begin; start.*
décagramme, s. m., *decagramme.*
décaler, v. a., *displace.*
décalitre, s. m., *decalitre.*
décamètre, s. m., *decametre.*
décanter, v. a., *decant.*
décaper, v. a., *cleanse* (surface of a metal).
déceler, v. a., *disclose.*
décharger, v. a., *discharge.*
dechiqueté, adj., *dissected.*
déchirer, v. a., *tear.*
décider, v. a. n., *decide.*
décigramme, s. n., *decigramme.*
décilitre, s. *m., decilitre.*
décimal, e, adj., *decimal.*
décimètre, s. m., *decimetre.*
déclarer, v. a., *declare.*
déclencher, v. a., *release; disconnect.*
déclinasion, s. f., *declination. See* boussole.
décomposer, v. a., *decompose; discompose.*
découvert, s m. Mettre à —, *lay bare; reveal.*
découverte, s. f., *discovery.*
découvrir, v. a., *discover.*
décrire, v. a., *describe.*
décroître, v. n., *decrease.*
dédoublement, s. m., *reduction by half.*
dédoubler, v. a., *reduce by half.*
déduire, v. a., *deduce.*
défaut, s. m. A — de, *if there is (are) no . . .* Faire —, *be missing.*
défense, s. f., *defence.*
défiler, v. n., *go past.*
défini, e, part. adj., *definite.*
définir, v. a., *define.*
définiti-f, -ve, adj., *final.*
définition, s. f., *definition.*
dégagé, e, part. adj., *free.*
dégagement, s. m., *release.* Tube à —, *escape tube.*

dégager, v. a., *release; liberate.*
dégât, s. m. (often pl.), *damage.*
degré, s. m., *degree.*
dehors, adv. Au —, *outside; out.*
déjà, adv., *already.*
delà, prep. adv. Au — de, *beyond.*
délayer, v. a., *dilute, dissolve; spin out.*
délicat, e, adj., *delicate.*
delta, s. m., *delta.*
demander, v. a. n., *ask.* Se —, *wonder.*
demeure, s. f., *habitat.*
demeurer, v. n., *remain.*
demi, e, adj. (invariable before s.), *half.* A —, (*by*) *half.*
démontrer, v. a., *demonstrate.*
dénominateur, s. m., *denominator.*
densité, s. f., *density.*
dent, s. f., *tooth.* — de lait, *milk tooth.* — de sagesse, *wisdom tooth.*
dentelle, s. f., *lacework.*
dentition, s. f., *denture.*
dénuder, v. a., *denude; lay bare.*
départ, s. m., *departure.*
dépasser, v. a. n., *surpass; exceed.*
dépendre, v. n., *depend* (+ de = *on*).
dépens, s. m. pl., *expense.*
dépense, s. f., *expenditure.*
déperdition, s. f., *loss.*
déplacement, s. m., *displacement; change of position; change of habitat.*
déplacer, v. a., *displace; remove.* Se —, *be moved; change one's position.*
dépoli, e, part. adj. Verre —, *ground glass.*
déposer, v. a., *lay; deposit.*
dépôt, s. m., *deposit.* En — sur le fer, *when iron is coated with it.*
dépourvu, e, part. adj., *devoid.*
dépression, s. f., *depression.*
depuis, prep., *since; for* (of time).
derni-er, -ère, adj. s., *latter; last; last-named.* Dans ces dernières années, *of recent years.*
déroulable, adj., *unrollable; capable of being unrolled.*
derrière, prep. adv., *behind.*
des (contraction of De + les, art.), *of the; some* (often untranslated).
dès, prep., *from the very . . .*
désagréable, adj., *unpleasant.*
désagréger, v. a., *disaggregate; disintegrate.*
désaturé, adj., *unsaturated.*
descendant, s. m., *descendant; offspring.*
descendre, v. n., *go down; come down; sink.* V. a., *let down; lower.*
désert, e, adj., *desert.*
désert, s. m., *desert.*
désexcitation, s. f., *de-energisation.*
déshydrater, v. a., *dehydrate.*
désigner, v. a., *designate; indicate.*
désinfecter, v. a., *disinfect.*
désirer, v. a., *desire; require.*
dessécher, v. a., *dry; dry up; desiccate.*
dessiccation, s. f., *desiccation.*
dessin, s. m., *design* (drawing); *pattern.*
dessiner, v. a., *design; form.*
dessous, adv. prep. Au— de, *below.*
dessus, adv. prep., *upon it* (*them*). Au— de, *above.* Par— de, *over.*
destinée, s. f., *destiny, fate.* Destiner, v. a., *intend.*
destiner, v. a., *intend.*
destruc-teur, -trice, adj., *destructive.*
destruction, s. f., *destruction.*
détacher, v. a. Se —, v. r., *break off.*
détail, s. m., *detail.*
détaillé, e, part. adj., *detailed.*
detente, s. f., *expansion* (*of gases*).
détermination, s. f., *determination.*
déterminé, e, part. adj., *fixed; specific.*
déterminer, v. a., *determine; cause, bring about.*
détonant, e, adj., *explosive.*
détoner, v. n., *detonate.*
détroit, s. m., *strait*(*s*).
détruire, v. a., *destroy; cancel.*
deux, adj., *two.* Tous (les) —, *both.*
devant, prep., *before; in front of.* S. m., *front.* Aller au— de, *go to meet.*
développement, *development.* Prendre un grand —, *develop*

greatly. Entrer dans de grands —s, *go into great detail.*
développer, v. a., *develop.*
devenir, v. n., *become.* Que devient . . .? *What becomes of . . .?*
déverser, v. a., *pour.*
déviation, s. f., *deviation.*
deviner, v. a., *guess.*
devoir, v. a. n., *owe; must; have to.* Il a dû (+ infin.), *he must have* (+ past part.).
dévonien, adj., as s. m. (geol.), *Devonian.*
dextrine, s. f., *dextrine.*

diaclase, s. f., *diaclasis; fracture.*
diagonale, s. f., *diagonal.*
diamant, s. m., *diamond.*
diamètre, s. m., *diameter.*
diapason, s. m., *tuning-fork.*
diaphane, adj., *diaphanous.*
diaphonie, s. f., *inductive disturbances.*
diastase, s. f., *diastase; enzyme.*
dicotylédone, s. f., adj., *dicotyledon, -ous.*
dièdre, adj., *dihedral.*
Dieu, s. m., *God; god.*
différence, s. f., *difference.*
différencier, v. a., *differentiate.*
différent, e, adj., *different.*
difficile, adj., *difficult.*
difficilement, adv. Très —, *with great difficulty.*
digérer, v. n. a., *digest.* Faire — (chem.), *digest.*
digesti-f, -ve, adj., *digestive.*
digne, adj., *worthy.*
dilatation, s. f., *expansion.*
dilater, v. a. Se —, v. r., *dilate; expand.*
diluer, v. a., *dilute.*
dimension, s. f., *dimension.*
dimère, adj., *dimerous.*
diminuer, v. a. n., *diminish.*
dire, v. a., *say; call.* C'est-à-—, *that is to say; i.e.* Pour ainsi —, *so to speak.*
directement, adv., *directly.*
direction, s. f., *direction.*
diriger, v. a., *direct; point.*
disciple, s. m., *pupil.*
discuter, v. a., *discuss; consider.*
disette, s. f., *dearth.*
disloquer, v. a., *dislocate.*
disparaître, v. n., *disappear.*
disparition, s. f., *disappearance.*
disparu, e, part. adj., *missing.*
disposer, v. a. n., *arrange.*
disposition, s. f., *arrangement; aptitude.*
disseminer, v. a., *spread; publicise; disseminate.*
dissiper, v. a., *dissipate.*
dissociation, s. f., *dissociation.*
dissolution, s. f., *dissolution; solution.*
dissolvant, s. m., *solvent.*
dissoudre, v. a., *dissolve.*
distance, s. f., *distance.* De — en —, *at certain intervals.*
distillation, s. f., *distillation.*
distillatoire, adj., *distillatory.*
distiller, v. a. n., *distil.*
distinct, e, adj., *distinct.*
distincti-f, -ve, adj., *distinctive.*
distinction, s. f., *distinction.*
distinguer, v. a. n., *distinguish.*
distribuer, v. a., *distribute.*
district, s. m., *district.*
dit, e, part. adj., *called; known as.*
diurne, adj., *diurnal.*
divergence, s. f., *divergence.*
diverger, v. n., *diverge.*
divers, e, adj., *divers; different.*
dividende, s. m., *dividend.*
diviser, v. a., *divide.*
diviseur, s. m., *divisor.*
divisibilité, s. f., *divisibility.*
divisible, adj., *divisible.*
division, s. f., *division.*
dix, adj., *ten.*
dixième, adj., *tenth.*

doigt, s. m., *finger.*
dolomies, s. f. pl., *dolomites.*
domaine, s. m., *estate; sphere.*
domesticité, s. f., *domesticated condition.*
domestique, adj., *domestic.*
dominer, v. a. n., *predominate.*
donc, adv., *therefore; accordingly.*
donnée, s. f., *given quantity.*
donner, v. a. n., *give.* Étant donné(e), *given.*

dont, pron., *whose, of whom (which), with whom (which),* etc.
dormir, v. n., *sleep.*
dorsal, e, adj., *dorsal.*
dos, s. m., *back.*
dosage, s. m., *estimation, titration; precise dosing.*
doser, v. a., *estimate quantitatively; dose.*
doubler, v. a. n., *double.*
douer, v. a., *endow.* Être doué de, *possess.*
douille, s. f., *socket; cartridge.*
doute, s. m. Sans —, *no doubt.*
douteu-x, -se, adj., *doubtful.*
dou-x, -ce, adj., *soft; gentle.* Eau douce, *fresh water.*
douzième, adj., *twelfth.*

drainage, s. m., *drainage, -ing.*
drap, s. m., *cloth.*
drapeau, s. m., *flag, colours.*
dresser, v. a., *draw up.*
droit, e, adj., *straight; right.*
droit, adv. Tout —, *straight (on).* Plonger tout —, *run straight down.*
droite, s. f., *right hand, right.* A —, to (*on*) *the right.*
drôle, adj., *comical.*

du (contraction of de + le, art.), *of the, of,* etc.; *some* (often untranslated).
dû, m., **due,** f., part. adj., *due.* — à ce que, *due to the fact that.*
ductilité, s. f., *ductility.*
dune, s. f., *dune; sandhill.*
duquel, contraction of de lequel, *whose, of which,* etc.
dur, e, adj., *hard.* Peu —, *soft.*
durcir, v. a. n., *harden.*
durée, s. f., *duration.*
durer, v. n., *last.*
dureté, adj., *hardness.*

dynamite, s. f., *dynamite.*
dynamo, s. m., *dynamo.*

E

eau, s. f., *water.*
ébauche, s. f., *basic plan.*
éblouir, v. a., *dazzle.*
éboulement, s. m., *falling in; landslide.*
ébouler, v. n. S'—, v. r., *slide down.*
ébranlement, s. m., *shaking.*
ébullition, s. f., *boiling.* Entrer en —, *reaching boiling-point.*

écart, s. m., *distance (apart).*
écarté, part. adj. Très —s les uns des autres, *very far apart.*
écarter, v. a., *deflect.* S'—, v. r., *deviate.*
échancrer, v. a., *indent.*
échange, s. m., *interchange.*
échantillon, s. m., *sample; pattern.*
échapper, v. n. a. S'—, v. r., *escape.*
échauffement, s. m., *heating.*
échauffer, v. a., *heat.*
échelle, s. f., *scale of map; ladder.*
échinoderme, s. m., *echinoderm.*
échoir, v. n., *occur; fall due, mature.*
éclairage, s. m., *lighting.*
éclairant, e, part. adj., *luminous.* Peu —, *not very luminous.*
éclaircir, v. a., *explain, illuminate* (fig.).
éclaire, s. f. Grande —, *greater celandine.*
éclairement, s. m., *illumination.*
éclairer, v. a., *illuminate.*
éclore, v. n., *hatch out, be hatched.*
école, s. f., *school.*
économie, s. f., *economy; saving.*
écorce, s. f., *bark; crust.* L'— terrestre, *the earth's crust.*
écouler, v. a. S'—, v. r., *flow out.*
écran, s. m., *screen.*
écrire, v. a. n., *write; state.*
écrivain, s. m., *writer.*
écrou, s. m., *screw-nut.*
écrouler. S'—, v. r., *fall in; collapse.*
écureuil, s. m., *squirrel.*

édenté, s. m., *edentate;* (zoöl.) *sloth.*
édifice, s. m., *building.*

effectivement, adv., *actually.*
effectuer, v. a., *carry out; bring about.*
effervescence, s. f., *effervescence.*
effet, s. m., *effect.* En —, *indeed; in fact.*
effilé, e, adj., *tapering.*
effondrer, v. a. S'—, v. r., *fall in; give way.*
efforcer, v. r. S'—, *try hard, strive.*
effort, s. m., *effort.*

égal, e, adj., *equal; even.*
egalement, adv., *equally; likewise.*
égaler, v. a., *equal.*
égalité, s. f., *equality; equation.*
égaré, e, part. adj., *strayed; having lost one's way.*
égayer, v. a., *make cheerful.*
égoutter, s', v. r., *to drain off.*
égyptien, ne, adj., *Egyptian.*

élancer. S'—, v. r., *shoot up.*
élasticité, s. f., *elasticity.*
élastique, adj., *elastic.*
électricité, s. f., *electricity.*
électrisation, s. f., *electrification.*
électriser, v. a., *electrify.*
électro-chimie, s. f., *electro-chemistry;* -statique, adj., *electrostatic;* -lytique, adj., *electrolytic;* -moteur, -trice, adj., *electromotive;* -scope, s. m., *electroscope* (à feuilles, d'or, *gold-leaf electroscope*).
électrode, s. m., *electrode.*
élégant, e, s. m., *smart; attractive.*
élément, s. m., *element.*
élémentaire, adj., *elementary.*
élevage, s. m., *breeding.*
élévation, s. f., *elevation; rise.*
élevé, e, adj., *high.*
élever, v. a., *raise; rear.* S'—, *rise; arise.*
éleveur, s. m., (*cattle-, etc.*) *breeder.*
éliminer, v. a., *eliminate.*
elle, pron. f., *she; her; it.* —s, *they.* —-même, *her-, it-self.*
ellipse, s. f., *ellipse.*
éloigné, e, part. adj., *remote.*
éloigner, v. a., *remove.* S'—, v. r., *go far away.*

émail, s. m., *enamel.*
émanée, e, part. adj., *emanating.*
émaner, v. n., *emanate.*
emboîter, v. a., *encase.*
embouchure, s. f., (*river-*) *mouth.*
embrasser, v. a., *embrace; include, comprise.*
embryon, s. m., *embryo.*
embryonnaire, adj., *embryonic.* Sac —, *embryo sac.*
émerger, v. n., *emerge.*
émersion, s. f., *emersion.*
émettre, v. a., *emit.*
emmagasinment, s. m., *storage; warehousing.*
empêcher, v. a., *prevent.*
empiéter, v. n. a., *encroach.*
emplacement, s. m., *site.*
emplir, v. a., *to fill.*
emploi, s. m., *use.*
employer, v. a., *use; spend.*
empoisonner, v. a., *envenom.*
emporte-pièce, s. m., *punch; die.*
emporter, v. a., *carry with one; carry down, off.* L'— sur, *prevail over.*
empreinte, s. f., *imprint.*
emprunter, v. a., *borrow* (+ à, *from*); *derive.*
émulsi-f, -ve, adj., *emulsive.*
émulsion, s. f., *emulsion.*
emulsionner, v. a., *emulsionise.*

en, prep., *in, into,* etc.; (+ -ant) *by,* etc.; *as, as a.*
en, pron., *of it, of them; from it, from them,* etc.; (+ plus *or* moins) *for that.* Il — est qui, *there are some who, that.* Il n'— est rien, *it is not so.*
énantiomorphe, adj., *enantiomorphous.*
encastrer, v. a., *embed.*
enceinte, adj., *pregnant;* s. f., *enclosure, precinct.*
enclore, v. a., *enclose.*
encore, adv., *still; also.*
encre, s. f., *ink.*
endroit, s. m., *place; district.*
enduire, v. a., *coat; smear.*
endurer, v. a., *put up with.*
énergie, s. f., *energy.*
énergique, adj., *energetic; vigorous.*
enfant, s. m. f., *child.*
enfermer, v. a., *shut up; enclose; contain.*
enfin, adv., *finally.*
enflammer, v. a., *ignite.*
enfoncer, v. a., *fix; drive in.* — trop fortement, *wedge too tightly.* S'—, v. n., *sink; go deep.*
enfouir, v. a., *bury.*
engager, v. a. S'— dans, *makes one's way into; get into.*
engendrer, v. a., *engender.* Les ravages engendrés, *the havoc wrought.*
englober, v. a., *unite.*
engrais, s. m., *manure.*
enlever, v. a., *remove* (+ à, *from*).

enmagasinment, s. m., *storing up.*
ennemi, s. m., *enemy.*
énoncé, s. m., *statement.*
énoncer, v. a., *enunciate; express.*
énorme, adj., *enormous.*
énormément, adv., *enormously.*
enrouler, v. a., *coil.*
ensemble, adv., *together; at the same time.* S. m., *mass; aggregate.*
ensuite, adv., *afterwards; then; next; in the next place.*
entacher, v. a., *taint.*
entailler, v. a., *cut; notch; mortise.*
entamer, v. a., *encroach upon.*
entasser, v. a. S'—, *be piled up.*
entendre, v. a. n., *hear.* — parler de, *hear of.*
entendu, e, part. adj. Bien —, *of course.*
enthousiasme, s. m., *enthusiasm.*
enti-er, -ère, adj., *whole; entire.* Tout —, *whole.* Nombre —, *integer.*
entièrement, adv., *entirely.*
entonnoir, s. m., *funnel.*
entourer, v. a., *surround.*
entraîner, v. a., *carry along; involve.*
entraver, v. a., *hinder.*
entre, prep., *between; among.* — eux, *together.* Certaines d'— elles, *some of them.*
entrée, s. f., *entr-ance, -y.*
entrer, v. n., *enter; go in.*
entretenir, v. a., *keep up; sus-, main-, tain.*
entretien, s. m., *upkeep.*
entr'ouvrir, v. a., *open a little; half-open.*
envaser, v. a., *silt up; sink in mud.*
enveloppe, s. f., *covering; integument.*
envergure, s. f., *span.*
environ, prep., adv., *about.* —s, s. m. pl., *environs.*
environner, v. a., *surround.*
envoyer, v. a., *send; throw.*

épais, se, adj., *thick.*
épaisseur, s. f., *thickness; density.*
épaissir, v. a., *thicken.*
épancher, v. a. S'—, *overflow.*
épanouir, v. a. S'—, v. r., *expand.*
épars, adj., *scattered, sparse, thin.*
épidémie, s. f., *epidemic.*
épiderme, s. m., *epidermis.*
épine, s. f., *thorn; spine; prickle.*
époque, s. f., *era; period; time.*
éprouver, v. a., *experience; undergo.*
éprouvette, s. f., *test-tube.* — à pied, *mounted test-tube.*
épuiser, v. a., *exhaust; wear out.*
épurer, v. a., *purify.*

équateur, s. m., *equator.*
équation, s. f., *equation.*
équidistant, e, adj., *equidistant.*
équilibre, s. m., *equilibrium; balance.* Se mettre en —, *attain equilibrium.*
équitablement, adv., *fairly.*
équivalent, e, adj., *equivalent.*
érable, s. m., *maple.*
ériger, v. a., *erect.*
erreur, s. f., *error.* Être dans l'—, *be mistaken.*
erroné, e, adj., *erroneous.*
érupti-f, -ve, adj., *eruptive.*
éruption, s. f., *eruption.* Faire —, *erupt.*

escargot, s. m., *snail.*
espace, s. m., *space.*
Espagnol, s. m., *Spaniard.*
espèce, s. f., *species.*
esprit, s. m., *intellect; spirit(s).* —de-vin, *spirits of wine.*
esquille, s. f., *fragment; splinter.*
esquisser, v. a., *to sketch.*
essai, s. m., *essay; test, experiment.*
essayer, v. a. n., *try.*
essence, s. f., *essence; petrol.*
essential, le, adj., *essential.*
essentiellement, adv., *for all essential purposes.*
essorer, v. a., *dry in the open air.*
est, s. m., *east.*
estimer, v. a. n., *estimate.*
estomac, s. m., *stomach.*
estomper, v. a., *to stump.*

et, conj., *and.*
établir, v. a., *establish; set up; lay down.* — que, *establish the fact that.*
étage, s. m., *storey, floor; rank, status.*
étain, s. m., *tin.*
étaler, v. a., *spread out; display.*

étalon, s. m., *standard* (of weight, etc.).
étamine, s. f., *stamen; coarse muslin; sieve, strainer.*
étanche, adj., *water-tight; impervious.*
étang, s. m., *pond.*
étape, s. f., *stage* (in progress, etc.).
état, s. m., *state; condition; stage.* En — de, *in a position to.*
étau, s. m., *vice* (tool).
été, s. m., *summer.*
éteindre, v. a., *extinguish.* S'—, *go out* (of light).
éteint, e, part. adj., *extinct.*
étendre, v. a., *extend; stretch (out); dilute.*
étendu, e, part. adj., *extensive; dilute.*
étendue, s. f., *extent; tract.*
éternellement, adv., *eternally.*
éther, s. m., *ether.*
éthylène, s. f., *ethylene.*
étincelle, s. f., *spark.*
étoffe, s. f., *stuff; fabric.*
étoile, s. f., *star.* — polaire, *Pole-star.*
étoilé, e, adj., *stellate.*
étrange, adj., *strange.*
étrang-er, -ère, adj., *foreign.*
étranglement, s. m., *strangulation; constriction.*
être, v. n., *be; exist.* C'est, *it is, that is.* C'est (donc) que, *because.* En —, *be.* Le plus . . . qui soit, *the most . . . in existence.*
être, s. m., *creature; being.*
étroit, e, adj., *narrow; close.*
étroitement, adv., *closely.*
étude, s. f., *study.*
étudier, v. a. n., *study.*
étui, s. m., *case; sheath.*
étymologie, s. f. Avoir pour —, *be derived from.*

eux, pron. m. pl., *they; them.* — -mêmes, *themselves.*

évaluer, v. a., *assess.*
évaporation, s. f., *evaporation.*
évaporer, v. a. S'—, *evaporate.*
évaser, v. a., *widen.*
éventail, s. m., *fan.*
évertuer. S'—, v. r., *strive.*
évidemment, adv., *evidently; obviously.*
évidence, s. f. Mettre en —, *make evident; prove.*
évident, e, adj., *evident.* —(e) par lui- (elle-) même, *self-evident.*
éviter, v. a., *avoid.*
évoluer, v. n., *evolve.*
évolution, s. f., *evolution.*
exact, e, adj., *exact; accurate.*
exactement, adv., *exactly; accurately.*
exagération, s. f., *exaggeration.*
examen, s. m., *examination.*
examiner, v. a. n., *examine; look at.*
excéder, v. a., *exceed.*
excellent, e, adj., *excellent.*
excepté, prep., *except.*
excepter, v. a., *except.*
exception, s. f., *exception.* A l'— de, *with the exception of.* Faire —, *be exceptional.*
exceptionel, le, adj., *exceptional.*
excès, s. m., *excess.*
excessivement, adj., *exceedingly.*
exclure, v. a., *exclude.*
exclusivement, adj., *exclusively; solely.*
excréteur, s. m. as adj., *excretory.*
excrétion, s. f., *excretion.*
excroissance, s. f., *excrescence.*
exécuter, v. a., *execute; carry out; perform; make.*
exempt, adj., *exempt; free from.*
exemple, s. m., *example.* Par —, *for instance.*
exercer, v. a. n., *exercise; exert.*
exhausser, v. a., *raise.*
exigence, s. f., *requirement.*
existence, s. f., *existence.*
exister, v. n., *exist.*
exode, s. m., *exodus.*
expérience, s. f., *experiment.*
expérimental, e, adj., *experimental.*
expérimentalement, adv., *experimentally.*
expliquer, v. a., *explain.*
exploiter, v. a., *exploit; work.*
explorateur, s. m., *explorer.*
explorer, v. a., *explore.*
explosion, s. f., *explosion.*
exposant, s. m. (math.), *exponent; index.*
exposer, v. a., *expose; set forth.*
exposition, s. f., *exposure.*
expression, s. f., *expression.*
exprimer, v. a., *express.*
expulser, v. a., *expel.*
extérieur, adj., *external.* S. m., *exterior.* A

l'—, *outside; outwardly.*
extérieurement, adv., *externally.*
externe, adj., *external; outer.*
extinction, s. f., *extinction.*
extraction, s. f., *extraction.*
extraire, v. a., *extract.*
extraordinaire, adj., *extraordinary.* -ment, adv., *extraordinarily.*
extrême, adj., *extreme.* -ment, adv., *exceedingly.*
extrêmité, s. f., *extremity; end.*

F

fa, s. m., *F* (music).
fabrication, s. f., *manufacture.*
fabrique, s. f., *factory.*
fabriquer, v. a., *manufacture.*
face, s. f., *face; surface.* En — (de), *opposite.*
Fâcheu-x, -se, adj., *unpleasant.*
facies, s. m., *aspect, appearance; cast of features.*
facile, adj., *easy; easily made.* —ment, adv., *easily; readily.*
facilité, s. f., *ease.*
faciliter, v. a., *facilitate.*
façon, s. f., *fashion; way.* A la — de, *after the manner of.* De — à, *in such a way as to.* De — que, *in such a way that.*
faculté, s. f., *faculty.*
faible, adj., *weak; slight; little.* Courant de — puissance, *low-powered current.*
faire, v. a. n., *make; do; cause; let; travel;* (of temperature, etc.) *be;* (+ inf.) *see that . . . (is done), cause (to be done).* C'est ce qui fait que, *that is why.* Se faire, *happen.*
faisceau, s. m., *bundle.* — lumineux, *pencil of rays.*
fait, s. m., *fact.* — d'observation, *observed fact.*
falaise, s. f., *cliff.*
falloir, v. n., *need.* Il faut, *it is necessary; (one) must.*
fameu-x, -se, *famous.*
famille, s. f., *family.*
fanon, s. m., *whalebone; baleen.*
fardeau, s. m., *load.*
fascicule, s. m., *fascicle; bundle; part* (of a work), *issue* (of a journal).
fasciculé, e, part. adj., *fasciculate.*
faune, s. f., *fauna.*
favorable, adj., *favourable.*
favori, te, adj., *favourite.*
fécondation, s. f., *fertilisation.*
fécondité, s. f., *fertility.*
fécule, s. f., *faecula.*
féculente, e, adj., *feculent.*
femelle, s. f., *female.*
fémur, s. m., *femur; thighbone.*
fente, s. f., *rift; slit.*
fer, s. m., *iron.* — de cheval, *horseshoe* (+ en, *in horseshoe fashion*).
ferme, adj., *firm; stiff.* S. f., *farm.*
ferment, s. m., *ferment; leaven.*
fermentation, s. f., *fermentation.*
fermenter, v. n., *ferment.*
fermer, v. a., *shut; close.*
fermeture, s. f., *closure.*
ferrer, v. a., *tire* (wheel).
ferreu-x, -se, adj., *ferrous.*
ferrique, adj., *ferric.*
ferrugineu-x, -se, adj., *ferruginous; chalybeate.*
fertile, adj., *fertile.*
feu, s. m., *fire.* Mettre le — à, *set fire to.*
feuillage, s. m., *foliage.*
feuille, s. f., *leaf; sheet.* A —s persistantes, *evergreen.* A —s caduques, *deciduous.*
feuillet, s. m., *leaf* (of book).
fève, s. f., *broad bean.*
février, s. m., *February.*

fibre, s. f., *fibre.*
fibreu-x, -se, adj., *fibrous.*
fibrille, s. f., *fibril.*
fibrine, s. f., *fibrine.*
fibroine, s. f., *fibroin.*
fiche, s. f., *plug; slip of paper, "chit"; small piece of material.*
figure, s. f., *figure; face.*
figuré, e, adj. Ferment —, *organised ferment.*
figurer, v. a. n., *figure.*
fil, s. m., *thread; wire.*
filament, s. m., *filament.*
filamenteu-x, -se, adj., *filamentous.*
filerie, s. f., *spinning house.*
filet, s. m., *thread; filament; ligament.*
filiforme, adj., *filiform.*
filon, s. m., *lode, vein of ore.*
fin, s. f., *end.* Mettre — à, *end, bring to an end.*

fin, e, adj., *fine; slender; acute.*
final, e, adj., *final.*
finalement, adv., *finally.*
finir, v. a. n., *finish; end.*
fiole, s. f., *phial.*
fixe, adj., *fixed.*
fixer, v. a., *fasten.*

flacon, s. m., *phial; bottle.*
flagellé, e, adj., *flagellate.*
flanc, s. m., *flank; side; slope* (of hill, etc.).
flécher, v. a., *flex, bend, soften, relent.*
fleur, s. f., *flower; blossom.* A — de terre, *on a level with the ground.*
fleurir, v. n., *blossom.*
fleuve, s. m., (*large*) *river.*
flexible, adj., *flexible.*
flotter, v. n., *float.*
flottille, f. s., *flotilla; fleet.*
fluide, adj. s. m., *fluid.*
fluidité, s. f., *fluidity.*
flux, s. m., *flux; flow.*

focal, e, adj., *focal.*
foie, s. m., *liver.*
foin, s. m., *hay.*
fois, s. f., *time* (*occasion*). A la —, *at the same time; both.* Chaque — que, toutes les — que, *whenever.* Une —, *once.* Une — que, *when once.* Une seule —, *only once.*
foncé, adj., *dark; dark-coloured.*
fonction, s. f., *function.*
fonctionnement, s. m., *working.*
fonctionner, v. n., *work; act.*
fond, s. m., *bottom.* De —, *fundamental.*
fondamental, e, adj., *fundamental.*
fondation, s. f., *foundation.*
fonder, v. a., *found; base.*
fondre, v. a., *melt; cast* (iron).
fondu, e, part. adj., *molten.*
foraminifères, s. m. pl., *foraminifera.*
force, s. f., *strength; force.*
forcément, adv., *necessarily.* — nécessaire, *absolutely necessary.*
forcer, v. a., *compel.*
forêt, s. f., *forest.*
formation, s. f., *formation.*
forme, s. f., *form; shape.*
former, v. a., *form.*
formule, s. f., *formula.*
formuler, v. a., *formulate; state.*
fort, e, adj., *strong; powerful; considerable.*
fort, adv., *very.*
fortement, adv., *strongly; powerfully; stoutly; tightly.*
fossette, s. f., *dimple; small pit.*
fossile, adj. s. m., *fossil.*
fougère, s. f., *fern.* — arborescente, *tree-fern.*
fouille, s. f., *excavation; search; digging.*
fouiller, v. a., *search; rummage.*
foule, s. f., *crowd; large number.*
fourchu, adj., *dichotomous.*
fournir, v. a., *furnish; supply.*
fovea, s. f., *fovea.*
foyer, s. m., *hearth; furnace; focus.*

fraction, s. f., *fraction.*
fracturer, v. a., *fracture.*
fra-is, -îche, adj., *fresh.*
frais, s. m. pl., *expense; cost.*
franc, s. m., *franc.*
franc, he, adj. *See* fusion.
français, e, adj., *French.*
franchir, v. a., *cross.*
frange, s. f., *fringe.*
frappant, e, part. adj., *striking; impressive.*
frapper, v. a. n., *strike.*
freinage, s. m., *braking.*
fréquemment, adv., *frequently.*
friction, s. f., *friction.*
frictionner, v. a. n., *rub.*
frisé, e, adj., *curled; curly.*
froid, e, adj., *cold.*
froid, s. m., *cold.* Les grands —s, *the very cold weather.*
fromage, s. m., *cheese.*
front, s. m., *front; forehead.*
frontal, e, adj., *forehead.*
frottement, s. m., *friction.*
frotter, v. a. n., *rub.*
fruit, s. m., *fruit.*

fuchsia, s. m., *fuchsia.*
fugiti-f, -ve, adj., *transient.*
fulgurant, adj., *flashing.*
fuligineu-x, -se, adj., *fuliginous; sooty.*
fumée, s. f., *smoke.*
fumerolle, s. f., *fumarole; smoke-hole.*
funéraire, adj., *funerary.*
fuseau, s. m., *spindle.*
fusible, adj., *fusible.*
fusion, s. f., *fusion; melting; transfusion.* — franche, *direct transfusion.* Eau de —, *water from melted ice.*

G

gagner, v. a. n., *gain; reach.*
gaine, s. f., *sheath;* (bot.) *leaf-base.*
galéopithèque, s. m., *galeopithecus; flying lemur.*
galet, s. m., *pebble.*
galvaniser, v. a., *galvanise.*
galvanoplastie, s. f., *electro-plating.*
gamètes, s. m. pl., *gametes.*
gamme, s. f., (musical) *scale.*
gamopétale, adj., *gamopetalous.*
gamosépale, adj., *gamosepalous.*
ganglion, s. m., *ganglion.*
ganglionnaire, adj., *ganglion-ary, -ic.*
garantir, v. a., *guarantee; protect.*
garder, v. a., *keep.*
garnir, v. a., *furnish; provide.*
gastéropode, s. m., *gasteropod.*
gastrique, adj., *gastric.*
gâteau, s. m., *cake.*
gauche, adj., *left.* **A —,** *to the left.*
gavial, s. m., *gavial* (species of crocodile).
gaz, s. m., *gas.* — ammoniac, *ammonia.*
gazelle, s. f., *gazelle.*
gazeu-x, -se, adj., *gaseous.*
gel, s. m., *gel; freezing; frosty weather.*
gélatineu-x, -se, adj., *gelatinous.*
gelé, e, part. adj., *frozen.*
gemmule, s. f., *gemmule; plumule.*
gêner, v. a., *disturb; trouble; inconvenience.*
général, e, adj., *general.* En —, *generally; as a rule.*
généralement, adv., *generally.*
généralité, s. f., *universality.* —s, *general features.*
généra-teur, -trice, adj., *generating.* S. m., *generator.*
génération, s. f., *generation.*
génitoires, s. f. pl., *genital organs.*
genre, s. m., *kind.*
gens, s. m. f. pl., *people.*
géocentrique, *geocentric.*
géograph-ic, s. f., *geography;* -ique, adj., *geographical.*
géologie, s. f., *geology.*
géologue, s. m., *geologist.*
géométr-ie, s. f., *geometry;* -ique, adj., *geometrical.*
géosynclinal, s. m., *syncline.*
géotropisme, s. m., *geotropism; tendency of plants to turn towards the earth.*
germe, s. m., *germ; bud.*
germé, e, part. adj., *germinated.*
germination, s. f., *germination.*
gésier, s. m., *gizzard.*
geyser, s. m., *geyser.*
gigantesque, adj., *gigantic.*
giroflée, s. f., *gillyflower.*
gisement, s. m., *bearing, direction; deposit of coal or ore.*

glace, s. f., *ice.*
glacer, v. a., *ice; freeze; chill; glaze.*
glacier, s. m., *glacier.*
glaïeul, s. m., *gladiolus.*
glande, s. f., *gland.*
glandulaire, adj., *glandular.*
glisser, adj., *slide; glide.*
globe, s. m., *globe.*
globule, s. m., *globule.*
gluant, adj., *sticky.*
glucose, s. f., *glucose.*
glucosique, adj., *glucosic.*

golfe, s. m., *gulf; bay.*
gomme, s. f., *gum.*
gommeu-x, -se, adj., *gummy.*
gorge, s. f., *throat.*
gorger, v. a., *cram.*
goudron, s. m., *tar.* — de houille, *coal-tar.*
goulot, s. m., *neck* (of bottle).
goût, s. m., *taste.* Sans —, *tasteless.*
goûter, v. a. n., *taste.*
goutte, s. f., *drop.* — à —, *drop by drop.*
gouttelette, s. f., *droplet.*
gouttière, s. f., *gutter.*

grâce, s. f. — à, *thanks to.*
graduation, s. f., *graduation.*
graduel, le, adj., *gradual.*
grain, s. m., *grain.* — d'orge, *barleycorn.*
graine, s. f., *seed.*
graisse, s. f., *fat; grease.*
gramme, s. m., *gramme.*
grand, e, adj., *large; great; grown-up; main.*
grandeur, s. f., *size.*
granit, s. m., *granite.*
graphique, adj., *graphic.* S. f., *drawing; graph.*
graphite, s. m., *graphite.*

gras, se, adj., *fat; fatty, greasy;* (bot.) *fleshy, pulpy.*
grauwacke, s. f., *greywacke.*
grave, adj., (of sound) *low.*
gravier, s. m., *gravel.*
gravité, s. f., *gravity.*
grec, que, adj., *Greek.*
grenadier, s. m., *pomegranate-tree.*
grès, s. m., *sandstone.*
griffe, s. f., *claw;* (bot.) *offset tuber.*
grille, s. f., *grating; railing; grid of radio valve.*
grimpeur, s. m. as adj., *climbing.*
gris, e, adj., *grey.*
gros, se, adj., *big; bulky; thick.*
grosseur, s. f., *bulk; thickness.* Avoir la — de, *be as big as.*
gross-ier, -ière, adj., *coarse.*
grossir, v. a. n., *swell.*
groupe, s. m., *group.*

guêpe, s. f., *wasp.*
guère, adv., (ne . . .), *hardly.* Ne — que, *hardly ever, except; hardly anything but.*
guerre, s. f., *war.*
gueule, s. f., *mouth; maw.* —-de-lion, *snapdragon.*
gui, s. m., *mistletoe.*

gymnosperme, s. m., *gymnosperm.*
gypse, s. m., *gypsum, plaster of Paris.*

H

habitant, s. m., *inhabitant.*
habitation, s. f., *housing.*
habiter, v. a. n., *inhabit; dwell (in).*
habitude, s. f., *habit.* D'—, *usually.*
habituel, le, adj., *habitual.*
habituellement, adv., *usually.*
hameçon, s. m., *fish-hook.*
hanche, s. f., *hip; haunch.*
hardi-e, adj., *bold.*
haricot, s. m., *kidney-bean; bean.*
harmoniser, v. a., *harmonise.*
hasard, s. m. Par —, *by chance.*
hâter, v. a., *hasten.*
haut, e, adj., *high; tall.* S. m., *height.* En —, *at the top.* Vers le —, *upward.*
hauteur, s. f., *height; pitch.* Avoir la — de, *be . . . in height.*
héberger, v. a., *harbour; lodge.*

hectogramme, s. m., *hectogramme.*
hectolitre, s. m., *hectolitre.*
hélice, s. f., *screw, propeller; helix.*
helminthe, s. m., *helminth.*
hémisphère, s. m., *hemisphere.*
herbacé, e, adj., *herbaceous.*
herbe, s. f., *grass.*
herbeu-x, -se, adj., *grassy.*
herbivore, adj., *herbivorous.*
hercynien, ne, adj., *Hercynian.*
hérisson, s. m., *hedgehog.*
hêtre, s. m., *beech tree.*
heure, s. f., *hour.*
heureu-x, -se, adj., *happy.*
hexagonal, e, adj., *hexagonal.*
hexatomique, adj., *hexatomic.*
hile, s. m., *hilum.*

himalayen, ne, adj., *Himalayan.*
hippomobile, adj., *horse-drawn.*
hirondelle, s. f., *swallow.*
histoire, s. f., *history; story.*
historique, adj., *historic, -al.*
hiver, s. m., *winter.*

holoturie, s. f., *holothurian.*
homme, s. m., *man.*
homogène, adj., *homogeneous.*
homologue, adj., *homologous.*
honneur, s. m., *honour.*
horizon, s. m., *horizon.*
horloge, s. f., *clock; public clock.*
houblon, s. m., *hop(s).*
houille, s. f., *(pit-) coal.*
houill-er, -ère, adj., *coal-.*

huile, s. f., *oil.*
huiler, adj., *oil.*
huit, adj., *eight.*
huître, s. f., *oyster.*
humain, e, adj., *human.*
humidité, adj., *dampness; moisture.*

hydratation, s. f., *hydration.*
hydrate, s. m., *hydrate.*
hydraté, e, adj., *hydrated.*
hydraulique, adj., *hydraulic.*
hydre, s. f., *hydra.*

hydro-carbure, s. f., *hydrocarbon.* -électrique, adj., *hydro-electric.* -gène, s. m., *hydrogen.* -gené, e, adj., *hydrogenated.* -statique, s. f., *hydrostatics.*
hyène, s. f., *hyena.*
hygiène, s. f., *hygiene.*
hygiénique, adj., *hygienic.*
hypoténuse, s. f., *hypotenuse.*
hypothèse, s. f., *hypothesis.*

I

ici, adv., *here.*

identique, adj., *identical.*

igné, e, adj., *igneous.*
ignition, s. f., *ignition.*
ignorer, v. a., *not know.*

il, pron., *he; it.* **—s,** *they.*
île, s. f., *island.*

image, s. f., *image.*
imaginer, v. a. n. S'—, v. r., *imagine.*
imbiber, v. a., *soak; steep.*
imbuvable, adj., *undrinkable.*
imiter, v. a., *imitate.*
immédiat, e, adj., *immediate.* -ement, adv., *immediately.*
immense, adj., *immense.*
immobile, adj., *motionless.*
immobilité, s. f., *immobility.*
impair, adj., *odd.*
imparfait, e, adj., *imperfect.*
impénétrabilité, s. f., *impenetrability.*
impénétrable, adj., *impenetrable; impervious.*
imperméable, adj., *impervious; waterproof.*
importance, s. f., *importance; size.*
important, e, adj., *important.*
importer, v. a., *import.* N'importe où, *anywhere;* — quel(le), *any.*
imposant, e, adj., *imposing.*
imposer, v. a. n., *impose.* S'—, v. r., *be essential.*
impossibilité, s. f., *impossibility.*
imprégner, v. a., *impregnate.*
impression, s. f., *impression.*
imprimer, v. a., (*im-*) *print; impart.*
impulsion, s. f., *impulse.*
impureté, s. f., *impurity.*

inacti-f, -ve, adj., *inactive.*
inactivité, s. f., *inactivity.*
inaltérable, adj., *durable.*
inanimé, e, adj., *inanimate.*
incandescence, s. f., *incandescence; white heat.*
incapable, adj., *incapable.*
incendie, s. m., *fire (conflagration).*
incessamment, adv., *incessantly.*
incessant, e, adj., *incessant.*
incidence, s. f., *incidence.*
incident, e, adj., *incident.*
incisive, s. f., *incisor.*
inclination, s. f. *See* boussole.
incliner, v. a. n. S'—, v. r., *incline; bend.*
incolore, adj., *colourless.*
incombustible, adj., *fireproof.*
incompl-et, -ète, adj., *incomplete.*
inconnu, e, adj., *unknown.* **—e,** s. f., *unknown quantity.*
indéfinie, e, adj., *indefinite.*
indéfiniment, adv., *indefinitely.*
indéterminé, e, adj., *indeterminate.*
indication, s. f., *indication; sign; hint.*
indice, s. m. (math., phys.), *index.*
indien, ne, adj., *Indian.*
indigène, s. m., *native.*
indigo, s. m., *indigo.*
indiquer, v. a., *indicate; show.*
indispensable, adj., *indispensable.*
indistinct, e, adj., *indistinct.*
individu, s. m., *individual.*
individuel, le, adj., *individual.*
indivisible, adj., *indivisible.*
industrie, s. f., *industry.*
industriel, -le, adj., *industrial.* -ellement, adv., *industrially.*
inégal, e, adj., *unequal.*
inégalement, adv., *unequally; unevenly.*
inerte, adj., *inert.*
inférieur, e, adj., *inferior; lower.*
infester, v. a., *infest.*
infiltration, s. f., *infiltration.*
infiltrer, v. a., *infiltrate.*
infini, e, adj., *infinite.* S. m., A l'—, *endlessly.*

infiniment, adv., *infinitely.*
infirmité, s. f., *infirmity.*
inflammation, s. f., *inflammation.*
influence, s. f., *influence; induction.*
influencé, e, part. adj., *induced.*
influencer, v. a., *influence.*
information, s. f., *information.*
infuser, v. a., *infuse; steep.*
infusoire, s. m. —s, *infusoria.*
ingénieur, s. m., *engineer.*
ingénieu-x, -se, adj., *ingenious.*
ininterrompu, e, adj., *uninterrupted.*
initial, e, adj., *initial.*
inodore, adj., *scentless.*
inscrire, v. a., *inscribe; write down; enter.*
insecte, s. m., *insect.*
insensible, adj., *imperceptible.* -ment, adv., *imperceptibly.*
insipide, adj., *tasteless.*
insister, v. n. — sur, *dwell upon.*
insolite, adj., *unusual.*
insoluble, adj., *insoluble.*
inspection, s. f. A l'— de, *on inspecting.*
instrument, s. m., *instrument.*
insuffisante, e, adj., *insufficient.*
intelligence, s. f., *intelligence.*
intendant, s. m., *steward, manager, surveyor, official.*
intense, adj., *intense.*
intensité, s. f., *intensity.*
intercaler, v. a., *insert.*
intercepter, v. a., *intercept.*
interdire, v. a., *forbid, suspend; astound.*
intéresser, v. a., *interest.*
intérêt, s. m., *interest.* —(s) composé(s), *compound interest.*
intérieur, e, adj., *internal.* S. m., *interior.*
intérieurement, adv., *internally.*
intermédiaire, adj., *intermediate.* S. m., *medium.* A l'— de, *by the medium of.*
intermitt-ence, s. f., *intermittence, intermittent character;* -ent, e, adj., *intermittent.*
international, e, adj., *international.*
interne, adj., *internal; inner.*
interposer, v. a., *interpose.*
interrompre, v. a., *interrupt.*
intersection, s. f., *intersection.*
intervenir, v. n., *occur.*
interversion, s. f., *inversion.*
intervertir, v. a., *invert.*
intestin, s. m., *intestine.*
intestinal, s. m., *intestinal.*
intime, adj., *intimate; close.*
intimider, v. a., *intimidate.*
introduire, v. a., *introduce; insert, put in.* S'—, *penetrate.*
introuvable, adj., *undiscoverable.*
intuiti-f, -ve, adj., *intuitive.*
inventer, v. a., *invent.*
inverse, adj., *inverse.* S. m., *reverse.* — -ment, adv., *in- (con-) versely, vice versa.*
invertebré, s. m., *invertebrate.*
invisible, adj., *invisible.*
involontaire, adj., *involuntary.*

iode, s. m., *iodine.*
iodisme, s. m., *iodism; treating with iodine.*
ionogène, adj., *producing ions.*

irrationel, le, adj., *irrational.*
irrégularité, s. f., *irregularity.*
irréguli-er, -ère, adj., *irregular.*
irrésistible, adj., *irresistible.*

islandais, e, adj. s., *Icelandic; Icelander.*
isogyre, s. f., *isogyre.*
isolant, s. m., *insulator.*
isolé, e, part. adj., *iso- (insu-) lated.*
isolément, adv., *separately.*
isoler, v. a., *iso- (insu-) late.* S'—, *separate.*
isotrope, adj., *isotropic.*
issue, s. f., *outlet.*

ivoire, s. m., *ivory; bone (of teeth).*

J

jabot, s. m., (of bird, etc.) *crop.*
jaillir, v. n., *gush forth; burst forth.*
jais, s. m. (geol.), *jet.*
jalon, s. m., *stake; landmark.*
jamais, adv., (ne . . .), *never.*
jambe, s. f., *leg;* (of bird, insect) *shank, tibia.*
jardin, s. m., *garden.*
jaspure, s. f., *marbling.*
jaunâtre, adj., *yellowish.*

jaune, adj., *yellow.*

jet, s. m., *spurt.*
jeter, v. a. n., *throw; spout.* Se —, v. r., *run.*
jeu, s. m., *play.*
jeune, adj., *young.* S. m., *young one.*

joindre, v. a., *join; connect.*
jointure, s. f., *joint.*
jouer, v. a. n., *play.*
jouir, v. n. (+ de), *enjoy; possess.*
jour, s. m., *day.* De nos —s, *nowadays.*
journée, s. f., *day.*
joyau, s. m., *jewel.*

jugement, s. m., *judgment.* Passer en —, *be brought up for trial.* Porter un —, *pass judgment.*
juillet, s. m., *July.*
jumeau, adj., *twin.*
jument, s. f., *mare.*
jurassique, adj., s. m., *Jurassic.*
jus, s. m., *juice.*
jusque, jusqu'à, prep., *until; up to; as much as.* Jusqu'à ce que, conj., *until.*
justesse, s. f., *justice; propriety.*
juxtaposer, v. a., *juxtapose, place side by side.*
juxtaposition, s. f., *juxtaposition.*

K

kilo-gramme, s. m., *kilogramme;* -litre, s. m., *kilolitre;* -mètre, s. m., *kilometre* ($\frac{5}{8}$ of a mile).

L

la, l', art., *the.*
la, s. m. (music), *A.*
là, adv., *there; that point.* De —, *from this (that); hence.*
laboratoire, s. m., *laboratory.*
labourer, v. a. n., *plough.*
lac, s. m., *lake.*
lâche, adj., *loose, lax; cowardly.*
lactifère, adj., *lactiferous.* S. m., *lactiferous vessel.*
lac-tique, adj., *lactic.* -tate, s. m., *lactate.*
lacune, s. f., *lacuna; gap.*
lacustre, adj. Habitation —, *lake-dwelling.*
lagune, s. f., *lagoon.*
laine, s. f., *wool.*
laisser, v. a., *leave; let; let alone; permit.* Se —, v. r. (+ active inf.), *may* (+ passive inf.).
lait, s. m., *milk.*
laiton, s. m., *brass.*
lame, s. f., *plate, sheet; blade; lamina.*
lamelle, s. f., *lamella; scale.*
lamellibranches, s. m. pl., *lamellibranchiates.*
lampe, s. f., *lamp; radio valve.*
lancer, v. a., *hurl; send forth.*
langage, s. m., *language* (style of talk).
langue, s. f., *tongue.*
lanterne, s. f., *lantern.* — de projection, *projector.*
lanthane, s. m., *lanthanium.*
lapidaire, s. m., *lapidary.* Adj., *pertaining to stones.*
lapin, s. m., *rabbit.*
laquelle, pron. f., *which.*
large, adj., *broad; wide.*
largeur, s. f., *breadth; width.*
larve, s. f., *larva.*
latéral, e, adj., *lateral.*
latex, s. m., *latex.*
latitude, s. f., *latitude.*
lavage, s. m., *washing; ablution.*
lave, s. f., *lava.*
laver, v. a., *wash.*

le, l', art. m., *the.* Pron. m., *him; it; so; to do so.*
lecture, s. f., *reading.*
léger, e, adj., *light; slight.*
légèrement, adv., *lightly; slightly.*
légèreté, s. f., *lightness.*
légume, s. m., *vegetable.*
légumineuses, s. f. pl., *leguminosae.*
lémuriens, s. m. pl., *lemuridae, lemurs.*
lent, e, adj., *slow.*
lentement, adv., *slowly.*
lenteur, s. f. Avec —, *slowly.*
lentille, s. f., *lens.*
lequel, pron. m. Lesquels, pl., *who, which,* etc.
les, art. pl., *the;* pron. pl., *them.*
lessive, s. f., *lye.*
lettre, s. f., *letter.*
leur, pron. pl. (*to*) *them.* Adj. (pl. leurs), *their.*
lever, v. a., *raise; remove.* Se —, v. r., *rise.*
lèvre, s. f., *lip.*
levure, s. f., *yeast.*

liaison, s. f., *connection, junction; collaboration.*

liber, s. m. (bot.), *bast.*
liberté, s. f., *liberty.* Mettre en —, *release.*
libre, adj., *free; unimpeded; unattached.*
librement, adv., *freely.*
liège, s. m., *cork.*
lien, s. m., *bond; link.*
lier, v. a., *bind.*
lierre, s. m., *ivy.*
lieu, s. m., *place;* (math.) *locus.* Au — de, *instead of.* Avoir —, *take place.* Il y a — de, *there is occasion to.* Comme cela a —, *as in the case.*
lièvre, s. m., *hare.*
ligne, s. f., *line.* En première —, *first and foremost.*
lignite, s. m., *lignite.*
lilas, s. m., *lilac.*
limaille, s. f., *filings.*
limbe, s. m., *limb; graduated edge.*
limite, s. f., *limit; boundary.*
limiter, v. a., *limit; circumscribe.*
limpide, adj., *limpid, clear.*
lin, s. m., *flax.*
lingules, s. f. pl., *lingulae.*
lion, s. m., *lion.*
liquéfaction, s. f., *liquefaction.*
liquéfier, v. a., *liquefy.*
liquide, s. m., *liquid.*
lis, s. m., *lily.*
lisière, s. f., *edge.*
lisse, adj., *smooth.*
lit, s. m., *bed.*
lithographique, adj., *lithographic.*
litre, s. m., *litre.*
littéral, e, adj., *literal.*
livre, s. f., *pound.*
livrer, v. a., *deliver; hand over.*
lobe, s. m., *lobe of the ear; lobe of a polar diagram.*
lobule, s. m., *lobule.*
local, e, adj., *local.*
localiser, v. a., *localise.*
locomo-teur, -trice, adj., *locomot-or, -ive.*
locomotion, s. f., *locomotion.*
logarithme, s. m., *logarithm.*
logis, s. m., *home.*
loi, s. f., *law.* — de Descartes, *Descartes' law.*
loin, adv., *far; away.* Plus —, *further; later.*
long, ue, adj., *long.* Avoir . . . de —, *be . . . in length.*
longitudinal, e, adj., *longitudinal.* -ement, adv., *longitudinally.*
longtemps, adv., (*for a*) *long* (*time*).
longueur, s. f., *length.* Avoir . . . de —, *be . . . in length.*
lors, adv., *then.* Dès —, *from that time, thenceforth.*
lorsque, conj., *when.*
loupe, s. f., *magnifying glass.*
lourd, e, adj., *heavy.*

lui, pron. m. f., *he; him; her; to, in,* etc., *him, her.*
lumière, s. f., *light.*
lumineu-x, -se, adj., *luminous; of light.*
lune, s. f., *moon.*
lutte, s. f., *struggle.*
luzerne, s. f., *lucern.*

M

macérer, v. a. Faire —, *steep.*
mâcher, v. a. n., *masticate.*
machine, s. f., *machine; engine.*
mâchoire, s. f., *jaw.*
mâcle, s. f., *twin-crystal.*
madréporique, adj., *coral.*
magma, s. m., *magma.*
magnésium, s. m., *magnesium.*
magnétique, adj., *magnetic.*
magnétisme, s. m., *magnetism.*
magnifique, adj., *magnificent.*
mai, s. m., *May.*
maille, s. f., *mesh.*
maillechort, s. m., *German silver.*
main, s. f., *hand.* A la —, *in the hand.*
maint, adj., *many a; several.*
maintenant, adv., *now.*
maintenir, v. a., *maintain; keep up.*
mais, adv., conj., *but.*
maïs, s. m., *maize.*
maison, s. f., *house.*
maîtresse, s. f., *mistress.* Adj., *main.*
majestueu-x, -se, adj., *majestic.*
majorité, s. f., *majority.*
maki, s. m., *maki.*
mal (pl. **maux**), s. m., *evil; ailment.*
malade, adj., *ill.* Les —s, *the sick.*
maladie, s. f., *disease.*
mâle, adj., *male.*
malgré, prep., *despite.*
malheureusement, adv., *unfortunately.*
malléable, adj., *malleable.*
malt, s. m., *malt.*
maltose, s. m., *maltose.*
mamelle, s. f., *breast.*
mammifère, s. m., *mammal.*
manche, s. m., *handle.*

manganèse, s. m., *manganese.*
manger, v. a. n., *eat.*
maniement, s. m., *handling.*
manier, v. a., *handle; manipulate.*
manière, s. f., *manner; way.* A la — de, *like.* De — à, *in such a way as to.*
manifestation, s. f., *manifestation.*
manifeste, adj., *obvious.*
mannite, s. f., *mannite.*
manque, s. m., *lack.*
manquer, v. a. n., *fail; miss.* Qui lui manquent, *which it lacks.*
manutention, s. f., *management.*
marais, s. m., *marsh.*
marbre, s. m., *marble.*
marche, s. f., *progress; course.*
marcher, v. n., *walk.*
marécageu-x, -se, adj., *marshy.*
marée, s. f., *tide.* Mer à —s, sans —s, *tidal, tideless, sea.*
marin, e, adj., *marine.*
marne, s. f., *marl.*
marquer, v. a., *mark; score, show.*
marronnier, s. m., *chestnut-tree.*
marsouin, s. m., *porpoise.*
marsupial, e, adj. s. m., *marsupial.*
masse, s. f., *mass; volume.*
Massif, s. m., *block; clump.*
mat, -e, adj., *dead colour.*
matelas, s. m., *mattress.*
mathématique, adj., *mathematical.* —s, s. f. pl., *mathematics.* -ment, adv., *mathematically.*
matière, s. f. (often pl.), *matter; substance.*
matras, s. m. (chem.), *receiver.*
maturité, s. f., *maturity.*
Maure, s. m., *Moor.*
mauvais, e, adj., *bad.*
maux. *See* mal.
maxima, s. m. pl., *maxima.*
maximum, s. m., *maximum.*

méat, s. m. (bot.), *meatus; air-cell.*
mécanique, s. f., *mechanics.*
mécanisme, s. m., *mechanism; machinery.*
médecinal, e, adj., *medicinal.*
médecine, s. f., *medicine.*
médian, e, adj., *med-ial, -ian.*
émdullaire, adj. Étui —, *medullary sheath.*
méduse, s. f., *jellyfish.*
meilleur, e, adj., *better; best.*
mélange, s. m., *mixture.*
mélanger, v. a., *mingle; blend.*
mélasse, s. f., *molasses.*
mêler, v. a., *mix.*
melon, s. m., *melon.*
membrane, s. f., *membrane.* — d'enveloppe, *covering membrane.*
membraneu-x, -se, adj., *membranous.*
membre, s. m., *member; limb.*
même, adj., *same.* Adv., *even.* De —, *in the same way.* Tout de —, *all the same, nevertheless.* Il en est de — pour, *the same applies to.*
mémoire, s. f., *memory.*
ménage, s. m., *household.*
mener, v. a. n., *lead; draw; convey.*
mentionner, v. a., *mention.*
menton, s. m., *chin.*
mer, s. f., *sea.* Sur —, *at sea.*
mercure, s. m., *mercury.*
mère, s. f., *mother.*
méridien, s. m., *meridian.*
mesure, s. f., *measure, measurement.* A — que, *in proportion as, according as, as.*
mesurer, v. a. n., *measure.*
métallique, adj., *metallic.*
métalloïde, s. m., *metalloid.*
métamorphose, s. f., *metamorphosis; transformation.*
méthode, s. f., *method.*
métier, s. m., *craft; profession; calling.*
mètre, s. m., *metre.*
métrique, adj., *metric.*
mets, s. m., *dish; viand.*
mettre, v. a., *put; place.* Se — en route, *start, set out.*

Mi, s. m. (music), *E.*
mica, s. m., *mica.*
micelle, s. f., *micella.*
micro-pyle, s. m., *micropile.* -scope, s. m., *microscope.* -scopique, adj., *microscopic.*
mieux, adv., *better.*
migrateur, adj., *migratory.*
migration, s. f., *migration.*
milieu, s. m., *middle, centre; medium; environment.*

mille, adj., *thousand.* S. m., *mile.*
millier, s. m., *thousand.*
milli-gramme, s. m., *milligramme.* -litre, s. m., *millilitre.*
million, s. m., *million.*
mince, adj., *thin; slender.*
mine, s. f., *mine.*
minérai, s. m., *ore.*
minéral, e, adj., *mineral.* S. m., *mineral.*
miroir, s. m., *mirror.*

mobile, adj., *movable; turning.* S. m., *body in motion.*
modalité, s. f., *details.*
mode, s. m., *mode.*
modéré, e, adj., *moderate.*
moderne, adj., *modern.*
modification, s. f., *change.*
modifier, v. a., *modify.*
module, s. m., *module; modulus.*
moelle, s. f., *marrow; pith.* — épinière, *spinal cord.*
moindre, adj., *less; lesser.*
moine, s. m., *monk.*
moins, adv. prep., *less; minus.* — grand, *smaller.* De — en —, *less and less.* Du —, *at least; at any rate.*
mois, s. m., *month.*
moitié, s. f., *half.*
molaire, s. f., *molar.*
môle, s. m., *mole; mass.*
moléculaire, adj., *molecular.*
molécule, s. f., *molecule.*
mollasse, s. f. (geol.), *molasse.*
mollusque, s. m., *mollusc; shellfish.*
moment, s. m., *moment; time.*
momentané, e, adj., *momentary; temporary.* -ment, adv., *temporarily.*
monde, s. m., *world.*
monnaie, s. f., *coin(age).*
mono-cotylédone, s. f. adj., *monocotyledon (ous).* -sépale, adj., *monosepalous.* -trèmes, s. m. pl., *monotremata.*
montage, s. m., *mounting, setting, fitting; electrical cross leakage.*
montagne, s. f., *mountain.*
montagneu-x, -se, adj., *mountainous.*
monter, v. n. a., *climb.*
montrer, v. a., *show.*
monture, s. f., *mount.*
moraine, s. f., *moraine.*
morceau, s. m., *piece; lump.*
morphologique, adj., *morphological.*
mort, s. f., *death.* S. m. pl., *(the) dead.*
morue, s. f., *cod.*
mot, s. m., *word.* En peu de —s, *in a few words, briefly.*
motilité, s. f., *motility.*
mou, molle, adj., *soft.*
moule, s. f., *mussel.*
moulure, s. f., *moulding.*
mourir, v. n., *die.*
mousse, s. f., *moss.*
moût, s. m., *must.*
mouvement, s. m., *movement.* Mettre en —, *set in motion.*
mouvoir, v. a., *move.*
moyen, s. m., *means;* (math.) *mean, middle term.* Au — de, *by means of.*
moyen, ne, adj., *medium, average.*
moyenne, s. f., *average, mean.*
mucique, adj., *music.*
multi-plicande, s. m., *multiplicand.* -plicateur, s. m., *multiplier.* -cation, s. f., *multiplication.* -plier, v. a. n., *multiply.* -tude, s. f., *multitude.*
munir, v. a., *provide.*
muqueu-x, -se, adj., *mucous.*
mur, s. m., *wall.*
mûr, e, adj., *ripe.*
mûrir, v. a. n., *ripen.*
muscle, s. m., *muscle.*
musculaire, adj., *muscular.*
museau, s. m., *muzzle; snout.*
musical, e, adj., *musical.*
mustang, s. m., *mustang.*
mutuel, le, adj., *mutual.*

mycoderme, s. m., *mycoderm.*
myriapode, s. m., *myriapod.*
mystère, s. m., *mystery.*

N

nageoire, s. f., *fin.*
nager, v. n., *swim.*
naguère, adv., *but lately.*
naissance, s. f., *birth.*
naître, v. n., *be born; arise.*
nappe, s. f., *sheet.*
naseau, s. m., *nostril of a horse.*
nation, s. f., *nation.*
naturaliste, s. m., *naturalist.*
nature, s. f., *nature.*
natur-el, le, adj., *natural.* S. m., *native.* -ellement, adv., *naturally.*
nautile, s. m., *nautilus.*
navet, s. m., *turnip.*
navigation, s. f., *navigation.*
navire, s. m., *ship.*

ne, n', adv., *not.* — pas, *not.* — que, *only.*
nécessaire, adj., *necessary.* -ment, adv., *necessarily; of course.*
necess-ité, s. f., *necessity.* -iter, v. a., *necessitate; require.*
négati-f, -ve, adj., *negative.*
négativement, adv., *negatively.*
négliger, v. a., *neglect; disregard.*
nègre, s. m., *negro.*
neige, s. f., *snow.*
nerf, s. m., *nerve; sinew.*
nerv-eux, -se, adj., *nervous.* -in, *nervine.* -ure, s. f. (bot.), *vein, rib;* (zoöl.) *nervure, rib* (of wing); *cooling fin* (of engine).
net, te, adj., *clear.*
nettement, adv., *cleanly; clearly; neatly.*
neutraliser, v. a., *neutralise.*
neutre, adj., *neutral.*
névé, s. m., *glacier-snow.*

ni, conj., *neither; nor.* — l'un — l'autre, pron., *neither.*
nickel, s. m., *nickel.*
nid, s. m., *nest.*
nitré, e, adj., *nitrated.*
nitrique, adj., *nitric.*
nitrophile, adj., *nitrophilous.*
niveau, s. m., *level.*

Noci-f, -ve, adj., *harmful; toxic.*
noctiluques, s. f. pl., *noctilucae.*
nocturne, adj., *nocturnal.*
nœud, s. m., *knot; node.*
noir, e, adj., *black; dark.*
nom, s. m., *name; kind.* Sous le — de, *by the name of.*
nomade, adj., *nomadic.*
nombre, s. m., *number; kind.* Être au — de, *be . . . in number.* Sans —, *countless.*
nombreu-x, -se, adj., *numerous.*
nommer, v. a., *name; call.*
non, adv., *no; not.* — plus, *no longer;* (+ negative) *either.*
nord, s. m., *north.*
normal, e, adj., *normal.* -ement, adv., *normally.*
notable, adj., *notable; considerable.* -ment, adv., *notably.*
notamment, adv., *especially.*
notation, s. f., *notation.*
note, s. f., *note.*
notion, s. f., *notion.*
notre, adj., *our.*
nôtre, pron., *ours, our own.*
nourrici-er, -ère, adj., *nutritive.*
nourrir, v. a., *feed.* Se —, *take nutriment.*
nourrissant, e, part. adj., *nourishing.*
nourriture, s. f., *food; diet.*
nous, pron., *we; us.*
nouv-eau, -el, m., **-elle**, f. adj., *new; fresh.* De —, *afresh, again.*
noyau, s. m., *kernel; nucleus.*

nu, e, adj., *naked.*
nucelle, s. f., *nucellus.*
nuit, s. f., *night.*
nul, le, adj., *no.* S. m., *nil.*
numérateur, s. m., *numerator.*
numérique, adj., *numerical.*
numéro, s. m., *number.*
nutriti-f, -ve, adj., *nutritive.*
nymphose, s. f., *pupation.*

O

oasis, s. f., *oasis.*

obéir, v. n. (+ à), *obey.*
objet, s. m., *object.*
obliger, v. a., *compel.*
obliquité, s. f., *obliqueness.*
obscurité, s. f., *darkness.* A l'—, *in the dark.*
observation, s. f., *observation.* Il est d'— vulgaire, *it is a matter of common observation.*
observer, v. a. n., *observe.*
obstacle, s. m., *obstacle.*
obstruer, v. a., *stop up.*
obtenir, v. a., *obtain; get.*
obtus, e, adj., *obtuse.*

occasion, s. f., *opportunity.* -ner, v. a., *cause.*
occidental, e, adj., *western.*
occuper, v. a., *occupy.* S'— de, *be concerned with.*
océan, s. m., *ocean.* -ique, adj., *oceanic.*
ocellé, adj., *ocellate.*
octave, s. f., *octave.*
octobre, s. m., *October.*

odeur, s. f., *odour.*
odorant, e, adj., *odorous.*
odorat, s. m., (*sense of*) *smell.*

œil (pl. **yeux**), s. m., *eye.*
œillet, s. m., *carnation.*
œsophage, s. m., *oesophagus.*
œsophagien, ne, adj., *oesophagal.*
œuf, s. m., *egg.*
œuvre, s. f., *work.* Mettre en —, *set to work; bring to bear.*

office, s. m., *office; function.*
offrir, v. a., *offer.*

oignon, ognon, s. m., *onion.*
oiseau, s. m., *bird.*
oiselet, s. m., *small bird.*

ombre, s. f., *shadow.*

on, pron., *one, people, they, we,* etc. (or paraphrase with verb in passive).
onde, s. f., *wave.*
ondée, s. f., *shower.*
ongle, s. m., *claw, talon, hoof, finger nail, toe nail.*
onglet, s. m., *unguis; claw* (of petal).
ontogenèse, s. f., *ontogenesis, history of individual from conception to adult state.*
onzième, adj., *eleventh.*

oolithique, adj., *oölitic.*
oosphère, s. f., *oösphere.*

opacité, s. f., *opaqueness.*
opaque, adj., *opaque.*
opérateur, s. m., *operator.*
opération, s. f., *operation.*
opinion, s. f., *opinion.*
opposé, e, part. adj., *opposite.*
opposer, v. a., *oppose.* — une résistance, *offer (a) resistance.*
optique, s. f., *optics.* Adj., *optical.*

or, conj., *now.*
or, s. m., *gold.*
orage, s. m., *(thunder-) storm.*
orangé, e, s. m., *orange (coloured).*
orbite, s. m., *orbit; socket.*
orchidées, s. f. pl., *orchidaceae.*
ordinaire, adj., *ordinary.*
ordonnée, s. f., *ordinate, y co-ordinate.*
ordonnance, s. f., *array, order, disposition; medical prescription.*
oreille, s. f., *ear.*
oreiller, s. m., *pillow.*
oreillette, s. f., *auricle.*
organe, s. m., *organ.* — de réserve, *reserve organ.*
organ-ique, adj., *organic.* -isme, s. m., *organism.*
organisation, s. f., *organisation; organism.*
orge, s. f., *barley.*
orgue, s. m. (f. in pl.), *organ (musical instrument), organ loft.*
orient, s. m., *East.*
oriental, e, adj., *eastern.*
orientation, s. f., *orientation; position.*
orientaux, s. m. pl., *Orientals.*
orienter, v. a., *orientate.*
orifice, s. m., *orifice.*
originaire, adj. Être — de, *to have originally come from.*
origine, s. f., *origin.* Avoir son —, *originate.*
ornement, s. m., *ornament.*
ornementer, v. a., *ornament.*

os, s. m., *bone.*
oscillatoire, adj., *oscillatory.*
osciller, v. n., *oscillate.*
oscule, s. m., *osculum; flue.*
oseille, s. f., *sorrel.* Sel d'—, *salt of sorrel* (binoxalate of potash).

ou, conj., *or; either.* — bien, *or else.*
où, adv., *where; on which, in which; when.* D'—, *whence.*
oui, adv., *yes.*
ouïe, s. f., *(sense of) hearing.* —s, *gills.*
oursin, s. m., *sea-urchin.*
outil, s. m., *tool; implement.*
outre, prep. adv. En —, *moreover, besides.*
ouverture, s. f., *opening; aperture.*
ouvrir, v. a., *open.*
ovaire, s. m., *ovary.*
ovipare, adj., *oviparous.*
ovule, s. m., *ovule.*

oxydable, adj., *oxidisable.*
oxydation, s. f., *oxidisation.*
oxyde, s. m., *oxide.*
oxygène, s. m., *oxygen.*

ozone, s. m., *ozone.*

P

Pacifique, s. m., *Pacific.*
paire, s. f., *pair.*
paître, v. a. n., *graze; feed* (animals).
palais, s. m., *palace, court, palate.*

pâle, adj., *pale; light-coloured.*
palette, s. f., *palette.*
palmier, s. m., *palm-tree.*
panache, s. m., *plume.*
pancréas, s. m., *pancreas.*
pancréatique, adj., *pancreatic.*
papier, s. m., *paper.*
par, prep., *by, through,* etc.; *multiplied* (or *divided*) *by.*
parachute, s. m., *parachute; patagium.* Mammifères à —, *flying* (or *patagial*) *mammals.*
paraffine, s. f., *paraffin.*
paragraphe, s. m., *paragraph.*
paraître, v. n., *appear; seem.*
parallèle, adj., *parallel.*
paramécie, s. f., *paramecium, slipper animalcule.*
parapluie, s. m., *umbrella.*
parasite, s. m., *parasite.* Adj., *parasitic.*
parcelle, s. f., *particle; portion; patch.*
parce que, conj., *because.*
parcourir, v. a., *travel over; traverse; cover.*
pareil, le, adj., *like; similar; equal.* -lement, adv., *similarly.*
parenchyme, s. m., *parenchyma; woof.*
parent, s. m., *relative.* —s, *parents.*
parer, v. a., *to adorn.*
parfait, e, adj., *perfect.* -ement, adv., *perfectly.*
parfois, adv., *sometimes.*
parler, v. a. n., *speak.*
parmi, prep., *among.*
paroi, s. f., *side; wall, partition.*
parsemer, v. a., *strew; sprinkle.*
part, s. f., *part; share; side.* A —, *apart from, besides.* D'autre —, *on the other hand.* De — et d'autre, *on both sides.* D'une —, *on the one hand.*
partager, v. a. n., *divide; distribute; share.* Se —, *share among themselves.*
particule, s. f., *particle.*
particuli-er, -ère, adj., *particular; peculiar; special.*
particulier, s. m. En —, *in particular.*
partie, s. f., *part; portion.* Faire — de, *form part of.*
partir, v. n., *start.* A — de, *from . . . onward, starting from.*
partisan, s. m., *partisan.*
partout, adv., *everywhere.*
parure, s. f., *adornment.*
parvenir, v. n., *arrive.*
pas, adv. (usually + ne), *not.*
passage, s. m., *passage.* Au —, *on one's way.*
passé, e, part. adj., *past; bygone; over.*
passer, v. m. a., *pass; spend* (time). Se —, *take place.* Se — de, *dispense with.*
pâte, s. f., *paste.*
patelle, s. f., *limpet.*
pâteu-x, -se, adj., *pasty.*
patiner, v. n., *skate.*
patte, s. f., *paw; foot; leg.* — antérieure, *front leg.*
pâturage, s. m., *pasture.*
paupière, s. f., *eyelid.*
pauvre, adj., *poor.*
pays, s. m. ,*country.*
peau, s. f., *skin, hide.*
pêche, s. f., *fishing.*
pêcher, v. n. a., *fish* (*for*). — à la ligne, *angle* (*for*).
pêcherie, s. f., *fishery.*
pêcheur, s. m., *fisherman.*
pédicelle, s. m., *pedicel, pedicle.*
pédieu-x, -se, adj., *pedal.*
pédoncule, s. m., *peduncle.*
peine, s. f., *difficulty.* A —, *hardly.*
pelage, s. m., *hairy coat.*
pelure, s. f., *peel; rind.* — d'oignon (adjectival), *tawny.*
pendant, prep., *during.* — que, conj., *while.*
pendule, s. m., *pendulum.*
pénétrer, v. a. n., *penetrate.* Dont il faut bien se —, *which we must thoroughly grasp.*
pénible, adj., *laborious; hard.*
penser, v. a. n., *think.* — à (+ infin.), *think of . . . ing.*
pente, s. f., *slope; gradient.*
pepsine, s. f., *pepsin.*
percer, v. a. n., *pierce; bore.*
percevoir, v. a., *perceive.*
percher, v. n. a., *perch.*
perdre, v. a. n., *lose.*
perfectionnement, s. m., *improvement.*
perforer, v. a., *bore through.*
péricarpe, s. m., *pericarp; seed-vessel.*
période, s. f., *period.*
périr, v. n., *perish.*
périsperme, s. m., *perisperm.*

permettre, v. a., *permit.* Permet de . . . (+ infin.), *allows us to.*
perpendiculaire, adj. s. f., *perpendicular* (sur, *to*).
persévérer, v. n., *persevere.*
persistance, s. f., *persistance.*
persistant, e, adj. *See* feuille.
personne, s. f., *person.* Pron., *nobody, no one,* (+ neg.) *anyone.*
persuader, v. a., *persuade.*
perturbation, s. f., *disturbance.*
pesant, e, adj., *weighty; ponderable.*
pesanteur, a. f., *weight.*
peser, v. a. n., *weigh.*
peste, s. f., *plague.*
pétale, s. f., *petal.*
pétiole, s. f., *leaf-stalk.*
petit, e, adj., *small.*
petitesse, s. f., *smallness.*
pétrir, v. a., *knead.*
pétrole, s. m., *petroleum; kerosene; paraffin oil.*
peu, adv. s. m., *little;* (+ de) *few.*
peupler, v. a., *people, populate.*
peut-être, adv., *perhaps.*

phanérogame, s. f., *phanerogam.*
phase, s. f., *phase.*
phén-ique, adj., *carbolic, phenic.* -ate, s. n., *phenate.* -ol, s. m., *phenol, carbolic acid;* — trinitré, *trinitrophenol.*
phénomène, s. m., *phenomenon.*
philosophal, e, adj. Pierre —e, *philosopher's stone.*
phosphore, s. m., *phosphorus.*
phosphorique, adj., *phosphoric.*
phys-icien, s. m., *physicist.* -iologie, s. f., *physiology.* -iologique, adj., *physiological.* -ique, s. f., *physics;* adj., *physical.*
phytozoaire, s. m., *phytozoon* (pl. *-zoa*).

pic, s. m., *peak.*
picr-ate, s. m., *picrate.* -ique, adj., *picric.*
pièce, s. f., *piece; portion.*
pied, s. m., *foot; base; stem; bottom.*
piège, s. m., *trap.*
pierre, s. f., *stone.* —s résistantes, *stone clamps.*
pierreu-x, -se, adj., *stony.*
pile, s. f., *pile; battery.* — de Volta, *Voltaic pile.*
pilotis, s. m., *pilework, piling.*
pin, s. m., *pine.*
piquant, s. m., *quill.*
pivot, s. m., *pivot; taproot.* En —, *pivotal.*
pivotant, e, adj. Racine —, *tap-root.*

place, s. f., *place.* Sur —, *on the spot.* Tenir une grande —, *play an important part.*
placement, s. m., *investment.*
placentaire, adj., *placental.*
placer, v. a., *place; situate; invest.*
plage, s. f., *beach.*
plaine, s. f., *plain.*
plan, e, adj., *plane.*
plan, s. m., *plane; plan.*
planète, s. f., *planet.*
plant, s. m., *sapling; haulm* (of potato).
plantain, s. m., *plantain.*
plante, s. f., *plant.*
planter, v. a., *plant; fix; drive in.*
plantule, s. f., *plantule.*
plaque, s. f., *slab.*
plastique, adj., *plastic.*
plat, e, adj., *flat; low.*
plateau, s. m., *tableland; plateau.*
platine, s. m., *platinum.*
pleine, adj., *full.* En — air, *in the open air.*
pli, s. m., *fold; wave.*
plissement, s. m., *wrinkling; corrugation.*
plisser, v. a., *corrugate.*
plomb, s. m., *lead.*
plonger, v. a. n., *plunge; immerse; dip; sink.*
plot, s. m., *plot; block.*
pluie, s. f., *rain.*
plumage, s. m., *plumage.*
plume, s. f., *feather.*
plupart, s. f., *most part; most; majority.* La — des, *most.*
plus, adv., *more; most;* (+ s.) *more of a; plus.* Un(e) (s.) des plus (adj.), *a most* (adj.) (s.). Au —, *at most.* De —, *moreover.* De — en —, *more and more.* Ne . . . —, *no more, no longer.* Ne . . . — que, *now only.* — . . . —, *the more . . . the more; the -er the -er.*
plusieurs, adj. pl., *several.*

pneumatique, adj., *pneumatic.* S. f., *pneumatics.* Machine —, *air-pump.*

poche, s. f., *pocket; pouch.*

poids, s. m., *weight.*
poil, s. m., *hair.*
poing, s. m., *fist.*
point, s. m., *point; extent.* — d'arrêt, *stationary point.* — de départ, *starting-point.* — de fusion, *melting-point.* — de vue, *point of view.* Au — que, *to such a degree that.*
point, adv., *not; no.*
pointe, s. f., *point; tip.*
pointu, e, adj., *pointed; sharp.*
poisson, s. m., *fish.*
poitrine, s. f., *chest; breast.*
polaire, adj., *polar.*
pôle, s. m., *pole.*
polir, v. a., *polish.*
pollinique, adj. Tube —, *pollen-tube.*
polype, s. m., *polyp.*
poly-pétale, adj., *polypetalous.* -sépale, adj., *polysepalous.*
polypier, s. m., *polypary.*
pomme, s. f., *apple.* — de terre, *potato.*
pommier, s. m., *apple-tree.*
pompe à palette, s. f., *vane pump.*
ponce, s. f. Pierre —, *pumice-stone.*
ponctuation, s. f., *punctuation; dotting.*
pondre, v. a., *lay* (eggs).
pont, s. m., *bridge.* — suspendu, *suspension bridge.*
ponte, s. f., *clutch* (of eggs).
populaire, adj., *popular.*
porcelaine, s. f., *porcelain.* En —, *porcelain* (adj.).
pore, s. m., *pore.*
porphyre, s. m., *porphyry.*
port, s. m., *port; harbour.*
portée, s. f. A sa —, *within his* (etc.) *reach.*
porter, v. a. n., *carry; bear; place; raise.* — sur, *be over; concern, affect.*
poser, v. a. n. *place; lay; lay on; lay down; assume.* Ceci posé, *this being assumed, granted.*
positi-f, -ve, adj., *positive.* -vement, adv., *positively.*
position, s. f., *position.*
posséder, v. a., *possess.*
possible, adj., *possible.*
postérieur, e, adj., *posterior; hind.*
potasse, s. f., *potash.*
potassium, s. m., *potassium.*
potatouche, s. m., *flying squirrel.*
potentiel, s. m., *potential.*
pouce, s. m., *thumb; inch.*
poudre, s. f., *powder; dust; gunpowder.* En —, *pulverised; ground.*
poule, s. f., *hen.* — couveuse, *broody hen.*
poumon, s. m., *lung.*
pour, prep., *for, to,* etc. — que, conj., *in order that; for . . . to . . . .* — cent, *per cent.*
pourquoi, adv., *why.*
poursuivre, v. a. n., *pursue.* Se —, *proceed.*
pousser, v. n., *grow.*
poussière, s. f., *dust.* —s, *specks of dust.*
poussin, s. m., *chick.*
poutre, s. f., *beam, girder.*
pouvoir, v. n. a., *be able; can; may.* Pour qu'il puisse, *to enable him to.*
pouvoir, s. m., *power.*

prairie, s. f., *prairie; meadow.*
pratique, s. f., *practice.* Dans la —, *in practice.* Adj., *practical.* -ment, adv., *practically; for practical purposes.* -r, v. a. n., *form; make; contrive.*
préalable, adj. s. m. Au —, *previously.*
précaution, s. f., *precaution.*
précédemment, adj., *previously.*
précédent, e, adj., *preceding; above* (*-mentioned*). Le —, la —e, *the preceding one.*
précéder, v. a. n., *precede.* Ce qui précède, *the foregoing remarks.*
précieux, adj., *precious.*
précipité, s. m., *precipitate.*
précis, adj., *precise.* -ément, adv., *precisely.*
préexistant, e, part. adj., *pre-existing.*
préférable, adj., *preferable.*
préférer, v. a., *prefer.*
préhension, s. f., *prehension.*
préhistorique, adj., *prehistoric.*
prèle, s. f. (bot.), *horsetail, equisetum.*
premi-er, -ère, adj., *first.*
prendre, v. a. n., *take; catch; assume.* Se —, *coagulate, solidify.*
préparatif, s. m., *preparation.*

préparation, s. f., *preparation (act of preparing).*
préparer, v. a., *prepare.*
prépondérant, e, adj., *preponderant.*
prés, adv., *near;* — de, prep., *near.* A peu —, *almost.* A une unité —, *to the nearest unit.*
présence, s. f., *presence.* En —, *face to face.*
présent, s. m., *present.* —, e, adj., *present.* -er, v. a., *show.* Se présenter, *appear.*
préserver, v. a., *preserve; protect.*
presque, adv., *almost; nearly.*
presqu'île, s. f., *peninsula.*
pression, s. f., *pressure.*
preuve, s. f., *prove.*
prévoir, v. a., *foresee.*
primaire, adj., *primary.*
primevère, s. f., *primrose.*
primine, s. f., *primine.*
primiti-f, -ve, adj., *original.* -vement, adv., *originally.*
principal, e, adj., *principal, chief.* -ement, adv., *chiefly.*
principe, s. m., *principle.*
printemps, s. m., *spring.* Au —, *in spring.*
privation, s. f., *privation; hardship.*
priver, v. a., *deprive.*
prix, s. m., *price.*
probable, adj., *probable; likely.* -ment, adv., *probably.*
problème, s.m., *problem.*
procédé, s. m., *process.*
procéder, v. n., *proceed; evolve.*
proche, adj., *near.* De — en —, *by degrees.*
proclamer, v. a., *proclaim.*
produire, v. a., *produce; yield.* Se —, *occur.*
produit, s. m., *product; proceeds.*
profond, e, adj., *deep.* Peu —, *shallow.*
profondeur, s. f., *depth.*
progrès, s. m., *progress.*
progresser, v. n., *progress.*
progressi-f, -ve, adj., *progressive.*
progression, s. f., *progression.*
proie, s. f., *prey.*
projection, s. f., *projection; splashing.*
projeter, v. a. n., *project; throw forward.* — au loin, *propel a long distance.*
prolongement, s. m., *extension.*
prolonger, v. a., *prolong;* (math.)*produce.* Se —, *stretch out.*
propagation, s. f., *propagation; diffusion.*
propager, v. a., *propagate.*
proportion, s. f., *proportion.*
proportionnel, le, adj., *proportion.*
propos, s. m. A — de, *with regard to; when dealing with.*
proposer, v. a. n., *propose.*
propre, adj., *own; appropriate; proper.* -ment, adv., *properly;* — dit, *properly so called;* à — parler, *strictly speaking.*
propreté, s. f., *cleanliness.*
propriété, s. f., *property.*
protéger, v. a., *protect.*
protéique, adj., *proteinic.*
prothalle, s. m., *prothallus.*
protoplasm-e, s. m., -ique, adj., *protoplasm, -ic.* Amas protoplasmique, *heap of protoplasm.*
protoxyde, s. m., *protoxide.*
protozoaire, s. m., *protozoon* (pl. *-zoa*).
prouver, v. a., *prove.*
provenir, v. n., *proceed; arise.*
provision, s. f., *store; supply.*
provisoire, adj., *provisional.*
provoquer, v. a., *provoke; produce.*

puis, adv., *then; next.*
puiser, v. a. n. — dans, *draw from . . .*
puisque, conj., *since; seeing that.*
puissance, s. f., *power.*
puissant, e, adj., *powerful.*
puits, s. m., *well; shaft.*
pulmonaire, adj., *pulmonary.*
pur, e, adj., *pure.* -ement, adv., *purely, merely.*
pureté, s. f., *purity.*
purifier, v. a., *purify.*

pyrénéen, ne, adj., *Pyrenean.*

Q

quadrupède, s. m. adj., *quadruped.*
qualité, s. f., *quality.*
quand, adv. conj., *when.*
quant à, prep., *as for, as to.*

quantité, s. f., *quantity.*
quart, s. m., *quarter.*
quaternaire, adj. s. m., *quaternary.*
quatre, adj., *four.*
que, qu', conj. adv., *that; when; than;* (following si), *if.*
que, qu', rel. pron., *whom; which; that.*
quel, le, adj., *what.* — que, *whatever.*
quelconque, adj., *any.* Un(e) . . . quelconque, *any . . . whatsoever.*
quelque, adj., adv., *some; a few.* — chose, *something.* — peu, *somewhat, slightly.*
quelquefois, adj., *sometimes.*
quelques-uns, -unes, pron. pl., *some; some people.*
question, s. f., *question; problem.* Il est — de . . ., *is (are) mentioned.*
queue, s. f., *tail.* — de la feuille, *leaf-stalk.*
qui, pron., *who; whom; which.*
quintal, s. m., *hundredweight.*
quinze, adj., *fifteen.*
quinzième, adj., *fifteenth.*
quitter, v. a., *leave.*
quoique, conj., *although.*
quotient, s. m., *quotient.*

R

rabattre, v. a., *bring down.*
raccourcir, v. a., *shorten.*
race, s. f., *race: breed.*
racine, s. f., *root.*
radicelle, radicule, s. f., *radicle; rootlet.*
radiculaire, adj. Poils —s, *root-hairs.*
radiolaires, s. m. pl., *radiolaria.*
raie, s. f., *stripe.*
raisin, s. m., *grape.*
raison, s. f., *reason; rate; ratio.* Il y n'y a pas de — pour que . . . ait . . ., *there is no reason why . . . should have . . .* En — de, *by reason of; in ratio to.*
raisonnement, s. m., *reasoning.*
ralenti, e, part. adj., *slow; passive.*
ralentir, v. a. n., *slacken.*
rallumer, v. a., *rekindle.*
rameau, s. m., *branch.*
ramener, v. a., *bring back.*
ramie, s. f., *ramie; China grass.*
ramification, s. f., *ramification.*
ramifier, v. a., *ramify; branch out.*
ramollir, v. a. n., *soften.*
ramper, v. n., *creep; crawl.*
rançon, s. f., *ransom; exhorbitant price; cost.*
rang, s. m., *row; rank; degree.*
rangée, s. f., *row.*
rapide, adj., *swift.* -ment, adv., *rapidly, quickly.*
rapidité, s. f., *speed.*
rappeller, v. a., *recall; remind; resemble.* Se —, *remember.*
rapport, s. m., *relation; ratio.* Être dans un — simple, *be simply related.* Mettre en — avec, *connect with.* Par — à, *in relation to.*
rapporter, v. a. n., *yield; bring in.* Se — à, *be connected with, relate to.*
rapproché, e, part. adj., *near.*
rapprocher, v. a., *bring near; compare.* Se —, *approach one another;* (+ de), *approach.*
rare, adj., *rare; sparse.* -ment, adv., *seldom.* -té, s. f., *scarcity.*
ras, s. m., *level.*
raser, v. a., *skim; level.*
rassembler, v. a., *collect.*
rationnel, le, adj., *rational.*
rattacher, v. a., *attach.*
ravage, s. m., *havoc.*
rayer, v. a., *scratch; erase.*
rayon, s. m., *ray; radius.* — lumineux, *light-ray.*
rayonné, e, part. adj., *radial.*
rayonnement, s. m., *radiation.*
rayure, s. f., *stripe; streak; groove.*

ré, s. m. (music), *D.*
réactif, s. m., *reagent.*
réaction, s. f., *reaction.*
réagir, v. n., *react.*
réaliser, v. a., *realise.*
réalité, s. f., *reality.*
recéder, v. a., *to return again.*
récepteur, s. m., *receiving apparatus.*
recevoir, v. a., *receive; get; admit of.*
réchauffer, v. a., *warm (up); heat.*
recherche, s. f. A la — de, *in search of.*
recherché, e, part. adj., Très —, *much in demand.*
rechercher, v. a., *seek after.*

récipient, s. m., *recipient; receiver.*
réciproquement, adv., *conversely.*
récolter, v. a., *gather in.*
reconnaissable, adj., *recognisable.*
reconnaître, v. a., *recognise; find out.*
reconstituant, s. m., *tonic.*
recourber, v. a., *bend round.*
recouvrir, v. a., *cover.*
rectangle, s. m., *rectangle.* Adj., *right-angled.*
rectangulaire, adj., *rectangular.*
rectifier, v. a., *rectify.*
rectiligne, adj., *rectilinear.*
recueillir, v. a., *collect; pick up.*
reculé, e, part. adj., *remote.*
reculer, v. n., *recede; retreat.* V. a., *move . . . back.*
redevenir, v. n., *become again.*
rediger, v. a., *to draw up; edit.*
redouter, v. a., *dread.*
redresser, v. a., *to straighten; set right.*
réduc-teur, -trice, adj., *reducing.*
réduire, v. a., *reduce.*
réduit, e, part. adj., *diminutive.*
réel, le, adj., *real.* -lement, adv., *really.*
réfléchir, v. a. n., *reflect.*
refléter, v. a. n., *reflect.*
réfracter, v. a., *refract.*
réfraction, s. f., *refraction.*
réfrigérant, s. m., *refriger-ant, -ator.* — ascendant, *reflux condenser.*
réfrigérant, e, adj., *cooling; freezing.*
refringent, e, adj., *refract-ing, -ive.*
refroidir, v. a. n., *cool.*
refroidissement, s. m., *cooling.*
refuge, s. m., *refuge.*
refuser, v. a. n., *refuse.*
regarder, v. a. n., *look (at, upon).*
regeler, v. a. n., *freeze again.*
régime, s. m., *diet.*
région, s. f., *region.*
règle, s. f., *rule.*
régler, v. a., *regulate; settle.*
règne, s. m., *reign; kingdom.*
régularité, s. f., *regularity.*
régul-ier, -ière, adj., *regular.*
reins, s. m. pl., *kidneys.*
réintégration, s. f., *reintegration.*
rejoindre, v. a., *join.*
relati-f, -ve, adj., *relative; relating.*
relation, s. f., *relation.* Mettre en — avec, *bring into touch with.*
relever, v. a. n., *set off.*
relief, s. m., *relief.*
relier, v. a., *bind; connect.*
remarquable, adj., *remarkable; noteworthy.*
remarque, s. f., *remark.*
remarquer, v. a., *observe, notice.*
remède, s. m., *remedy; medicine.*
remettre, v. a. S'en — à, *leave (it, things) to.*
remonter, v. a. n., *ascend; go up; go back.*
remorquer, v. a., *tug.*
remplacer, v. a., *replace; take the place of.*
remplir, v. a., *fill; fulfil; stuff.*
remuer, v. a. n., *move; stir.*
rencontre, s. f., *meeting.*
rencontrer, v. a., *meet; meet with.*
rendement, s. m., *yield, product.*
rendre, v. a., *render; make; give out.* Se —, *proceed; yield.* —domestique, *domesticate.*
renfermer, v. a., *enclose; contain.*
renflé, e, part. adj., *swelling.*
renflement, s. m., *swelling.*
renforcement, s. m., *increase in numbers.*
réniforme, adj., *reniform, kidney-shaped.*
renne, s. m., *reindeer.*
renommé, e, part. adj., *renowned.*
renoncer, v. n., (+ à) *give up;* (+ à + infin.) *give up the attempt to.*
renouveler, v. a. n., *renew.*
renseigner, v. a., *inform.*
renversé, e, part. adj., *upside down.*
renvoyer, v. a., *send out.*
répandre, v. a., *spill; shed; scatter; diffuse.*
répandu, e, part. adj. Fort —, *widely distributed.*
reparaître, v. n., *reappear.*
réparat-eur, -rice, adj., *restoring.*
réparer, v. n., *repair; recruit.*
répartir, v. a., *divide; distribute.*

répartition, s. f., *distribution.*
repasser, v. a. n., *pass back.*
repérer, v. a., *take bearings of, from.* On repère cette direction AD, *this direction is found to be AD.*
repli, s. m., *fold.*
replier, v. a., *fold up.*
repos, s. m., *rest.* En —, *at rest; steadily.*
repousser, v. a., *repel.* Se —, *repel each other.*
reprécipiter, v. a., *reprecipitate.*
reprendre, v. a. n., *resume; regain; take up; attack . . . again.*
représentant, s. m., *representative.*
représenter, v. a. n., *represent.*
reprise, s. f., *renewal.* A plusieurs —s, *several times.*
reproduc-teur, adj., *reproductive;* -tion, s. f., *reproduction.*
reproduire, *reproduce.* Se —, *propagate one's species.*
reptile, s. m., *reptile.*
répulsion, s. f., *repulsion.*
requin, m. s., *shark.*
reseau, s. m., *network.*
réserve, s. f., *reserve.* Le (etc.) . . . en —, *the reserve supply of . . .*
réservoir, s. m., *reservoir; tank; bulb.*
résidu, s. m., *residue.*
résineu-x, -se, adj., *resinous.*
résist-ance, s. f., *resistance.* -ant, e, adj., *resistant; tough.* -er, v. n., *resist.* -ivité, s. f., (*power of*) *resistance.*
résolu, e, adj., *resolved.*
résolution, s. f., *solution.*
résonance, s. f., *resonance.*
résonner, v. n., *resound.*
résorber, v. a., *reabsorb.*
résoudre, v. a. n., *resolve; solve.*
respectivement, adv., *respectively.*
respir-able, adj., *breathable.* -ation, s. f., *breathing.* -er, v. a. n., *breathe.*
ressaut, s. m., *projection; break in the level.*
ressembl-ance, s. f., *resemblance, similarity;* -er, v. n. (+ à), *resemble.*
ressentir, v. a., *feel; perceive.*
resserrer, v. a., *put by, stow away; narrow; strengthen.*
reste, s. n., *rest; remainder; remnant.*
rester, v. a., *remain; stay.* . . . resté, . . . *that had remained.*
restituer, v. a., *restore.*
restreint, adj., *restricted, particular, special.*
résultat, s. m., *result.*
résulter, v. n., *result.*
résumé, s. m. En —, *to sum up.*
résumer, v. a., *sum up.*
retarder, v. a., *retard.*
retenir, v. a., *retain; attach.*
rétin-e, s. f., *retina.* -ien, ne, adj., *retinal.*
retirer, v. a., *withdraw; draw in; extract.* Se —, *recede.*
retomber, v. n., *fall back.*
retour, s. m., *return.*
retourner, v. n., *return* (*go back*).
retracter, v. a. Se —, *retract.*
retraite, s. f., *retreat.*
rétroaction, s. f., *retrospective action; interaction; reaction.*
retrousser, se, v. r., *to curl* or *turn up.*
retrouver, v. a., *find.*
réunion, s. f., *combination.*
réunir, v. a. Se —, *unite, join.*
revanche, s. f. En —, *by way of compensation.*
réveil, s. m., *awakening.*
révélateur, s. m., *developer.*
revenir, v. n., *return* (*come back*); *revert, cost.* Cela revient à dire, *this amounts to saying.*
revêtir, v. a., *coat.*
revient, s. m., *net cost.* Prix net de —, *net cost price.*

rhizome, s. m., *rhizome, underground stem of plants.*
rhizopodes, s. m. pl., *rhizopoda.*

riche, adj., *rich.*
ricin, s. m., *castor-oil plant.*
rien, s. m., *nothing;* (+ ne, *not*) *anything.* Il n'en est —, *it is not so; nothing of the sort.*
rigide, adj., *rigid; stiff.*
rigoureusement, adv., *strictly.*
rivière, s. f., *river.*

roc, s. m., *rock.*
roche, s. f., *rock.*
rocher, s. m., *rock.*

rôle, s. m., *part.* Jouer le même —, *fulfil the same function.*
rond, e, adj., *round.*
rosacées, s. f. pl., *rosaceae.*
rose, s. f., *rose.*
rosier, s. m., *rose-tree.*
rotation, s. f., *rotation.* De —, *rotatory.*
roue, s. f., *wheel.*
rouge, adj., *red.* Chauffé au —, *red-hot.* -brun, *reddish-brown.* -vermeil, *vermilion.*
rouge-gorge, s. m., *robin redbreast.*
rougeâtre, adj., *reddish.*
rougir, v. a. n., *redden.*
rouille, s. f., *rust.*
routier, adj., *connected with roads.*
royal, e, adj., *royal.*
royaume, s. m., *kingdom.*

rudéral, adj., *ruderary; growing in rubbish.*
rudimentaire, adj., *rudimentary.*
rue, s. f., *street.*
ruisellement, s. m., *running* (of water).
ruminant, s. m., *ruminant.*
rupture, s. f., *breakage.*
rutilant, e, adj. *See* vapeur.

rythmer, v. n., *harmonise.*

S

sa, possessive f. adj., *his; her; its.*
sable, s. m., *sand.*
sac, s. m., *bag; sac.*
saccharifier, v. a., *saccharify.*
saccharose, s. m., *saccharose.*
sacciforme, adj., *sacciform.*
saillant, e, part. adj., *projecting.*
saillie, s. f., *projection.*
saillir, v. n., *project.*
sain, e, adj., *healthy.*
saisir, v. a., *seize; grasp.*
saison, s. f., *season.*
salé, e, adj., *salt, salty.* Eau —e, *salt water.*
saler, v. a., *salt.*
saleté, s. f., *dirt.*
salin, e, adj., *saline.*
salivaire, adj., *salivary.*
salive, s. f., *saliva.*
salpêtre, s. m., *saltpetre.*
sang, s. m., *blood.* Avoir le — chaud, *be warm-blooded.*
sangsue, s. f., *leech.*
sans, prep., *without; but for.*
saponifier, v. a., *saponify.*
sarigue, s. m. f., *sarigue, South American opossum.*
satellite, s. m., *satellite.*
satisfaisant, e, adj., *satisfactory.*
saturer, v. a., *saturate.*
sauf-conduit, s. m., *pass; safe-conduct.*
saumâtre, adj., *briny, brackish.*
saut, s. m., *leap.*
sauter, v. n., *leap.*
sauvage, adj., *wild.*
savane, s. f., *savannah.*
savant, s. m., *scientist.*
saveur, s. f., *flavour.*
savoir, v. a. n., *know; be able;* (adverbially) *to wit, namely.*
savon, s. m., *soap.*

scaphandre, s. m., *diving apparatus; cork life-jacket.*
sceller, v. a., *seal; imbed.*
schéma, s. m., *circuit diagram.*
schiste, s. m., *schist.*
scie, s. f., *saw.*
science, s. f., *science.*
scier, v. a. n., *saw.*
scinder, v. a., *divide.*

se, s', pers. pron., *oneself, him-, her, it-self, themselves; each other, one another.* (Forms reflexive verbs, often to be rendered by Eng. intransitive or passive verbs.)
seau, s. m., *pail; bucket.*
sec, sèche, adj., *dry.*
sécante, s. f., *secant.*
sécher, v. a. n., *dry.* Faire —, *dry.*
sécheresse, s. f., *dryness; drought.*
secondaire, adj., *secondary.*
seconde, s. f., *second.*
secondine, s. f., *secundine.*
secousse, s. f., *shake; shock.*
secréter, v. a., *secrete.*
séculaire, adj., *centurial.*
sédentaire, adj., *sedentary.*
sédimentaire, adj., *sedimentary.*
séduire, v. a., *to entice; to charm.*
segment, s. m., *segment.*
seiche, s. f., *cuttle-fish.*
sein, s. m., *bosom; midst.*
séism, s. m., *seism, earthquake.*
séjourner, v. n., *remain.*
sel, s. m., *salt.*
selon, prep., *according to; in.*
semblable, adj., *like; similar; such.*

sembler, v. n., *seem.*
semence, s. f., *seed, seed-corn; semen; tin tacks.*
semer, v. a. n., *sow.*
sens, s. m., *sense; direction.* En — inverse, *in the opposite direction.*
sensation, s. f., *sensation; feeling.*
sensib-ilité, s. f., *sensibility.* -le, adj., *perceptible; endowed with feeling.* -lement, adv., *perceptibly; obviously.*
sensitive, s. f., *sensitive plant.*
sentir, v. a. n., *feel.* Se faire —, *be felt.*
sépale, s. m. *sepal.*
séparément, adv., *separately.*
séparer, v. a., *separate.*
série, s. f., *series.*
sérieu-x, -se, adj., *earnest.*
serré, e, part. adj., *compact.*
serrer, v. a., *tighten; clamp.*
servir, v. a. n., *serve.* — de, *serve as.* Se — de, *use.*
ses, possessive pl. adj., *his, her, its.*
sessile, adj., *sessile.*
seul, e, adj., *alone; only; single; sole.* A lui —, à elle —e, *by him-, her-, it-self.*
seulement, adv., *only.*
sève, s. f., *sap.*
sévir, v. n., *rage.*

si, s', conj., *if.*
si (music), *B.*
siècle, s. m., *century; age.*
siège, s. m., *seat, bench; see, bishopric; head office.*
sien, ne, possessive pron., *his, hers, its own.*
siffler, v. a. n., *whistle; hiss.*
signal, s. m., *signal.*
signe, s. m., *sign; mark.*
signifier, v. a., *signify; mean.*
silencieu-x, -se, adj., *silent.*
silex, s. m., *silex, flint.*
silicate, s. m., *silicate.*
silice, s. f., *silica.*
siliceu-x, -se, adj., *siliceous.*
sillon, s. m., *furrow.*
silloné, adj., *furrowed.*
silurien, adj. as s. m., *Silurian.*
simpl-e, adj., *simple.* -ement, adv., *simply.* -icité, adj., *simplicity.* -ifier, v. a., *simplify.*
simultanément, adv., *simultaneously.*
singe, s. m., *monkey; ape.*
singul-ier, -ière, adj., *singular; curious.*
sinistre, adj., *sinister; evil.*
sinueux, adj., *sinuous, winding.*
sinus, s. m., *sine.*
siphonophores, s. m. pl., *siphonophora.*
situé, e, part. adj., *situated.*
sixième, adj., *sixth.*

société, s. f., *company.*
socle, s. m., *plinth, base, pedestal.*
soie, *silk.* —s, *silky hairs.*
soigneusement, adv., *carefully.*
soin, s. m., *care.* Avoir — de (+ infin.), *take care to.*
soit, subjunctive of être, *let . . . be.* — à comparer, *let us suppose that . . . have to be compared.* Adv., *namely.* Conj., *either, or.*
soixantaine, s. f. Une — de, *about sixty.*
soixante, adj., *sixty.*
sol, s. m., *soil, ground.*
sol, s. m. (music), *G.*
soleil, s. m., *sun; sunshine.*
solfatare (Italian), s., *solfatare, sulphur-mine.*
solid-e, adj., *solid; strong.* S. m., *solid.* -ifier, v. a., *solidify.* -ité, s. f., *strength; stability.*
solipédiation, s. f., *having whole hoofs.*
solliciter, v. a. n., *demand; induce.*
solvant, s. m., *solvent.*
solu-bilité, s. f., *solubility.* -ble, adj., *soluble.* -tion, s. f., *solution.*
sombre, adj., *dark.*
somme, s. f., *sum.* En —, *finally; to sum up.*
sommet, s. m., *summit; peak.*
sommier, s. m., *undermattress.*
son, possessive adj., *his, her, its.* S. m., *sound.*
sondage, s. m., *sounding* (of depth).
sonnerie, s. f., (*set of*) *bells.*
sonneur, s. m., *bell-ringer.*
sonore, adj., *sonorous; sounding.*
sort, s. m., *fate.*
sorte, s. f., *sort, species.* De — que, *so that.*
sortie, s. f., *emergence.*

sortir, v. n., *go, come, out; emerge.* V. a., *take out.*
souche, s. f., *root-stock.*
souci, s. m., *care, concern; marigold.*
soudan, s. m., *sultan.*
soude, s. f., *soda.*
soud-er, v. a., *weld, join.* -ure, s. f., *welding.*
souffler, v. a. n., *blow.*
soufre, s. m., *sulphur.*
soulèvement, s. m., *upheaval.*
soulever, v. a. n., *raise.*
souligner, v. a., *underline.*
soumettre, v. a., *subject.*
source, s. f., *source; spring.* Eau de —, *spring water.*
sous, prep., *under, in,* etc.
sous-classe, s. f., *subclass; subdivision.*
sous-marin, e, adj., *submarine.*
sous-tendre, v. a., *subtend.* Sous-tendu, e, part. adj., *subtended.*
soustraire, v. a., *subtract; withdraw; shelter.*
soutenir, v. a. n., *support; maintain.*
souterrain, adj., *underground.*
soutien, s. m., *support.*
souvent, adv., *often.* Le plus —, *most frequently.*

spécial, e, adj., *special.* -ement, adv., *especially.*
spécifique, adj., *specific.*
spectacle, s. m., *sight.*
spectre, s. m., *ghost; spectrum.*
sphèr-e, s. f., *sphere.* -ique, adj., *spherical.*
spicule, s. m., *spiculum.*
spirale, s. f., *spiral.* En —, *spiral* (adj.).
spirifère, s. m., *spirifer.*
spongiaire, s. m., *sponge.*
spongieu-x, -se, adj., *spongy.*
spontanément, adv., *spontaneously.*
sporange, s. m., *sporangium.*
spore, s. f., *spore.*

squelette, s. m., *skeleton.*

stable, adj., *stable.* Peu —, *unstable.*
stade, s. m., *period; phase.*
steppe, s. f., *steppe.*
stérile, adj., *barren.*
stigmate, s. m., *stigma* (pl. *-ata*).
stipule, s. f., *stipule.*
stomacal, e, adj., *stomachic.*
stomate, s. m., *stoma, breathing-pore.*
strie, s. f., *stria* (pl. *striae*).
strié, e, adj., *striated.*
strontium, s. m., *strontium.*
structure, s. f., *structure.*
style, s. m., *style.*

subdiviser, v. a., *subdivide.*
subéreu-x, -se, adj., *suberous.*
subir, v. a., *undergo.*
submerger, v. a., *submerge.* Plante submergée, *aquatic plant.*
subsister, s. n., *remain.*
substance, s. f., *substance.*
substituer, v. a., *substitute* (à, *for*). Se — à, *replace.*
suc, s. m., *juice.*
succéder, v. n. (+ à), *succeed* (*come after*); *follow.*
successi-f, -ve, adj., *successive.*
successivement, adv., *successively.*
succulent, e, *succulent; juicy.*
suceur, adj., *suctorial.*
sucre, s. m., *sugar.*
sucré, e, adj., *sugary.*
sud, s. m., *south.* Plus au —, *further south.*
suffire, v. n., *suffice.* Il suffit de, . . . *suffices.*
suffisant, e, adj., *sufficient.*
suif, s. m., *suet.*
suisse, adj., *Swiss.*
suite, s. f., *series.* A la — de, *after;* (also par — de), *as a result of.* De —, *directly.* Par —, *in consequence.*
suivant, prep., *according to;* — que, *according as.* —, e, adj., *following.* Sont les —s, *are as follows.*
suivre, v. a. n., *follow.*
sujet, te, adj., *subject.* S. m., *subject.* Au — de, *concerning.*
sulfate, s. m., *sulphate.*
sulfhydrique, adj. Acide —, *sulphydric acid, sulphuretted hydrogen.*
sulfuré, e, part. adj., *sulphuretted.*
sulfureu-x, -se, adj., *sulphureous.*
sulfurique, adj., *sulphuric.*
superficiel, le, adj., *superficial.* -lement, adv., *superficially; at the surface.*
supérieur, e, adj., *upper; higher.* — à, *greater than.*
superposer, v. a., *superimpose.*

suppléer, v. a. n. Y —, *make up for it.*
supplementaire, adj., *supplementary.*
support, s. m., *support; prop.*
supporter, v. a., *sustain.*
supposer, v. a., *suppose; assume.*
supprimer, v. a., *suppress; remove; leave out.*
suprématie, s. f., *supremacy.*
suprême, adj., *supreme;* -ment, adv., *supremely.*
sur, prep., *on, to, concerning,* etc.
sûr, e, adj., *sure.*
sureau, s. m., *elder (tree).*
surface, s. f., *surface.*
surfondu, e, part. adj., *superfused.*
surfusion, s. f., *superfusion.*
suroxyder, v. a., *superoxidise.*
surplomber, v. a., *overhang.*
sursaturation, s. f., *supersaturation.*
surtout, adv., *especially.*
susceptible, adj., *susceptible; capable.*
suspen-dre, v. a., *suspend.* -u, e, part. adj., *suspended; hanging.* -seur, adj. s. m., *suspensor.* -sion, s. f., *suspension.*

symétr-ie, s. f., *symmetry.* -ique, adj., *symmetrical.*
sympathique, adj. s. m., *sympathetic.*
syncipital, adj., *syncipital.*
synclinal, adj., *synclinal* (of a valley).
synthèse, s. f., *synthesis.*
système, s. m., *system; group of objects.*

T

table, s. f., *table* (article of furniture).
tableau, s. m., *table* (tabulation).
tâche, s. f., *task, duty.*
tache, s. f., *spot; stain.*
tacher, v. a., *stain.*
tacheter, v. a., *speckle.*
taille, s. f., *size.*
tailler, v. a. n., *cut; hew.*
tamis, s. m., *sieve; flour-dredge.*
tampon, s. m., *stopper; pad; buffer.*
tandis que, conj., *whereas.*
tangent, e, adj., *tangential.* —e, s. f., *tangent.*
tant, adv., *so much; so.*
tantôt, adv. — . . . — . . . , *now . . . now . . . ; in some cases . . . in others.*
tapis, s. m., *carpet, covering of floor or ground.*
tarage, s. m., *calibration.*
tard, adv., *late.* Plus —, *later.*
tarder, v. n., *delay.* Ne pas tarder à (+ inf.), *very soon be* (+ past part.).
tardivement, adv., *tardily; slowly.*
taré, adj., *tared.*
tarse, s. m., *tarsus.*
Tartare, s. m., *Tartar.*
taux, s. m., *rate; rate of interest.*

tegmen, s. m., *tegmen.*
tégument, s. m., *integument.*
teindre, v. a., *dye.*
teinte, s. f., *tint.*
teinture, s. f., *dye; tincture.*
tel, le, adj., *such.* — ou —, *this or that.* -lement, adv., *so.*
télé-graphie, s. f., *telegraphy.* -phonie, s. f., *telephony.*
température, s. f., *temperature.*
tempéré, e, adj., *temperate.*
temple, s. m., *temple.*
temps, s. m., *time; days; weather;* pl. *times, ages.* De tout —, *at all times.* En même —, *at the same time.* Quelque —, *for some time.*
ténacité, s. f., *tenacity.*
tendance, s. f., *tendency.*
tendre, v. a., *stretch.* V. n., *tend.*
tendu, e, part. adj., *stretched; taut.*
teneur (en), s. m., *amount (of).*
ténia, s. m., *taenia, tapeworm.*
tenir, v. a. n., *hold; keep.* — de, *take after, resemble.*
tentacule, s. m., *tentacle.*
tenter, v. n., *attempt; try; tempt.*
terme, s. m., *term; limit.* En d'autres —, *in other words.*
terminer, v. a., *end; terminate.*
terne, adj., *dull (lack-lustre).*
terrain, s. m., *ground; soil.*
terre, s. f., *earth; ground; land.*
terreau, s. n., *mould; compost.*

terrestre, adj., *terrestrial; of the land, of the earth; land-.*
terreu-x, -se, adj. *earthy.*
terrible, adj., *terrible.*
territoire, s. m., *territory.*
tertiaire, adj. as s. m., *tertiary.*
testa, s. f., *testa; seed-coat.*
testicule, s. m., *testicle.*
tête, s. f., *head.*

théorème, s. m., *theorem.*
théorie, s. f., *theory.*
thermal, e, adj., *thermal.* Sources —es, *hot springs.*
thermomètre, s. m., *thermometer.*
thèse, s. f., *thesis; argument.*
thoracique, adj., *thoracic.*
thorax, s. m., *thorax.*

tiède, adj., *lukewarm, tepid.*
tiers, s. m., *third.*
tige, s. f., *stem; stalk; shaft.* — métallique, *metal rod.*
tigelle, s. f., *tigel, embryonic axis.*
timbre, s. m., *bell.*
tirage, s. m., *drawing; pulling.*
tiraillement, s. m., *pull; stress.*
tirailler, v. a. n., *pull about.*
tirer, v. a. n., *draw; extract; deduce.*
tisser, v. a., *weave.*
tissu, s. m., *tissue.*
titre, s. m. (chem.), *titre; standard.*

toit, s. m., *roof.*
tôle, s. f., *sheet-iron.*
tomber, v. n., *fall.*
tonne, s. f., *ton.*
tordre, se, v. r., *twist.*
torpille, s. f., *torpedo.*
torsion, s. f. Balance de —, *torsion-balance.*
total, e, adj., *total.*
toucher, v. n., *touch; draw* (money).
touffe, s. f., *cluster; clump.*
touffu, e, adj., *tufted; bushy.*
toujours, adv., *always.*
toundra, s. m., *tundra.*
tour, s. m., *turn.* A son —, *in his (her, its) turn.* — du soleil, *revolution round the sun.*
tourbe, s. f., *peat.*
tourie, s. f., *carboy.*
tourner, v. a. n., *turn.*
tournesol, s. m., *litmus.*
tout, e (pl. tous, toutes), adj., *all; whole; every.*
tout, s. m., *everything.* Pas du —, *not at all, by no means.* Le —, *the whole.* Adv., *quite, right.* — à fait, *quite, altogether.* — en -ant, *while -ing.*
toutefois, adv., *nevertheless.*

traçant, e, part. adj. (bot.), *running, creeping.*
trace, s. f., *trace.*
tracer, v. a., *trace, trace out.*
trachée, s. f., *trachea.*
traduction, s. f., *translation.*
traduire, v. a., *translate.*
trait, s. m., *feature.*
traité, s. m., *treatise; treaty.*
traitement, s. n., *treatment.*
traiter, v. a. n., *treat.*
trajet, s. m., *journey; course.*
tranchée, s. f., *trench, ditch; colic; labour pains.*
trancher, v. a. n., *cut; carve; slice.*
tranquille, adj., *quiet.*
transfert, s. m., *transfer.*
transform-er, v. a., *transform;* -ation, s. f., *transformation.*
transformisme, s. m., *transformism.*
transition, s. f., *transition.*
translation, s. f., *transfer; translatory movement.*
translucide, adj., *translucent.*
transmettre, v. a., *transmit.*
transparent, e, adj., *transparent.*
transporter, v. a., *transport.*
transversal, e, adj. *transversal.*
travail, s. m., *work; labour.*
travailler, v. a. n., *work.*
travailleur, s. n., *worker.*
travers, s. m. A —, *across; through.* Au — de, *through.*
traverser, v. a. n., *cross; go through; run through; penetrate.*
traversin, s. m., *bolster.*
trèfle, s. m., *clover.*
tremper, v. a., *soak; temper.* Acier trempé, *tempered steel, hard steel.*
trente, adj., *thirty.*
très, adv., *very.*
triangle, s. m., *triangle.*
trias, s. m., *Triassic.*
trièdre, adj., *trihedral.*
tripler, v. a. n., *treble.*
trois, adj., *three.*

troisième, adj., *third.*
tronc, s. m., (*tree-*) *trunk.*
tronçon, s. m., *fragment, stump; branch railway line; spur.*
trop, adj., *too; too much; too many; too often.*
tropical, e, adj., *tropical.*
trou, s. m., *hole.*
troupe, s. f., *herd.*
trouver, v. a., *find.* On n'en trouvera plus, *they will be extinct.* Se —, *be found, be present, be.*

tube, s. m., *tube.* En forme de —s, *tubular.*
tubercule, s. m., *tuber.*
tubéreux, adj., *tuberous.*
tubulure, s. f., *tubulure; pipe.*
tuer, v. a. n., *kill.*
tulipe, s. f., *tulip.*
tunnel, s. m., *tunnel.*
tuyau, s. m., *tube; pipe.* — sonore, *resonance tube.*

type, s. m., *type.*

U

ultérieur, adj., *ulterior; further; advanced; subsequent; later.*
un, e, adj. art., *one; a, an.* L'un(e) et l'autre, *both.* Les un(e)s et les autres, *every one of them.* L'un de l'autre, *of (from) one another.* Les uns, *some.* Les uns . . . les autres . . . , *some . . . others.*
unicellulaire, adj., *unicellular.*
uniforme, adj., *uniform.*
uniformément, adv., *uniformly.*
unique, adj., *sole; single.* -ment, adv., *solely.*
unir, v. a., *join; combine.*
unisson, s. m. A l'—, *in unison.*
unité, s. f., *unity; unit; digit.*
univalve, adj. s. m., *univalve.*
univers, s. m., *universe; world.*
universel, le, adj., *universal.*

urinaire, adj., *urinary.*
urne, s. f., *urn.*
urticant, e, adj., *urticarial.*

usage, s. m., *use.* Faire — de, *make use of.* D'— courant, *in current use.*
user, v. a., *wear away.* S'—, *wear out.*
ustensile, s. m., *utensil.*
usuel, le, adj., *usual; customary.*
usure, s. f., *wearing out; wastage.*

ut (music), *C.*
utile, adj., *useful.*
utilis-able, adj., *utilisable.* —er, v. a., *utilise; use.*
utilité, s. f., *utility.*

V

va-et-vient, s. m. Mouvement de —, *reciprocating movement.*
vache, s. f., *cow.*
vacuole, s. f., *vacuole; vesicle; protoplasm cavity.*
vagin, s. m., *vagina.*
vague, s. f., *wave.*
vague, adj., *vague.*
vaisseau, s. m., *vessel.*
valable, adj., *valid, available.*
valenciel, adj., *pertaining to valency.*
valeur, s. f., *value; application.*
vallée, s. f., *valley.*
vallonment, s. m., *foothill.*
valoir, v. a. n., *be worth; be equivalent to.*
valve, valvule, s. f., *valve.*
vapeur, s. f., *vapour; steam; fume.* —s rutilantes, *coloured fumes.*
vaporiser, v. a., *vaporise.*
variable, adj., *variable; varying.*
vari-é, e, part. adj., *various.* -er, v. a. n., *vary.* -été, s. f., *variety.*

vasculaire, adj., *vascular.*
vase, s. m., *vessel.* — distillatoire, *distilling flask.*
vaste, adj., *vast; spacious.*

végét-al, e, adj., *vegetable.* Terre —e, *vegetal soil.* -al, s. m., *plant.* -ati-f, -ve, adj., *vegetative.* -ation, s. f., *vegetation.*
veine, s. f., *vein.*
veineu-x, -se, adj., *venous.*
vendre, v. a. n., *sell.*
vénéneu-x, -se, adj., *poisonous.*
vénimeu-x, -se, adj., *venomous.*
venir, v. n., *come.* — en, *coincide with.* — de (+ inf.), *have just* (+ part.). Il vient, *we get.* (+ inf.) often untranslated; *e.g.* viennent converges, *converge.*

vent, s. m., *wind.*
ventouse, s. f., *air-hole.*
ventral, e, adj., *ventral.*
ventre, s. m., *belly; abdomen.* — de vibration, *vibrating segment* (*loop*), *ventral segment.*
ventricule, s. m., *ventricle.*
venu, e, part. —s, *who have come.*
ver, s. m., *worm.* — de terre, *earthworm.*
verdure, s. f., *verdure.*
vérifier, v. a., *verify; confirm.*
véritable, adj., *true; regular.*
vérité, s. f., *truth.*
verjus, s. m., *acid fruit juice.*
vernis, s. m., *varnish.*
verre, s. m., *glass.*
verrue, s. f., *wart.*
vers, prep., *towards; about.*
versant, s. m., *slope; watershed.*
vers à soire, s. m., *silk worm.*
verser, v. a. n., *pour.*
vert, e, adj., *green.*
vertèbre, s. f., *vertebra.*
vertébré, s. m., *vertebrate.*
vertical, e, adj., *vertical.*
vésicule, s. f., *vesicle; bladder.*
vessie, s. f., *bladder; blister.*

viande, s. f., *meat.*
vibra-tile, adj., *vibratile.* -tion, s. f., *vibration.* -toire, adj., *vibratory.*
vibrer, v. n., *vibrate.*
vide, adj., *empty.* S. m., *vacuum.*
vie, s. f., *life.* En —, *alive.*
vieux, vieil, vieille, adj., *old; ancient.*
vi-f, -ve, adj., *lively; vigorous.*
vignoble, s. m., *vineyard.*
ville, s. f., *town; city.*
vinaigre, s. m., *vinegar.*
vingt, adj., *twenty.*
virgule, s. f., *comma; decimal point* (French use comma for this).
virtuel, le, adj., *virtual.*
vis-à-vis, adv. — de, prep., *opposite; respecting.*
viscères, s. m. pl., *viscera.*
visible, adj., *visible.*
vision, s. f., *vision.*
visite, s. f., *visit.*
visqueu-x, -se, adj., *viscous.*
visuel, le, adj., *visual.*
vital, e, adj., *vital. See* concurrence.
vite, adv., *quickly.*
vitesse, s. f., *speed; velocity.*
vitreu-x, -se, adj., *vitreous; glassy.*
vivant, e, adj., *living; extant.*
vivement, adv., *vigorously.*
vivifiant, e, part. adj., *life-giving.*
vivre, v. n., *live.*

voici, prep., *here is.*
voie, s. f., *way.* Par — de déduction, *by deductive methods.*
voilette, s. f., *small veil.*
voir, v. a. n., *see.* Faire —, *show.* Se —, *find oneself.*
voisin, e, adj., *neighbouring; near.*
voisinage, s. m., *neighbourhood.*
voiture, s. f., *vehicle.*
volatil, e, adj., *volatile.*
volatiliser, v. a., *volatilise.*
volcan, s. m., *volcano.*
volcanique, adj., *volcanic.*
voler, v. n., *fly.*
volontaire, adj., *voluntary.*
volonté, s. f., *will; will-power.*
volume, s. m., *volume.*
volumineu-x, -se, adj., *voluminous; bulky.* Peu —, *slender.*
vorace, adj., *voracious.*
voracité, s. f., *voracity.*
vorticelles, s. f. pl., *vorticellae.*
vouloir, v. a., *wish;* (imperfect) *would.* — dire, *mean.* — parler de, *refer to.*
vous, pers. pron., *you.*
voûte, s. f., *vault; ceiling, roof.*
voyage, s. m., *journey; voyage.* — de circumnavigation, *voyage round the world.*
voyageur, s. m., *traveller.*
vrai, e, adj., *true.* A — dire, *as a matter of fact.*
vraiment, adv., *really.*
vraisemblable, adj., *likely, probable.*
vue, s. f., *sight.*
vulgaire, adj., *common.*

Y

y, adv., *there.* Pers. pron., *to it, in it* (*them*).
yeux, pl. of Œil.

Z

zèbre, s. m., *zebra.*
zélé, e, adj., *zealous.*
zéro, s. m., *nought; zero.*
zinc, s. m., *zinc.* — hydraté, *zinc hydrate.*
zone, s. f., *zone.*
zoologie, s. f., *zoology.*

PROPER NAMES

Afrique, f., *Africa.*
Allemagne, f., *Germany.*
Alpes, f. pl., *Alps.*
Amérique, f., *America.* L'— du Nord, du Sud, *North, South America.*
Angleterre, f., *England.*
Antilles, f. pl., *West Indies.*
Apennins, m. pl., *Apennines.*
Arabie, f., *Arabia.*
Aristote, *Aristotle.*
Asie, f., *Asia.*
Atlantique, m., *Atlantic.*
Australie, f., *Australia.*
Autriche, f., *Austria.*

Belgique, f., *Belgium.*
Bretagne, f., *Brittany.*

Carpathes, m. pl., *Carpathians.*
Caspienne, f., *Caspian Sea.*
Caucase, m., *Caucasus.*
Copernic, *Copernicus.*
Côte d'Or (range of hills in Burgundy).

Dauphiné (former French province).
Dunkerque, m., *Dunkirk.*

Écosse, f., *Scotland.* La Nouvelle —, *Nova Scotia.*
Egée (La Mer), f., *Aegean.*
Egypte, f., *Egypt.*
Espagne, f., *Spain.*
Etats-Unis, m. pl., *United States.*
Europe, f., *Europe.*

France, f., *France.*

Galilée, *Galileo.*
Gênes, f., *Genoa.*
Groënland, m., *Greenland.*
Guyane, f., *Guiana.*

Hongrie, f., *Hungary.*
Hudson. La Mer d'—, *Hudson Bay.*

Inde, f., *India.* Les —s, *the Indies; India.*
Irlande, f., *Ireland.*
Islande, f., *Iceland.*
Italie, f., *Italy.*

Jacques, *James.*

Londres, m., *London.*
Lyon, m., *Lyons.*

Méditerranée, f., *Mediterranean.*
Mexique, m., *Mexico.*

Nil, m., *Nile.*
Normandie, f., *Normandy.*

Nouvelle-Zemble, f., *Nova Zembla.*
Nouvelle-Guinée, f., *New Guinea.*

Océanie, f., *Oceania.*

Paracelse, *Paracelsus.*
Perse, f., *Persia.*
Philippines, f. pl., *Philippines.*
Polynésie, f., *Polynesia.*
Pouzzoles, m., *Pozzuoli.*
Pythagore, *Pythagoras.*

Sibérie, f., *Siberia.*
Sonde, f. Les Iles de la —, *the Sunda Islands.*
Spitzberg, m., *Spitzbergen.*
Suisse, f., *Switzerland.*

Tartarie, f., *Tartary.*
Terre. La — de Feu, *Tierra del Fuego.* — Neuve, *Newfoundland.*
Transylvanie, f., *Transylvania.*

Vienne, f., *Vienna.*
Virginie, f., *Virginia.*

Zélande. La Nouvelle —, *New Zealand.*

INDEX

The numbers refer to the sections of the book

I.—GENERAL INDEX

II.—FRENCH WORDS AND SUFFIXES

PRINTED IN GREAT BRITAIN BY UNIVERSITY TUTORIAL PRESS LTD, FOXTON NEAR CAMBRIDGE

9 780820 600383

Printed by Libri Plureos GmbH in Hamburg,
Germany